APPLIED
FOURIER
ANALYSIS

HARCOURT BRACE JOVANOVICH COLLEGE OUTLINE SERIES

APPLIED FOURIER ANALYSIS

Hwei P. Hsu

Fairleigh Dickinson University

Books for Professionals
Harcourt Brace Jovanovich, Publishers
San Diego   New York   London

Requests for permission to make copies of any part of the work should be mailed to:

Permissions Department
Harcourt Brace Jovanovich, Publishers
8th Floor
Orlando, Florida 32887

Printed in the United States of America

Library of Congress Cataloging in Publication Data

Hsu, Hwei P. (Hwei Piao), 1930-
    Applied Fourier analysis.

    (College outline series) (Books for professionals)
    Includes index.
    1. Fourier analysis.   I. Title.   II. Series: College
outline series (San Diego, Calif.)   III. Series: Books for professionals.
QA403.5.H78 1984   515′.2433   83-22732

ISBN 0-15-601609-5

First edition

D E F G

PREFACE

Jean-Baptiste-Joseph Fourier's *Théorie analytique de la chaleur* [The Mathematical Theory of Heat] inaugurated simple methods for the solution of boundary-value problems occurring in the conduction of heat. But this "great mathematical poem," as Fourier analysis was called by Lord Kelvin, has extended far beyond the physical applications for which it was originally intended. In fact it has become "an indispensable instrument" in the treatment of nearly every recondite question in modern physics, communication theory, linear systems, and other fields.

The intention of the author in writing this book is to develop classical Fourier analysis fully and to show the link between it and its modern applications.

This book is designed for students in mathematics, physics, and the various disciplines of engineering to be used in a formal course in Fourier analysis and throughout the numerous related courses that introduce and employ Fourier techniques. It combines the advantages of both the textbook and the so-called review book. And in the direct way characteristic of the review book, it gives hundreds of completely solved problems that use essential theory and techniques. The solved problems constitute an integral part of the text, illustrating and amplifying the fundamental concepts and developing the techniques of Fourier analysis. The supplementary exercises are designed not only for practice but also to strengthen the skill and insight necessary for the practical use of Fourier techniques.

The only formal prerequisite is an eight-semester-hour course (or its equivalent) in elementary calculus; however, Chapters 8 through 10 assume a basic familiarity with advanced calculus and applied mathematics.

The first three chapters deal with Fourier series and the concept of frequency spectra. Chapters 4 through 6 cover generalized functions, Fourier transforms, and generalized Fourier transforms. Chapter 7 deals with convolution and correlation. The remaining three chapters discuss the applications of Fourier analysis to signal theory, linear systems, and boundary-value problems.

The author is grateful to his daughter Diana for helping with the typing and to his wife Daisy, whose understanding and constant supportiveness were necessary factors to the completion of this book.

Hwei P. Hsu

CONTENTS

1 FOURIER SERIES

THIS CHAPTER IS ABOUT

☑ **Periodic Functions**
☑ **Fourier Series**
☑ **Properties of Sine and Cosine: Orthogonal Functions**
☑ **Evaluation of Fourier Coefficients**
☑ **Approximation by Finite Fourier Series**

1-1. Periodic Functions

A. Definition of periodic functions

A **periodic function** is any function for which

Periodic function
$$f(t) = f(t + T) \qquad \text{(1.1)}$$

for all t. The smallest constant T that satisfies (1.1) is called the **period** of the function. By iteration of (1.1), we obtain

$$f(t) = f(t + nT), \qquad n = 0, \pm 1, \pm 2, \ldots \qquad \text{(1.2)}$$

Figure 1-1 shows an example of a periodic function. Note that a constant is a periodic function of a period T for any value of T.

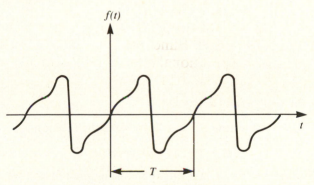

Figure 1-1   A periodic function.

EXAMPLE 1-1: Find the period of the function $f(t) = \cos 2t$.

Solution: From the trigonometric formula

$$\cos(\theta + 2\pi m) = \cos\theta \cos 2\pi m - \sin\theta \sin 2\pi m$$

we obtain $\cos(\theta + 2\pi m) = \cos\theta$ for any integer m since $\cos 2\pi m = 1$ and $\sin 2\pi m = 0$ for any integer m. Thus, we see from definition (1.1) that $\cos\theta$ is a periodic function with a period 2π since the smallest value of $2\pi m$ is 2π when $m = 1$. Now

$$\cos 2(t + T) = \cos 2t$$

or

$$\cos(2t + 2T) = \cos 2t$$

Since we know that $\cos(\theta + 2\pi m) = \cos\theta$ for any integer m, we have $2T = 2\pi m$. Therefore, $T = \pi m$. When $m = 1$, we obtain the smallest value of T. Hence, the period $T = \pi$.

In general, if the function

$$f(t) = \cos\omega_1 t + \cos\omega_2 t$$

is periodic with a period T, it must be possible to find two integers m and n such that

$$\omega_1 T = 2\pi m \quad \text{and} \quad \omega_2 T = 2\pi n$$

or

$$\omega_1/\omega_2 = m/n$$

That is, the ratio ω_1/ω_2 must be a rational number.

EXAMPLE 1-2: Find the period of the function $f(t) = \cos(t/3) + \cos(t/4)$.

Solution: From definition (1.1),

$$\cos\frac{1}{3}(t + T) + \cos\frac{1}{4}(t + T) = \cos\frac{t}{3} + \cos\frac{t}{4}$$

Since $\cos(\theta + 2\pi m) = \cos\theta$ for any integer m, we see that

$$\frac{1}{3}T = 2\pi m \quad \text{and} \quad \frac{1}{4}T = 2\pi n$$

where m and n are integers. Therefore, $T = 6\pi m = 8\pi n$. When $m = 4$ and $n = 3$, we obtain the smallest value of T. (You can do this by trial-and-error.) Hence, $T = 24\pi$.

B. Properties of periodic functions

If $f(t + T) = f(t)$, we have

**Properties of
periodic functions**

$$\int_\alpha^\beta f(t)\,dt = \int_{\alpha+T}^{\beta+T} f(t)\,dt \tag{1.3}$$

$$\int_0^T f(t)\,dt = \int_a^{a+T} f(t)\,dt \tag{1.4}$$

for any α, β, and a.

EXAMPLE 1-3: Verify the properties (1.3) and (1.4).

Solution: If $f(t + T) = f(t)$, then letting $t = \tau - T$, we have the relation

$$f(\tau - T + T) = f(\tau) = f(\tau - T)$$

Now, making the substitution $t = \tau - T$ in $\int_\alpha^\beta f(t)\,dt$ and using the relation we just obtained, we have

$$\int_\alpha^\beta f(t)\,dt = \int_{\alpha+T}^{\beta+T} f(\tau - T)\,d\tau = \int_{\alpha+T}^{\beta+T} f(\tau)\,d\tau = \int_{\alpha+T}^{\beta+T} f(t)\,dt$$

Next, the right-hand side of (1.4) can be written as

$$\int_a^{a+T} f(t)\,dt = \int_a^0 f(t)\,dt + \int_0^{a+T} f(t)\,dt$$

Now, by (1.3) we have

$$\int_a^0 f(t)\,dt = \int_{a+T}^T f(t)\,dt$$

Thus,

$$\int_a^{a+T} f(t)\,dt = \int_{a+T}^T f(t)\,dt + \int_0^{a+T} f(t)\,dt = \int_0^{a+T} f(t)\,dt + \int_{a+T}^T f(t)\,dt = \int_0^T f(t)\,dt$$

1-2. Fourier Series

Let the function $f(t)$ be periodic with period T. This function can then be represented by the trigonometric series

Trigonometric
Fourier series

$$f(t) = \frac{1}{2}a_0 + a_1\cos\omega_0 t + a_2\cos 2\omega_0 t + \cdots + b_1\sin\omega_0 t + b_2\sin 2\omega_0 t + \cdots$$

$$= \frac{1}{2}a_0 + \sum_{n=1}^{\infty}(a_n\cos n\omega_0 t + b_n\sin n\omega_0 t) \tag{1.5}$$

where $\omega_0 = 2\pi/T$.

A series such as (1.5) is called a **trigonometric Fourier series**, where a_n and b_n are the **Fourier coefficients** of $f(t)$. Series (1.5) can be rewritten as

Trigonometric
Fourier series
(harmonics form)

$$f(t) = C_0 + \sum_{n=1}^{\infty} C_n\cos(n\omega_0 t - \theta_n) \tag{1.6}$$

From (1.6), you can see that the Fourier series representation of a periodic function describes a periodic function as a *sum of sinusoidal components having different frequencies*. The sinusoidal component of frequency $\omega_n = n\omega_0$ is called the **nth harmonic** of the periodic function. The first harmonic $C_1\cos(\omega_0 t - \theta_1)$ is commonly called the **fundamental component** (because it has the same period as the function $f(t)$), so $\omega_0 = 2\pi f_0 = 2\pi/T$ is called the **fundamental angular frequency** and $f_0 = 1/T$ is the **fundamental frequency**. The coefficients C_n and the angles θ_n are the **harmonic amplitudes** and **phase angles**, respectively.

EXAMPLE 1-4: Derive the harmonics form (1.6) from the trigonometric Fourier series (1.5), and express the harmonic amplitudes C_n and the phase angles θ_n in terms of the Fourier coefficients a_n and b_n.

Solution: Multiplying and dividing by $\sqrt{a_n^2 + b_n^2}$, we can write

$$a_n\cos n\omega_0 t + b_n\sin n\omega_0 t = \sqrt{a_n^2 + b_n^2}\left(\frac{a_n}{\sqrt{a_n^2 + b_n^2}}\cos n\omega_0 t + \frac{b_n}{\sqrt{a_n^2 + b_n^2}}\sin n\omega_0 t\right)$$

Applying a trigonometric identity, we obtain

$$a_n\cos n\omega_0 t + b_n\sin n\omega_0 t = C_n(\cos\theta_n\cos n\omega_0 t + \sin\theta_n\sin n\omega_0 t)$$
$$= C_n\cos(n\omega_0 t - \theta_n)$$

where

$$C_n = \sqrt{a_n^2 + b_n^2}, \qquad \cos\theta_n = \frac{a_n}{\sqrt{a_n^2 + b_n^2}}, \qquad \sin\theta_n = \frac{b_n}{\sqrt{a_n^2 + b_n^2}}$$

Hence

$$\tan\theta_n = \frac{b_n}{a_n} \quad \text{or} \quad \theta_n = \tan^{-1}\left(\frac{b_n}{a_n}\right)$$

Also, letting $C_0 = \frac{1}{2}a_0$, we obtain

$$f(t) = \frac{1}{2}a_0 + \sum_{n=1}^{\infty}(a_n\cos n\omega_0 t + b_n\sin n\omega_0 t) = C_0 + \sum_{n=1}^{\infty} C_n\cos(n\omega_0 t - \theta_n)$$

1-3. Properties of Sine and Cosine: Orthogonal Functions

A. Definition of orthogonality

A set of functions $\{\phi_k(t)\}$ is **orthogonal** on an interval $a < t < b$ if, for any two functions $\phi_m(t)$ and $\phi_n(t)$ in the set $\{\phi_k(t)\}$, the relation (1.7) holds.

Orthogonality relation

$$\int_a^b \phi_m(t)\phi_n(t)\,dt = \begin{cases} 0 & \text{for } m \neq n \\ r_n & \text{for } m = n \end{cases} \tag{1.7}$$

B. Orthogonality relations for sine and cosine functions

Let's consider a set of sinusoidal functions, for example. Using elementary calculus, we can show that

$$\int_{-T/2}^{T/2} \cos(m\omega_0 t)\, dt = 0 \qquad \text{for } m \neq 0 \tag{1.8a}$$

Orthogonality relations for sine and cosine functions

$$\int_{-T/2}^{T/2} \sin(m\omega_0 t)\, dt = 0 \qquad \text{for all } m \tag{1.8b}$$

$$\int_{-T/2}^{T/2} \cos(m\omega_0 t)\cos(n\omega_0 t)\, dt = \begin{cases} 0, & m \neq n \\ T/2, & m = n \neq 0 \end{cases} \tag{1.8c}$$

$$\int_{-T/2}^{T/2} \sin(m\omega_0 t)\sin(n\omega_0 t)\, dt = \begin{cases} 0, & m \neq n \\ T/2, & m = n \neq 0 \end{cases} \tag{1.8d}$$

$$\int_{-T/2}^{T/2} \sin(m\omega_0 t)\cos(n\omega_0 t)\, dt = 0 \qquad \text{for all } m \text{ and } n \tag{1.8e}$$

where $\omega_0 = 2\pi/T$.

These relations (1.8a–e) show that the functions $\{1, \cos \omega_0 t, \cos 2\omega_0 t, ..., \cos n\omega_0 t, ..., \sin \omega_0 t, \sin 2\omega_0 t, ..., \sin n\omega_0 t, ...\}$ form an orthogonal set of functions on an interval $-T/2 < t < T/2$.

EXAMPLE 1-5: Verify the orthogonality relation (1.8c).

Solution: In view of the trigonometric identity

$$\cos A \cos B = \frac{1}{2}[\cos(A + B) + \cos(A - B)]$$

and

$$\omega_0 t \Big|_{t = \pm T/2} = \frac{2\pi}{T}\left(\pm \frac{T}{2}\right) = \pm \pi$$

we have

$$\int_{-T/2}^{T/2} \cos(m\omega_0 t)\cos(n\omega_0 t)\, dt = \frac{1}{2}\int_{-T/2}^{T/2} \{\cos[(m + n)\omega_0 t] + \cos[(m - n)\omega_0 t]\}\, dt$$

$$= \frac{1}{2}\frac{1}{(m + n)\omega_0}\sin[(m + n)\omega_0 t]\Big|_{-T/2}^{T/2}$$

$$+ \frac{1}{2}\frac{1}{(m - n)\omega_0}\sin[(m - n)\omega_0 t]\Big|_{-T/2}^{T/2}$$

$$= \frac{1}{2}\frac{1}{(m + n)\omega_0}\{\sin[(m + n)\pi] + \sin[(m + n)\pi]\}$$

$$+ \frac{1}{2}\frac{1}{(m - n)\omega_0}\{\sin[(m - n)\pi] + \sin[(m - n)\pi]\}$$

$$= 0 \qquad \text{if } m \neq n$$

since $\sin k\pi = 0$ for any k.

If $m = n \neq 0$, by using the trigonometric identity $\cos^2\theta = \frac{1}{2}(1 + \cos 2\theta)$, we have

$$\int_{-T/2}^{T/2} \cos(m\omega_0 t)\cos(n\omega_0 t)\, dt = \int_{-T/2}^{T/2} \cos^2(m\omega_0 t)\, dt$$

$$= \frac{1}{2}\int_{-T/2}^{T/2} [1 + \cos 2m\omega_0 t]\, dt$$

$$= \frac{1}{2}t\Big|_{-T/2}^{T/2} + \frac{1}{4m\omega_0}\sin 2m\omega_0 t\Big|_{-T/2}^{T/2}$$

$$= \frac{T}{2}$$

1-4. Evaluation of Fourier Coefficients

Using the orthogonality relations (1.8a–e), we can now evaluate the coefficients a_n and b_n of the Fourier series $f(t) = \frac{1}{2}a_0 + \sum_{n=1}^{\infty}(a_n\cos n\omega_0 t + b_n\sin n\omega_0 t)$ (1.5):

$$a_n = \frac{2}{T}\int_{-T/2}^{T/2} f(t)\cos(n\omega_0 t)\,dt, \qquad n = 0, 1, 2,\ldots \tag{1.9}$$

Fourier coefficients

$$b_n = \frac{2}{T}\int_{-T/2}^{T/2} f(t)\sin(n\omega_0 t)\,dt, \qquad n = 1, 2,\ldots \tag{1.10}$$

$$\frac{1}{2}a_0 = \frac{1}{T}\int_{-T/2}^{T/2} f(t)\,dt \tag{1.11}$$

EXAMPLE 1-6: Verify (1.9)–(1.11).

Solution: Multiplying both sides of (1.5) by $\cos m\omega_0 t$ and integrating over $(-T/2, T/2)$, we get

$$\int_{-T/2}^{T/2} f(t)\cos(m\omega_0 t)\,dt = \frac{1}{2}a_0\int_{-T/2}^{T/2}\cos(m\omega_0 t)\,dt$$
$$+ \int_{-T/2}^{T/2}\left[\sum_{n=1}^{\infty} a_n\cos(n\omega_0 t)\right]\cos(m\omega_0 t)\,dt$$
$$+ \int_{-T/2}^{T/2}\left[\sum_{n=1}^{\infty} b_n\sin(n\omega_0 t)\right]\cos(m\omega_0 t)\,dt$$

Interchanging the order of integration and summation, we obtain

$$\int_{-T/2}^{T/2} f(t)\cos(m\omega_0 t)\,dt = \frac{1}{2}a_0\int_{-T/2}^{T/2}\cos(m\omega_0 t)\,dt$$
$$+ \sum_{n=1}^{\infty} a_n\int_{-T/2}^{T/2}\cos(n\omega_0 t)\cos(m\omega_0 t)\,dt$$
$$+ \sum_{n=1}^{\infty} b_n\int_{-T/2}^{T/2}\sin(n\omega_0 t)\cos(m\omega_0 t)\,dt$$

In view of the orthogonality relations (1.8a, c, and e), we see that

$$\int_{-T/2}^{T/2} f(t)\cos(m\omega_0 t)\,dt = \frac{T}{2}a_m$$

Hence,

$$a_m = \frac{2}{T}\int_{-T/2}^{T/2} f(t)\cos(m\omega_0 t)\,dt \tag{1.12}$$

Similarly, if (1.5) is multiplied by $\sin m\omega_0 t$ and integrated term-by-term over $(-T/2, T/2)$, we get

$$\int_{-T/2}^{T/2} f(t)\sin(m\omega_0 t)\,dt = \frac{1}{2}a_0\int_{-T/2}^{T/2}\sin(m\omega_0 t)\,dt$$
$$+ \sum_{n=1}^{\infty} a_n\int_{-T/2}^{T/2}\cos(n\omega_0 t)\sin(m\omega_0 t)\,dt$$
$$+ \sum_{n=1}^{\infty} b_n\int_{-T/2}^{T/2}\sin(n\omega_0 t)\sin(m\omega_0 t)\,dt$$

Here the orthogonality relations (1.8b, d, and e) yield

$$\int_{-T/2}^{T/2} f(t)\sin(m\omega_0 t)\,dt = \frac{T}{2}b_m$$

Hence,

$$b_m = \frac{2}{T}\int_{-T/2}^{T/2} f(t)\sin(m\omega_0 t)\,dt \tag{1.13}$$

If we integrate (1.5) over $(-T/2, T/2)$ and use (1.8a and b), we obtain

$$\int_{-T/2}^{T/2} f(t)\,dt = \frac{1}{2} a_0 \int_{-T/2}^{T/2} dt + \int_{-T/2}^{T/2} \left[\sum_{n=1}^{\infty} (a_n \cos n\omega_0 t + b_n \sin n\omega_0 t) \right] dt$$

$$= \frac{1}{2} a_0 T + \sum_{n=1}^{\infty} a_n \int_{-T/2}^{T/2} \cos(n\omega_0 t)\,dt + \sum_{n=1}^{\infty} b_n \int_{-T/2}^{T/2} \sin(n\omega_0 t)\,dt$$

$$= \frac{1}{2} a_0 T$$

Hence,

$$\frac{1}{2} a_0 = \frac{1}{T} \int_{-T/2}^{T/2} f(t)\,dt$$

or

$$a_0 = \frac{2}{T} \int_{-T/2}^{T/2} f(t)\,dt \tag{1.14}$$

Note that $a_0/2$ is the average value of $f(t)$ over a period.

Equation (1.14) indicates that (1.9), which evaluates the coefficients of the cosine series, also gives the coefficient a_0 correctly since $\cos m\omega_0 t |_{m=0} = 1$.

Replacing m by n, we can rewrite (1.12) and (1.13) as

$$a_n = \frac{2}{T} \int_{-T/2}^{T/2} f(t)\cos(n\omega_0 t)\,dt, \qquad n = 0, 1, 2, \ldots$$

and

$$b_n = \frac{2}{T} \int_{-T/2}^{T/2} f(t)\sin(n\omega_0 t)\,dt, \qquad n = 1, 2, \ldots$$

In general, there is no necessity that the interval of integration of (1.9)–(1.11) be symmetric about the origin. In view of (1.4), the only requirement is that the integral be taken over a complete period.

EXAMPLE 1-7: Find the Fourier series for the function $f(t)$ defined by

$$f(t) = \begin{cases} -1, & -T/2 < t < 0 \\ 1, & 0 < t < T/2 \end{cases}$$

and $f(t + T) = f(t)$, as shown in Figure 1-2.

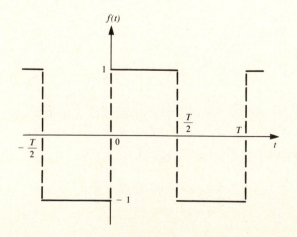

Figure 1-2

Solution: From (1.9) and

$$\omega_0 t \Big|_{t = \pm T/2} = \frac{2\pi}{T} \left(\pm \frac{T}{2} \right) = \pm \pi$$

we obtain

$$a_n = \frac{2}{T} \int_{-T/2}^{T/2} f(t) \cos(n\omega_0 t)\, dt$$

$$= \frac{2}{T} \left[\int_{-T/2}^{0} -\cos(n\omega_0 t)\, dt + \int_{0}^{T/2} \cos(n\omega_0 t)\, dt \right]$$

$$= \frac{2}{T} \left(\frac{-1}{n\omega_0} \sin n\omega_0 t \Big|_{-T/2}^{0} + \frac{1}{n\omega_0} \sin n\omega_0 t \Big|_{0}^{T/2} \right)$$

$$= \frac{2}{T} \left\{ \frac{-1}{n\omega_0} [\sin 0 - \sin(-n\pi)] + \frac{1}{n\omega_0} [\sin(n\pi) - \sin 0] \right\}$$

$$= 0 \quad \text{for } n \neq 0$$

since $\sin 0 = \sin(n\pi) = 0$.

For $n = 0$, from (1.11),

$$\frac{1}{2} a_0 = \frac{1}{T} \int_{-T/2}^{T/2} f(t)\, dt = 0$$

since the average value of $f(t)$ over a period is zero.

From (1.10) and $\omega_0 T = (2\pi/T)T = 2\pi$,

$$b_n = \frac{2}{T} \int_{-T/2}^{T/2} f(t) \sin(n\omega_0 t)\, dt$$

$$= \frac{2}{T} \left[\int_{-T/2}^{0} -\sin(n\omega_0 t)\, dt + \int_{0}^{T/2} \sin(n\omega_0 t)\, dt \right]$$

$$= \frac{2}{T} \left[\frac{1}{n\omega_0} \cos(n\omega_0 t) \Big|_{-T/2}^{0} + \frac{-1}{n\omega_0} \cos(n\omega_0 t) \Big|_{0}^{T/2} \right]$$

$$= \frac{2}{n\omega_0 T} \{ [1 - \cos(-n\pi)] - [\cos(n\pi) - 1] \}$$

$$= \frac{2}{n\pi} (1 - \cos n\pi)$$

Since $\cos n\pi = (-1)^n$,

$$b_n = \begin{cases} 0, & n \text{ even} \\ \dfrac{4}{n\pi}, & n \text{ odd} \end{cases}$$

Hence,

$$f(t) = \frac{4}{\pi} \sum_{n=\text{odd}}^{\infty} \frac{1}{n} \sin n\omega_0 t$$

$$= \frac{4}{\pi} \left(\sin \omega_0 t + \frac{1}{3} \sin 3\omega_0 t + \frac{1}{5} \sin 5\omega_0 t + \cdots \right)$$

EXAMPLE 1-8: Find the Fourier series for the function $f(t)$ defined by

$$f(t) = \begin{cases} 0, & -\pi < t < 0 \\ \dfrac{1}{\pi} t, & 0 < t < \pi \end{cases}$$

and $f(t + 2\pi) = f(t)$, as shown in Figure 1-3.

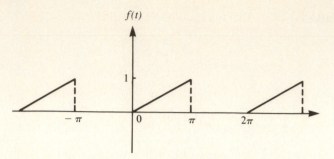

Figure 1-3

Solution: If $T = 2\pi$ and $\omega_0 = 2\pi/T = 1$, from (1.9),

$$a_n = \frac{1}{\pi} \int_{-\pi}^{\pi} f(t)\cos(nt)\, dt$$

$$= \frac{1}{\pi} \int_{0}^{\pi} \frac{1}{\pi} t \cos(nt)\, dt$$

$$= \frac{1}{\pi^2} \left[\frac{t}{n} \sin(nt) \Big|_{0}^{\pi} - \frac{1}{n} \int_{0}^{\pi} \sin(nt)\, dt \right]$$

$$= \frac{1}{\pi^2 n^2} \left[\cos(nt) \Big|_{0}^{\pi} \right]$$

$$= \frac{1}{\pi^2 n^2} (\cos n\pi - 1) = \frac{1}{\pi^2 n^2} [(-1)^n - 1]$$

so that

$$a_n = \begin{cases} 0, & n \text{ even} \\ -\dfrac{2}{\pi^2 n^2}, & n \text{ odd} \end{cases}$$

From (1.10),

$$b_n = \frac{1}{\pi} \int_{-\pi}^{\pi} f(t)\sin(nt)\, dt$$

$$= \frac{1}{\pi} \int_{0}^{\pi} \frac{1}{\pi} t \sin(nt)\, dt$$

$$= \frac{1}{\pi^2} \left[-\frac{t}{n} \cos(nt) \Big|_{0}^{\pi} + \frac{1}{n} \int_{0}^{\pi} \cos(nt)\, dt \right]$$

$$= -\frac{1}{n\pi} \cos(n\pi) = -\frac{1}{n\pi} (-1)^n$$

From (1.11),

$$\frac{1}{2} a_0 = \frac{1}{2\pi} \int_{-\pi}^{\pi} f(t)\, dt = \frac{1}{2\pi} \int_{0}^{\pi} \frac{1}{\pi} t\, dt = \frac{1}{4}$$

Hence,

$$f(t) = \frac{1}{4} - \frac{2}{\pi^2} \sum_{n=\text{odd}}^{\infty} \frac{1}{n^2} \cos nt - \frac{1}{\pi} \sum_{n=1}^{\infty} \frac{(-1)^n}{n} \sin nt$$

$$= \frac{1}{4} - \frac{2}{\pi^2} \left(\cos t + \frac{1}{9} \cos 3t + \frac{1}{25} \cos 5t + \cdots \right)$$

$$- \frac{1}{\pi} \left(-\sin t + \frac{1}{2} \sin 2t - \frac{1}{3} \sin 3t + \cdots \right)$$

1-5. Approximation by Finite Fourier Series

A. Mean-square error

Let

Finite Fourier series
$$S_k(t) = \frac{a_0}{2} + \sum_{n=1}^{k} (a_n \cos n\omega_0 t + b_n \sin n\omega_0 t) \tag{1.15}$$

be the sum of the first $(2k + 1)$ terms of a Fourier series that represents $f(t)$ on $-T/2 < t < T/2$. Then

$$f(t) = \frac{a_0}{2} + \sum_{n=1}^{k} (a_n \cos n\omega_0 t + b_n \sin n\omega_0 t) + \varepsilon_k(t) \tag{1.16}$$

and

Error
$$\varepsilon_k(t) = f(t) - S_k(t) \tag{1.17}$$

where $\varepsilon_k(t)$ is the difference or **error** between $f(t)$ and its approximation. Then the **mean-square error** E_k is defined by

Mean-square error
$$E_k = \frac{1}{T} \int_{-T/2}^{T/2} [\varepsilon_k(t)]^2 \, dt = \frac{1}{T} \int_{-T/2}^{T/2} [f(t) - S_k(t)]^2 \, dt \tag{1.18}$$

and E_k is given by

$$E_k = \frac{1}{T} \int_{-T/2}^{T/2} [f(t)]^2 \, dt - \frac{a_0^2}{4} - \frac{1}{2} \sum_{n=1}^{k} (a_n^2 + b_n^2) \tag{1.19}$$

EXAMPLE 1-9: If we approximate a function $f(t)$ by the finite Fourier series $S_k(t)$, show that this approximation has the least mean-square error property.

Solution: Substituting (1.15) into (1.18),

$$E_k = \frac{1}{T} \int_{-T/2}^{T/2} \left[f(t) - \frac{a_0}{2} - \sum_{n=1}^{k} (a_n \cos n\omega_0 t + b_n \sin n\omega_0 t) \right]^2 \, dt \tag{1.20}$$

Consider E_k as a function of a_0, a_n, and b_n. Then, in order to have a minimum for the mean-square error E_k, its partial derivatives with respect to a_0, a_n, and b_n must be zero; that is,

$$\frac{\partial E_k}{\partial a_0} = 0, \qquad \frac{\partial E_k}{\partial a_n} = 0, \qquad \frac{\partial E_k}{\partial b_n} = 0 \qquad (n = 1, 2, \dots)$$

Interchanging the order of differentiation and integration,

$$\frac{\partial E_k}{\partial a_0} = -\frac{1}{T} \int_{-T/2}^{T/2} \left[f(t) - \frac{a_0}{2} - \sum_{n=1}^{k} (a_n \cos n\omega_0 t + b_n \sin n\omega_0 t) \right] \, dt \tag{1.21}$$

$$\frac{\partial E_k}{\partial a_n} = -\frac{2}{T} \int_{-T/2}^{T/2} \left[f(t) - \frac{a_0}{2} - \sum_{n=1}^{k} (a_n \cos n\omega_0 t + b_n \sin n\omega_0 t) \right] \cos(n\omega_0 t) \, dt \tag{1.22}$$

$$\frac{\partial E_k}{\partial b_n} = -\frac{2}{T} \int_{-T/2}^{T/2} \left[f(t) - \frac{a_0}{2} - \sum_{n=1}^{k} (a_n \cos n\omega_0 t + b_n \sin n\omega_0 t) \right] \sin(n\omega_0 t) \, dt \tag{1.23}$$

Using the orthogonality properties (1.8) and in view of (1.9)–(1.11), the integrals (1.21)–(1.23) reduce to

$$\frac{\partial E_k}{\partial a_0} = \frac{a_0}{2} - \frac{1}{T} \int_{-T/2}^{T/2} f(t) \, dt = 0$$

$$\frac{\partial E_k}{\partial a_n} = a_n - \frac{2}{T} \int_{-T/2}^{T/2} f(t) \cos(n\omega_0 t) \, dt = 0$$

$$\frac{\partial E_k}{\partial b_n} = b_n - \frac{2}{T} \int_{-T/2}^{T/2} f(t) \sin(n\omega_0 t) \, dt = 0$$

EXAMPLE 1-10: Verify (1.19).

Solution: From (1.18),

$$E_k = \frac{1}{T} \int_{-T/2}^{T/2} [f(t) - S_k(t)]^2 \, dt$$

$$= \frac{1}{T} \int_{-T/2}^{T/2} \{[f(t)]^2 - 2f(t)S_k(t) + [S_k(t)]^2\} \, dt$$

$$= \frac{1}{T} \int_{-T/2}^{T/2} [f(t)]^2 \, dt - \frac{2}{T} \int_{-T/2}^{T/2} f(t)S_k(t) \, dt + \frac{1}{T} \int_{-T/2}^{T/2} [S_k(t)]^2 \, dt \qquad \textbf{(1.24)}$$

Now

$$\frac{2}{T} \int_{-T/2}^{T/2} f(t)S_k(t) \, dt = \frac{2}{T} \frac{a_0}{2} \int_{-T/2}^{T/2} f(t) \, dt + \frac{2}{T} \sum_{n=1}^{k} a_n \int_{-T/2}^{T/2} f(t)\cos(n\omega_0 t) \, dt$$

$$+ \frac{2}{T} \sum_{n=1}^{k} b_n \int_{-T/2}^{T/2} f(t)\sin(n\omega_0 t) \, dt$$

In view of (1.9) and (1.10),

$$\frac{2}{T} \int_{-T/2}^{T/2} f(t)S_k(t) \, dt = \frac{a_0^2}{2} + \sum_{n=1}^{k} (a_n^2 + b_n^2) \qquad \textbf{(1.25)}$$

Using the orthogonality relations (1.8),

$$\frac{1}{T} \int_{-T/2}^{T/2} [S_k(t)]^2 \, dt = \frac{1}{T} \int_{-T/2}^{T/2} \left[\frac{a_0}{2} + \sum_{n=1}^{k} (a_n\cos n\omega_0 t + b_n\sin n\omega_0 t) \right]^2 dt$$

$$= \frac{a_0^2}{4} + \frac{1}{2} \sum_{n=1}^{k} (a_n^2 + b_n^2) \qquad \textbf{(1.26)}$$

Substituting (1.25) and (1.26) into (1.24),

$$E_k = \frac{1}{T} \int_{-T/2}^{T/2} [f(t)]^2 \, dt - \frac{a_0^2}{2} - \sum_{n=1}^{k} (a_n^2 + b_n^2) + \frac{a_0^2}{4} + \frac{1}{2} \sum_{n=1}^{k} (a_n^2 + b_n^2)$$

$$= \frac{1}{T} \int_{-T/2}^{T/2} [f(t)]^2 \, dt - \frac{a_0^2}{4} - \frac{1}{2} \sum_{n=1}^{k} (a_n^2 + b_n^2)$$

B. Parseval's identity

If a_0, a_n, and b_n are the coefficients in the Fourier expansion of a periodic function $f(t)$ with period T, then

Parseval's identity
$$\frac{1}{T} \int_{-T/2}^{T/2} [f(t)]^2 \, dt = \frac{a_0^2}{4} + \frac{1}{2} \sum_{n=1}^{\infty} (a_n^2 + b_n^2) \qquad \textbf{(1.27)}$$

which is known as **Parseval's identity**.

EXAMPLE 1-11: From definition (1.18) the mean-square error E_k is nonnegative, and from (1.19) we obtain

$$E_{k+1} = E_k - \frac{1}{2}(a_{k+1}^2 + b_{k+1}^2) \qquad \textbf{(1.28)}$$

Thus, the sequence $\{E_k\}$ contains only nonnegative terms and is nonincreasing. The sequence therefore converges. Also from (1.17),

$$\lim_{k \to \infty} \varepsilon_k(t) = f(t) - \lim_{k \to \infty} S_k(t) = 0$$

Hence,

$$\lim_{k \to \infty} E_k = 0$$

Consequently, from (1.19) we conclude that

$$\frac{1}{T} \int_{-T/2}^{T/2} [f(t)]^2 \, dt = \frac{a_0^2}{4} + \frac{1}{2} \sum_{n=1}^{\infty} (a_n^2 + b_n^2)$$

SUMMARY

1. A function $f(t)$ satisfying $f(t + T) = f(t)$ is called a periodic function, and the smallest T is called its period.
2. A periodic function $f(t)$ can be represented by the trigonometric Fourier series

$$f(t) = \frac{1}{2}a_0 + \sum_{n=1}^{\infty}(a_n\cos n\omega_0 t + b_n\sin n\omega_0 t)$$

or

$$f(t) = C_0 + \sum_{n=1}^{\infty}C_n\cos(n\omega_0 t - \theta_n)$$

where $\omega_0 = 2\pi/T$.
3. The coefficients of the Fourier series are found by using the orthogonality properties of sine and cosine functions over a period:

$$\left\{\begin{matrix}a_n\\b_n\end{matrix}\right\} = \frac{2}{T}\int_{-T/2}^{T/2}f(t)\left\{\begin{matrix}\cos n\omega_0 t\\\sin n\omega_0 t\end{matrix}\right\}dt$$

$$C_0 = \frac{1}{2}a_0, \qquad C_n = \sqrt{a_n^2 + b_n^2}, \qquad \theta_n = \tan^{-1}\left(\frac{b_n}{a_n}\right)$$

4. A set of functions $\{\phi_k(t)\}$ is called an orthogonal set on an interval $a < t < b$ if, for any two functions $\phi_m(t)$ and $\phi_n(t)$ in the set, the relation

$$\int_a^b \phi_m(t)\,\phi_n(t)\,dt = \begin{cases}0 & \text{for } m \neq n\\r_n & \text{for } m = n\end{cases}$$

holds. The functions $\{1, \cos\omega_0 t, \cos 2\omega_0 t, \ldots, \cos n\omega_0 t, \ldots, \sin\omega_0 t, \sin 2\omega_0 t, \ldots, \sin n\omega_0 t, \ldots\}$ form an orthogonal set of functions on an interval $-T/2 < t < T/2$.
5. If $f(t)$ is approximated by the finite Fourier series $S_k(t)$, this approximation has the least mean-square error property.

RAISE YOUR GRADES

Can you explain...?

☑ what a periodic function is and how you find its period
☑ what the orthogonality properties of sine and cosine functions over a period are
☑ why the constant term of the Fourier series expansion of $f(t)$ is equal to the average values of $f(t)$ over a period
☑ why the interval of integration need not be symmetric about the origin in the evaluation of the coefficients of the Fourier series, and why the only requirement is that the integral be taken over a complete period

SOLVED PROBLEMS

Periodic Functions

PROBLEM 1-1 Find the period of the function $f(t) = (10\cos t)^2$.

Solution: Using the trigonometric identity $\cos^2\theta = \frac{1}{2}(1 + \cos 2\theta)$, we see that

$$f(t) = (10\cos t)^2 = 100\cos^2 t = 100\tfrac{1}{2}(1 + \cos 2t) = 50 + 50\cos 2t$$

Since a constant is a periodic function of period T for any value of T and the period of $\cos 2t$ is π, we conclude that the period of $f(t)$ is π.

PROBLEM 1-2   Is the function $f(t) = \cos 10t + \cos(10 + \pi)t$ periodic?

Solution: Here $\omega_1 = 10$ and $\omega_2 = 10 + \pi$. Since $\omega_1/\omega_2 = 10/(10 + \pi)$ is not a rational number, it is impossible to find a value T for which $f(t) = f(t + T)$ (1.1) is satisfied. Hence, $f(t)$ is not periodic.

PROBLEM 1-3   Find the period of the function $f(t) = |\sin \omega_1 t|$ and its fundamental angular frequency, where $\omega_1 = 2\pi/T_1$ and T_1 is the period of $\sin \omega_1 t$.

Solution: $\sin \omega_1 t$ and $|\sin \omega_1 t|$ are plotted, respectively, in Figure 1-4a and b. From Figure 1-4b, we can see that the period of $f(t) = |\sin \omega_1 t|$ is $T_1/2$ and its fundamental angular frequency is $2\pi/(T_1/2) = 4\pi/T_1 = 2\omega_1$.

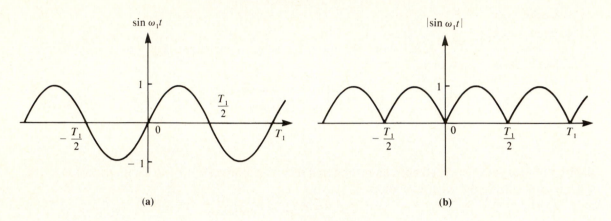

(a)   (b)

Figure 1-4

PROBLEM 1-4   Let $f(t + T) = f(t)$ and $F(t) = \int_0^t f(\tau)\,d\tau - \frac{1}{2}a_0 t$, where $a_0 = (2/T)\int_{-T/2}^{T/2} f(t)\,dt$. Show that $F(t + T) = F(t)$.

Solution: Since

$$F(t) = \int_0^t f(\tau)\,d\tau - \frac{1}{2}a_0 t$$

then

$$F(t + T) = \int_0^{t+T} f(\tau)\,d\tau - \frac{1}{2}a_0(t + T) = \int_0^T f(\tau)\,d\tau + \int_T^{T+t} f(\tau)\,d\tau - \frac{1}{2}a_0 t - \frac{1}{2}a_0 T$$

From $\int_\alpha^\beta f(t)\,dt = \int_{\alpha+T}^{\beta+T} f(t)\,dt$ (1.3) and $\int_0^T f(t)\,dt = \int_a^{a+T} f(t)\,dt$ (1.4),

$$\int_0^T f(\tau)\,d\tau = \int_{-T/2}^{T/2} f(\tau)\,d\tau = \frac{1}{2}a_0 T$$

$$\int_T^{T+t} f(\tau)\,d\tau = \int_0^t f(\tau)\,d\tau$$

Hence,

$$F(t + T) = \frac{1}{2}a_0 T + \int_0^t f(\tau)\,d\tau - \frac{1}{2}a_0 t - \frac{1}{2}a_0 T$$

$$= \int_0^t f(\tau)\,d\tau - \frac{1}{2}a_0 t = F(t)$$

PROBLEM 1-5   Let $f(t + T) = f(t)$ and $g(t) = \int_0^t f(\tau)d\tau$. Show that

$$g(t + T) = g(t)$$

if $\int_{-T/2}^{T/2} f(t)\,dt = 0$.

Solution: Since $g(t) = \int_0^t f(\tau)\,d\tau$,

$$g(t + T) = \int_0^{t+T} f(\tau)\,d\tau = \int_0^T f(\tau)\,d\tau + \int_T^{T+t} f(\tau)\,d\tau$$

From (1.3) and (1.4),

$$\int_0^T f(\tau)\,d\tau = \int_{-T/2}^{T/2} f(\tau)\,d\tau = \int_{-T/2}^{T/2} f(t)\,dt \qquad \int_T^{T+t} f(t)\,dt = \int_0^t f(t)\,dt$$

Hence,

$$g(t+T) = \int_{-T/2}^{T/2} f(t)\,dt + \int_0^t f(t)\,dt$$

and $g(t+T) = g(t)$ if $\int_{-T/2}^{T/2} f(t)\,dt = 0$.

Orthogonal functions

PROBLEM 1-6 Verify $\int_{-T/2}^{T/2} \sin(m\omega_0 t)\cos(n\omega_0 t)\,dt = 0$ for all m and n (1.8e).

Solution: With the trigonometric identity $\sin A \cos B = \frac{1}{2}[\sin(A+B) + \sin(A-B)]$, we have

$$\int_{-T/2}^{T/2} \sin(m\omega_0 t)\cos(n\omega_0 t)\,dt = \frac{1}{2}\int_{-T/2}^{T/2} \{\sin[(m+n)\omega_0 t] + \sin[(m-n)\omega_0 t]\}\,dt$$

$$= \frac{1}{2}\frac{-1}{(m+n)\omega_0}\cos[(m+n)\omega_0 t]\Big|_{-T/2}^{T/2} + \frac{1}{2}\frac{-1}{(m-n)\omega_0}\cos[(m-n)\omega_0 t]\Big|_{-T/2}^{T/2}$$

$$= 0 \quad \text{if} \quad m \neq n$$

If $m = n \neq 0$, by using the trigonometric identity $\sin 2\theta = 2\sin\theta\cos\theta$, we get

$$\int_{-T/2}^{T/2} \sin(m\omega_0 t)\cos(n\omega_0 t)\,dt = \int_{-T/2}^{T/2} \sin(m\omega_0 t)\cos(m\omega_0 t)\,dt$$

$$= \frac{1}{2}\int_{-T/2}^{T/2} \sin(2m\omega_0 t)\,dt$$

$$= -\frac{1}{4m\omega_0}\cos(2m\omega_0 t)\Big|_{-T/2}^{T/2}$$

$$= 0$$

Certainly for $m = n = 0$, the integral equals zero.

PROBLEM 1-7 An infinite set of real functions $\{\phi_n(t)\}$, where $n = 1, 2, \ldots$, is said to be an **orthonormal** set on the interval (a, b) if $\int_a^b \phi_n(t)\phi_m(t)\,dt = \delta_{mn}$, where δ_{mn} is Kronecker's delta, defined by

$$\delta_{mn} = \begin{cases} 1, & m = n \\ 0, & m \neq n \end{cases}$$

Let $f(t)$ be a function defined on the interval (a, b) and suppose that $f(t)$ can be represented as

$$f(t) = c_1\phi_1(t) + c_2\phi_2(t) + \cdots + c_n\phi_n(t) + \cdots = \sum_{n=1}^{\infty} c_n\phi_n(t)$$

on (a, b) everywhere, where c_n are constants. Show that

$$c_n = \int_a^b f(t)\phi_n(t)\,dt, \qquad n = 1, 2, \ldots$$

Solution: Multiplying $f(t)$ by $\phi_m(t)$ and integrating over the interval (a, b), we obtain

$$\int_a^b f(t)\phi_m(t)\,dt = \int_a^b \left[\sum_{n=1}^{\infty} c_n\phi_n(t)\right]\phi_m(t)\,dt$$

$$= \sum_{n=1}^{\infty} c_n \int_a^b \phi_n(t)\phi_m(t)\,dt$$

$$= \sum_{n=1}^{\infty} c_n\delta_{nm} = c_m$$

Changing m to n, we obtain

$$c_n = \int_a^b f(t)\phi_n(t)\,dt$$

Evaluation of Fourier Coefficients

PROBLEM 1-8 Find the Fourier series for the function whose waveform is shown in Figure 1-5.

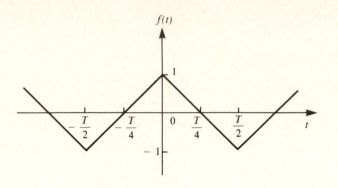

Figure 1-5

Solution: The function $f(t)$ can be expressed analytically as

$$f(t) = \begin{cases} 1 + \dfrac{4t}{T}, & -\dfrac{T}{2} \leqslant t < 0 \\[2ex] 1 - \dfrac{4t}{T}, & 0 \leqslant t < \dfrac{T}{2} \end{cases}$$

Since the average value of $f(t)$ over a period is zero,

$$\frac{1}{2} a_0 = \frac{1}{T} \int_{-T/2}^{T/2} f(t)\,dt = 0$$

From (1.9),

$$a_n = \frac{2}{T} \int_{-T/2}^{T/2} f(t)\cos(n\omega_0 t)\,dt$$

$$= \frac{2}{T} \int_{-T/2}^{T/2} \cos(n\omega_0 t)\,dt + \frac{2}{T} \int_{-T/2}^{0} \frac{4}{T} t \cos(n\omega_0 t)\,dt + \frac{2}{T} \int_{0}^{T/2} -\frac{4}{T} t \cos(n\omega_0 t)\,dt$$

The first integral on the right-hand side equals zero.

Letting $t = -\tau$ in the second integral, we get

$$a_n = \frac{8}{T^2} \int_{T/2}^{0} (-\tau)\cos[n\omega_0(-\tau)](-d\tau) - \frac{8}{T^2} \int_{0}^{T/2} t \cos(n\omega_0 t)\,dt$$

$$= \frac{8}{T^2} \int_{T/2}^{0} \tau \cos(n\omega_0 \tau)\,d\tau - \frac{8}{T^2} \int_{0}^{T/2} t \cos(n\omega_0 t)\,dt$$

$$= -\frac{8}{T^2} \int_{0}^{T/2} \tau \cos(n\omega_0 \tau)\,d\tau - \frac{8}{T^2} \int_{0}^{T/2} t \cos(n\omega_0 t)\,dt$$

$$= -\frac{16}{T^2} \int_{0}^{T/2} t \cos(n\omega_0 t)\,dt$$

Now integrating by parts,

$$\int_{0}^{T/2} t \cos(n\omega_0 t)\,dt = \frac{1}{n\omega_0} t \sin(n\omega_0 t)\Big|_{0}^{T/2} - \frac{1}{n\omega_0} \int_{0}^{T/2} \sin(n\omega_0 t)\,dt$$

$$= \frac{1}{(n\omega_0)^2} \cos(n\omega_0 t)\Big|_{0}^{T/2}$$

$$= \frac{1}{(n2\pi/T)^2} (\cos n\pi - 1)$$

Hence,

$$a_n = -\frac{16}{T^2}\frac{1}{(n2\pi/T)^2}(\cos n\pi - 1)$$

$$= \frac{4}{n^2\pi^2}(1 - \cos n\pi)$$

Since $\cos n\pi = (-1)^n$,

$$a_n = \begin{cases} 0, & n \text{ even} \\ \dfrac{8}{n^2\pi^2}, & n \text{ odd} \end{cases}$$

Similarly, from (1.10),

$$b_n = \frac{2}{T}\int_{-T/2}^{T/2} f(t)\sin(n\omega_0 t)\,dt$$

$$= \frac{2}{T}\int_{-T/2}^{T/2}\sin(n\omega_0 t)\,dt + \frac{2}{T}\int_{-T/2}^{0}\frac{4}{T}t\sin(n\omega_0 t)\,dt + \frac{2}{T}\int_{0}^{T/2}-\frac{4}{T}t\sin(n\omega_0 t)\,dt$$

$$= \frac{8}{T^2}\int_{T/2}^{0}(-\tau)\sin[n\omega_0(-\tau)](-d\tau) - \frac{8}{T^2}\int_{0}^{T/2}t\sin(n\omega_0 t)\,dt$$

$$= \frac{8}{T^2}\int_{0}^{T/2}t\sin(n\omega_0 t)\,dt - \frac{8}{T^2}\int_{0}^{T/2}t\sin(n\omega_0 t)\,dt$$

$$= 0$$

Hence,

$$f(t) = \frac{8}{\pi^2}\left(\cos\omega_0 t + \frac{1}{3^2}\cos 3\omega_0 t + \frac{1}{5^2}\cos 5\omega_0 t + \cdots\right)$$

PROBLEM 1-9 Find the Fourier series for the function $f(t)$ defined by

$$f(t) = \begin{cases} 0, & -\dfrac{T}{2} < t < 0 \\ A\sin\omega_0 t, & 0 < t < \dfrac{T}{2} \end{cases}$$

and $f(t + T) = f(t)$, $\omega_0 = 2\pi/T$. (See Figure 1-6.)

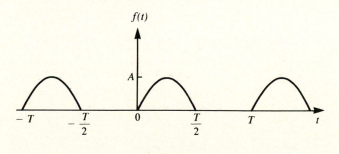

Figure 1-6

Solution: Since $f(t) = 0$ when $-T/2 < t < 0$, from (1.14) and (1.9),

$$a_0 = \frac{2}{T}\int_{0}^{T/2} A\sin(\omega_0 t)\,dt = \frac{2A}{T\omega_0}(-\cos\omega_0 t)\Big|_{0}^{T/2}$$

$$= \frac{A}{\pi}(1 - \cos\pi)$$

$$= \frac{2A}{\pi}$$

$$a_n = \frac{2}{T} \int_0^{T/2} A \sin(\omega_0 t) \cos(n\omega_0 t)\, dt$$

$$= \frac{A}{T} \int_0^{T/2} \{ \sin[(1+n)\omega_0 t] + \sin[(1-n)\omega_0 t] \}\, dt$$

When $n = 1$,

$$a_1 = \frac{A}{T} \int_0^{T/2} \sin(2\omega_0 t)\, dt = \frac{A}{T} \left(-\frac{1}{2\omega_0} \cos 2\omega_0 t \right)\Big|_0^{T/2} = \frac{A}{4\pi} [1 - \cos(2\pi)]$$

$$= \frac{A}{4\pi} (1 - 1)$$

$$= 0$$

When $n = 2, 3, \ldots$,

$$a_n = \frac{A}{T} \left\{ -\frac{\cos[(1+n)\omega_0 t]}{(1+n)\omega_0} - \frac{\cos[(1-n)\omega_0 t]}{(1-n)\omega_0} \right\}\Big|_0^{T/2}$$

$$= \frac{A}{2\pi} \left\{ \frac{1 - [\cos(1+n)\pi]}{1+n} + \frac{1 - \cos[(1-n)\pi]}{1-n} \right\}$$

$$= \begin{cases} 0, & n \text{ odd} \\ \dfrac{A}{2\pi}\left(\dfrac{2}{1+n} + \dfrac{2}{1-n} \right) = -\dfrac{2A}{(n-1)(n+1)\pi}, & n \text{ even} \end{cases}$$

Similarly, from (1.10),

$$b_n = \frac{2}{T} \int_0^{T/2} A \sin(\omega_0 t) \sin(n\omega_0 t)\, dt$$

$$= \frac{A}{T} \int_0^{T/2} \{ \cos[(1-n)\omega_0 t] - \cos[(1+n)\omega_0 t] \}\, dt$$

When $n = 1$,

$$b_1 = \frac{A}{T} \int_0^{T/2} dt - \frac{A}{T} \int_0^{T/2} \cos(2\omega_0 t)\, dt = \frac{A}{2} - \frac{A}{T} \frac{\sin 2\omega_0 t}{2\omega_0}\Big|_0^{T/2} = \frac{A}{2}$$

When $n = 2, 3, \ldots$,

$$b_n = \frac{A}{T} \left\{ \frac{\sin[(1-n)\omega_0 t]}{(1-n)\omega_0} - \frac{\sin[(1+n)\omega_0 t]}{(1+n)\omega_0} \right\}\Big|_0^{T/2}$$

$$= \frac{A}{2\pi} \left\{ \frac{\sin[(1-n)\pi] - \sin 0}{1-n} - \frac{\sin[(1+n)\pi] - \sin 0}{1+n} \right\}$$

$$= 0$$

Hence,

$$f(t) = \frac{A}{\pi} + \frac{A}{2} \sin \omega_0 t - \frac{2A}{\pi} \left(\frac{1}{1\cdot 3} \cos 2\omega_0 t + \frac{1}{3\cdot 5} \cos 4\omega_0 t + \cdots \right)$$

PROBLEM 1-10   Expand $f(t) = \sin^5 t$ in Fourier series.

Solution: Rather than proceed as in Problem 1-9, we shall make use of the identities

$$e^{\pm jn\theta} = \cos n\theta \pm j \sin n\theta$$

$$\cos n\theta = \frac{e^{jn\theta} + e^{-jn\theta}}{2}$$

$$\sin n\theta = \frac{e^{jn\theta} - e^{-jn\theta}}{2j}$$

We write

$$\sin^5 t = \left(\frac{e^{jt} - e^{-jt}}{2j}\right)^5 = \frac{1}{32j}\left(e^{j5t} - 5e^{j3t} + 10e^{jt} - 10e^{-jt} + 5e^{-j3t} - e^{-j5t}\right)$$

$$= \frac{5}{8}\sin t - \frac{5}{16}\sin 3t + \frac{1}{16}\sin 5t$$

Here the Fourier series has only three terms.

Approximation by Finite Fourier Series

PROBLEM 1-11 Approximate the function $f(t)$ defined by

$$f(t) = \begin{cases} -1, & -\pi < t < 0 \\ 1, & 0 < t < \pi \end{cases}$$

with a finite Fourier series to three nonvanishing terms. Also calculate the mean-square error in the approximation.

Solution: From the result of Example 1-7 and noting $T = 2\pi$, $\omega_0 = 2\pi/T = 1$, the three nonvanishing terms of the Fourier series of $f(t)$ are given by

$$S_5(t) = \frac{4}{\pi}\left(\sin t + \frac{1}{3}\sin 3t + \frac{1}{5}\sin 5t\right)$$

Now, from (1.19) the mean-square error E_5 is given by

$$E_5 = \frac{1}{2\pi}\int_{-\pi}^{\pi} [f(t)]^2\, dt - \frac{1}{2}\left(\frac{16}{\pi^2}\right)\left(1 + \frac{1}{3^2} + \frac{1}{5^2}\right)$$

$$= 1 - \frac{8}{\pi^2}\left(1 + \frac{1}{9} + \frac{1}{25}\right) = 0.067$$

PROBLEM 1-12 Establish the following inequality, which is known as **Bessel's inequality:**

$$\frac{2}{T}\int_{-T/2}^{T/2} [f(t)]^2\, dt \geqslant \frac{a_0^2}{2} + \sum_{n=1}^{k} (a_n^2 + b_n^2)$$

Solution: From definition (1.18),

$$E_k = \frac{1}{T}\int_{-T/2}^{T/2} [f(t) - S_k(t)]^2\, dt \geqslant 0$$

And from (1.19) we deduce that

$$\frac{2}{T}\int_{-T/2}^{T/2} [f(t)]^2\, dt \geqslant \frac{a_0^2}{2} + \sum_{n=1}^{k} (a_n^2 + b_n^2)$$

PROBLEM 1-13 If $\{a_n\}$ and $\{b_n\}$ are the sequence of the Fourier coefficients of $f(t)$, show that

$$\lim_{n\to\infty} a_n = \lim_{n\to\infty} b_n = 0$$

Solution: From the result of Problem 1-12,

$$\frac{1}{2}a_0^2 + \sum_{n=1}^{\infty} (a_n^2 + b_n^2) \leqslant \frac{2}{T}\int_{-T/2}^{T/2} [f(t)]^2\, dt$$

Since the series on the left-hand side is convergent, it is necessary that

$$\lim_{n\to\infty} (a_n^2 + b_n^2) = 0$$

which implies

$$\lim_{n\to\infty} a_n = \lim_{n\to\infty} b_n = 0$$

PROBLEM 1-14 Multiply

$$f(t) = \frac{1}{2}a_0 + \sum_{n=1}^{\infty} (a_n \cos n\omega_0 t + b_n \sin n\omega_0 t)$$

by $f(t)$, integrate term-by-term, and show that

$$\frac{1}{T}\int_{-T/2}^{T/2} [f(t)]^2 \, dt = \frac{1}{4}a_0^2 + \frac{1}{2}\sum_{n=1}^{\infty}(a_n^2 + b_n^2)$$

(*Hint:* See the proof of Parseval's identity in Example 1-11.)

Solution: Using (1.9) and (1.10),

$$\int_{-T/2}^{T/2} [f(t)]^2 \, dt = \frac{1}{2}a_0 \int_{-T/2}^{T/2} f(t) \, dt + \sum_{n=1}^{\infty}\left[a_n \int_{-T/2}^{T/2} f(t)\cos(n\omega_0 t) \, dt \right.$$

$$\left. + b_n \int_{-T/2}^{T/2} f(t)\sin(n\omega_0 t) \, dt \right]$$

$$= \frac{1}{4}a_0^2 T + \frac{T}{2}\left[\sum_{n=1}^{\infty}(a_n^2 + b_n^2) \right]$$

Thus,

$$\frac{1}{T}\int_{-T/2}^{T/2} [f(t)]^2 \, dt = \frac{1}{4}a_0^2 + \frac{1}{2}\sum_{n=1}^{\infty}(a_n^2 + b_n^2)$$

PROBLEM 1-15 If c_n are the Fourier coefficients of $f(t)$ with respect to the orthonormal set $\{\phi_n(t)\}$, show that

$$\int_a^b [f(t)]^2 \, dt = \sum_{n=1}^{\infty} c_n^2$$

Solution: From Problem 1-7,

$$f(t) = \sum_{n=1}^{\infty} c_n \phi_n(t)$$

which also can be written as

$$f(t) = \sum_{m=1}^{\infty} c_m \phi_m(t)$$

Thus,

$$\int_a^b [f(t)]^2 \, dt = \int_a^b \left[\sum_{n=1}^{\infty} c_n \phi_n(t) \right]\left[\sum_{m=1}^{\infty} c_m \phi_m(t) \right] dt$$

$$= \sum_{n=1}^{\infty} c_n \left[\sum_{m=1}^{\infty} c_m \int_a^b \phi_n(t)\phi_m(t) \, dt \right]$$

$$= \sum_{n=1}^{\infty} c_n \left[\sum_{m=1}^{\infty} c_m \delta_{nm} \right]$$

$$= \sum_{n=1}^{\infty} c_n c_n = \sum_{n=1}^{\infty} c_n^2$$

PROBLEM 1-16 Use Parseval's identity (1.27) to prove that

$$\sum_{n=1}^{\infty} \frac{1}{(2n-1)^2} = \frac{\pi^2}{8}$$

Solution: Parseval's identity (1.27) is given by

$$\frac{1}{T}\int_{-T/2}^{T/2} [f(t)]^2 \, dt = \frac{1}{4}a_0^2 + \frac{1}{2}\sum_{n=1}^{\infty}(a_n^2 + b_n^2)$$

Now, from Example 1-7, we have

$$f(t) = \frac{4}{\pi}\sum_{n=\text{odd}}^{\infty} \frac{1}{n}\sin n\omega_0 t$$

which can be rewritten as

$$f(t) = \sum_{n=1}^{\infty} \frac{4}{\pi(2n-1)} \sin(2n-1)\omega_0 t$$

Now

$$\frac{1}{T} \int_{-T/2}^{T/2} [f(t)]^2 \, dt = \frac{1}{T} \int_{-T/2}^{T/2} dt = 1$$

Thus,

$$1 = \frac{1}{2} \sum_{n=1}^{\infty} \frac{16}{\pi^2 (2n-1)^2} = \frac{8}{\pi^2} \sum_{n=1}^{\infty} \frac{1}{(2n-1)^2}$$

and

$$\sum_{n=1}^{\infty} \frac{1}{(2n-1)^2} = \frac{\pi^2}{8}$$

PROBLEM 1-17   Rework Problem 1-16 by using the Fourier series expansion of Problem 1-8.

Solution: From Problem 1-8, we have

$$f(t) = \frac{8}{\pi^2} \left(\cos \omega_0 t + \frac{1}{3^2} \cos 3\omega_0 t + \frac{1}{5^2} \cos 5\omega_0 t + \cdots \right)$$

$$= \frac{8}{\pi^2} \sum_{n=1}^{\infty} \frac{1}{(2n-1)^2} \cos(2n-1)\omega_0 t$$

Setting $t = 0$ and $f(0) = 1$, we obtain

$$1 = \frac{8}{\pi^2} \sum_{n=1}^{\infty} \frac{1}{(2n-1)^2}$$

Thus,

$$\sum_{n=1}^{\infty} \frac{1}{(2n-1)^2} = \frac{\pi^2}{8}$$

Supplementary Exercises

PROBLEM 1-18   Find the periods of the following functions: **(a)** $\cos nt$,   **(b)** $\cos 2\pi t$, **(c)** $\sin(2\pi t/k)$,   **(d)** $\sin t + \sin(t/3) + \sin(t/5)$,   **(e)** $|\sin \omega_0 t|$.

Answer:
(a) $2\pi/n$,   **(b)** 1,   **(c)** k,   **(d)** 30π,   **(e)** π/ω_0.

PROBLEM 1-19   If $f(t)$ is a periodic function of t with period T, show that $f(at)$ for $a \neq 0$ is a periodic function of t with period T/a.

PROBLEM 1-20   Find the Fourier series for the function $f(t)$ defined by $f(t) = 1$ for $-\pi < t < 0$, $f(t) = 0$ for $0 < t < \pi$, and $f(t + 2\pi) = f(t)$. (See Figure 1-7.)

Answer: $\dfrac{1}{2} - \dfrac{2}{\pi} \sum_{n=1}^{\infty} \dfrac{\sin(2n-1)t}{2n-1}$

Figure 1-7

PROBLEM 1-21 Find the Fourier series of the function $f(t)$ defined by $f(t) = t$ for the interval $(-\pi, \pi)$ and $f(t + 2\pi) = f(t)$. (See Figure 1-8.)

Answer: $2 \sum_{n=1}^{\infty} \dfrac{(-1)^{n-1}}{n} \sin nt$

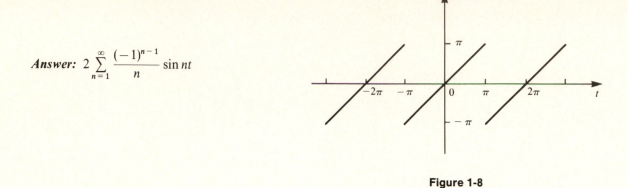

Figure 1-8

PROBLEM 1-22 Find the Fourier series for the function $f(t)$ defined by $f(t) = t^2$ for the interval $(-\pi, \pi)$ and $f(t + 2\pi) = f(t)$. (See Figure 1-9.)

Answer: $\dfrac{1}{3} \pi^2 + 4 \sum_{n=1}^{\infty} \dfrac{(-1)^n}{n^2} \cos nt$

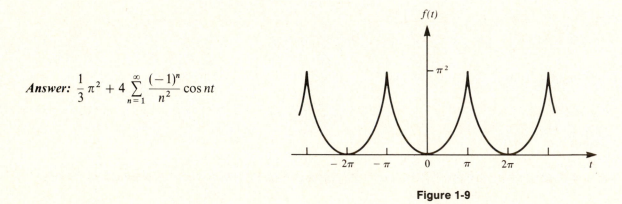

Figure 1-9

PROBLEM 1-23 Find the Fourier series for the function $f(t) = |A \sin \omega_0 t|$. (See Figure 1-10.)

Answer: $\dfrac{2A}{\pi} + \dfrac{4A}{n} \sum_{n=1}^{\infty} \dfrac{1}{1 - 4n^2} \cos(2n\omega_0 t)$

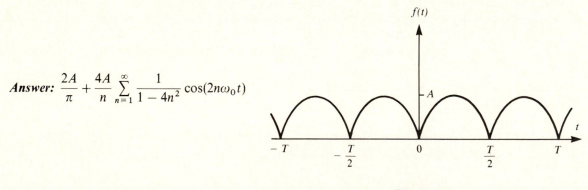

Figure 1-10

PROBLEM 1-24 Expand $f(t) = \sin^2 t \cos^3 t$ in Fourier series.

Answer: $\dfrac{1}{16}(2 \cos t - \cos 3t - \cos 5t)$

PROBLEM 1-25 Expand $f(t) = e^{r \cos t} \cos(r \sin t)$ in Fourier series. [*Hint:* Use the power series for e^z when $z = re^{jt}$.]

Answer: $1 + \sum_{n=1}^{\infty} \dfrac{r^n}{n!} \cos nt$

PROBLEM 1-26 Approximate the function $f(t) = t$ in the interval $(-\pi, \pi)$ with a finite Fourier series to five nonvanishing terms. Also, calculate the mean-square error in the approximation.

Answer: $2 \sum\limits_{n=1}^{5} \left[\dfrac{(-1)^{n-1}}{n} \sin nt \right]$, $\quad E_5 = 0.363$

PROBLEM 1-27 Using the Fourier series expansion of Example 1-7, show that

$$\frac{\pi}{4} = 1 - \frac{1}{3} + \frac{1}{5} - \frac{1}{7} + \cdots$$

[*Hint:* Set $t = T/4$ in the result of Example 1-7.]

PROBLEM 1-28 Prove that

$$\sum_{n=1}^{\infty} \frac{1}{n^2} = 1 + \frac{1}{4} + \frac{1}{9} + \frac{1}{16} + \cdots = \frac{\pi^2}{6}$$

[*Hint:* Set $t = \pi$ in the result of Problem 1-22.]

PROBLEM 1-29 Find the sum $\sum_{n=1}^{\infty} [1/(2n-1)^2]$. [*Hint:* Set $t = 0$ in the result of Problem 1-8.]

Answer: $\pi^2/8$

PROBLEM 1-30 Let $f(t)$ and $g(t)$ be piecewise continuous with period T, and let a_n, b_n and α_n, β_n be the respective Fourier coefficients of $f(t)$ and $g(t)$. Show that

$$\frac{2}{T} \int_{-T/2}^{T/2} f(t)g(t)\, dt = \frac{1}{2} a_0 \alpha_0 + \sum_{n=1}^{\infty} (a_n \alpha_n + b_n \beta_n)$$

PROBLEM 1-31 If $f(t)$ of Problem 1-7 is approximated by $f_k(t) = \sum_{n=1}^{k} c_n \phi_n(t)$, show that the mean-square error $[1/(b-a)] \int_a^b [f(t) - f_k(t)]^2\, dt$ is a minimum.

ANALYSIS OF PERIODIC WAVEFORMS

THIS CHAPTER IS ABOUT

☑ **Waveform Symmetry**
☑ **Fourier Coefficients of Symmetric Waveforms**
☑ **Fourier Expansion of a Function over a Finite Interval**

2-1. Waveform Symmetry

In this chapter we'll discuss the effect of waveform symmetry in a Fourier series.

A. Even and odd functions

A function $f(t)$ is **even** when it satisfies the condition

Even symmetry $$f(-t) = f(t) \qquad \text{(2.1)}$$

and it is **odd** if

Odd symmetry $$f(-t) = -f(t) \qquad \text{(2.2)}$$

Examples of even and odd functions are shown in Figure 2-1. Note that an even function is symmetric about the vertical axis at the origin. On the other hand, an odd function is antisymmetric about the vertical axis at the origin.

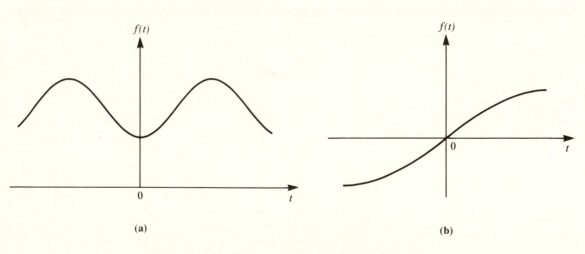

Figure 2-1   (a) An even function; (b) an odd function.

B. Half-wave symmetry

If a function $f(t)$ is periodic with period T, the periodic function $f(t)$ has **half-wave symmetry** when it satisfies the condition

Half-wave symmetry $$f(t) = -f\left(t + \frac{1}{2}T\right) \qquad \text{(2.3)}$$

A waveform with half-wave symmetry is shown in Figure 2-2. Note that the negative portion of the waveform is the mirror image of the positive portion of the waveform, displaced horizontally by a half period.

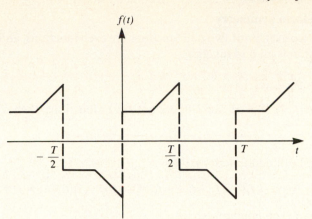

Figure 2-2   Half-wave symmetry.

C. Quarter-wave symmetry

If a periodic function $f(t)$ has half-wave symmetry and, in addition, is either an even or odd function, then $f(t)$ is said to have **even** or **odd quarter-wave symmetry**. Figure 2-3 illustrates waveforms with quarter-wave symmetry.

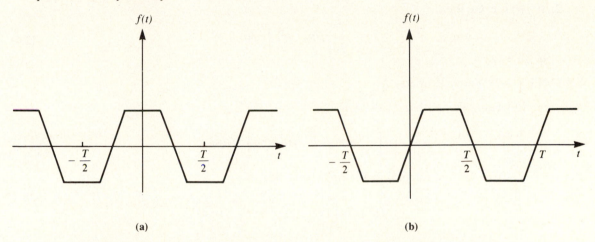

(a)   (b)

Figure 2-3   (a) Even quarter-wave symmetry; (b) odd quarter-wave symmetry.

D. Hidden symmetry

Often the symmetry of a periodic function is obscured by a constant term. Figure 2-4a shows that the subtraction of a constant $A/2$ from $f(t)$ merely shifts the horizontal axis upward by the amount $A/2$. As Figure 2-4b shows, the new function, $g(t) = f(t) - A/2$, is an odd function.

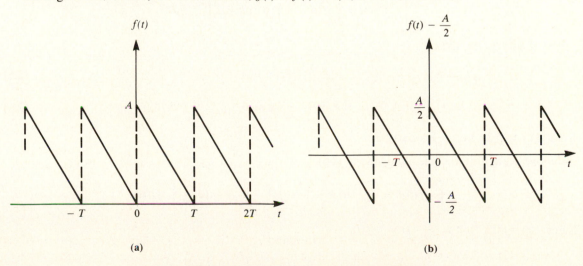

(a)   (b)

Figure 2-4   (a) Hidden symmetry; (b) odd symmetry.

E. Properties of waveform symmetry

1. The products of two even or of two odd functions are even functions, and the product of an even and an odd function is an odd function.

EXAMPLE 2-1: Verify Property 1.

Solution: Let $f(t) = f_1(t)f_2(t)$. If $f_1(t)$ and $f_2(t)$ are both even functions, then

$$f(-t) = f_1(-t)f_2(-t) = f_1(t)f_2(t) = f(t)$$

and if $f_1(t)$ and $f_2(t)$ are both odd functions, then

$$f(-t) = f_1(-t)f_2(-t) = -f_1(t)[-f_2(t)] = f_1(t)f_2(t) = f(t)$$

This proves that $f(t)$ is an even function.

Similarly, if $f_1(t)$ is even and $f_2(t)$ is odd, then

$$f(-t) = f_1(-t)f_2(-t) = f_1(t)[-f_2(t)] = -f_1(t)f_2(t) = -f(t)$$

This proves that $f(t)$ is an odd function.

2. If $f(t)$ is even, then

$$\int_{-a}^{a} f(t)\, dt = 2\int_{0}^{a} f(t)\, dt \tag{2.4}$$

EXAMPLE 2-2: Verify Property 2.

Solution: Rewriting the left-hand side of (2.4),

$$\int_{-a}^{a} f(t)\, dt = \int_{-a}^{0} f(t)\, dt + \int_{0}^{a} f(t)\, dt$$

Letting $t = -x$ in the first integral of the right-hand side,

$$\int_{-a}^{0} f(t)\, dt = \int_{a}^{0} f(-x)(-dx) = \int_{0}^{a} f(-x)\, dx$$

Since $f(t)$ is even, that is, $f(-x) = f(x)$,

$$\int_{0}^{a} f(-x)\, dx = \int_{0}^{a} f(x)\, dx = \int_{0}^{a} f(t)\, dt$$

Hence,

$$\int_{-a}^{a} f(t)\, dt = \int_{0}^{a} f(t)\, dt + \int_{0}^{a} f(t)\, dt = 2\int_{0}^{a} f(t)\, dt$$

3. If $f(t)$ is odd, then

$$\int_{-a}^{a} f(t)\, dt = 0 \tag{2.5}$$

$$f(0) = 0 \tag{2.6}$$

EXAMPLE 2-3: Verify Property 3.

Solution: Rewriting the left-hand side of (2.5),

$$\int_{-a}^{a} f(t)\, dt = \int_{-a}^{0} f(t)\, dt + \int_{0}^{a} f(t)\, dt = \int_{0}^{a} f(-t)\, dt + \int_{0}^{a} f(t)\, dt$$

Since $f(t)$ is odd, that is, $f(-t) = -f(t)$,

$$\int_{-a}^{a} f(t)\, dt = -\int_{0}^{a} f(t)\, dt + \int_{0}^{a} f(t)\, dt = 0$$

In particular, $f(-0) = -f(0)$ and $f(-0) = f(0)$; hence,

$$f(0) = 0$$

4. If a periodic function $f(t)$ is half-wave symmetric, then

$$f(t) = -f\left(t - \frac{1}{2}T\right) \tag{2.7}$$

EXAMPLE 2-4 Verify Property 4.

Solution: If $f(t)$ is half-wave symmetric, from (2.3)

$$f(t) = -f\left(t + \frac{1}{2}T\right)$$

Since $f(t)$ is periodic with period T,

$$f\left(t - \frac{1}{2}T\right) = f\left(t + T - \frac{1}{2}T\right) = f\left(t + \frac{1}{2}T\right)$$

Hence,

$$f(t) = -f\left(t + \frac{1}{2}T\right) = -f\left(t - \frac{1}{2}T\right)$$

2-2. Fourier Coefficients of Symmetric Waveforms

The use of symmetry properties simplifies the calculation of Fourier coefficients.

A. Even periodic functions

If $f(t)$ is an even periodic function with period T, its Fourier series consists of a *constant and cosine terms only*. That is,

$$f(t) = \frac{1}{2}a_0 + \sum_{n=1}^{\infty} a_n \cos n\omega_0 t \tag{2.8}$$

where $\omega_0 = 2\pi/T$, and a_n is given by

$$a_n = \frac{4}{T}\int_0^{T/2} f(t)\cos(n\omega_0 t)\,dt \tag{2.9}$$

EXAMPLE 2-5: Verify (2.8) and (2.9).

Solution: Fourier series expansion of $f(t)$ is

$$f(t) = \frac{1}{2}a_0 + \sum_{n=1}^{\infty}(a_n \cos n\omega_0 t + b_n \sin n\omega_0 t)$$

From our evaluation of Fourier coefficients in Chapter 1, we use (1.9) and (1.10):

$$a_n = \frac{2}{T}\int_{-T/2}^{T/2} f(t)\cos(n\omega_0 t)\,dt, \qquad n = 0, 1, 2, \ldots$$

$$b_n = \frac{2}{T}\int_{-T/2}^{T/2} f(t)\sin(n\omega_0 t)\,dt, \qquad n = 1, 2, \ldots$$

Since $\sin n\omega_0 t$ is odd and $f(t)$ is even, the product $f(t)\sin n\omega_0 t$ is an odd function. Hence, according to (2.5), $b_n = 0$. Also, since $\cos n\omega_0 t$ is an even function, the product $f(t)\cos n\omega_0 t$ is an even function. Hence, from (2.4),

$$a_n = \frac{4}{T}\int_0^{T/2} f(t)\cos(n\omega_0 t)\,dt$$

B. Odd periodic functions

If $f(t)$ is an odd periodic function with period T, its Fourier series consists of *sine terms only*. That is,

$$f(t) = \sum_{n=1}^{\infty} b_n \sin n\omega_0 t \tag{2.10}$$

where $\omega_0 = 2\pi/T$, and b_n is given by

$$b_n = \frac{4}{T} \int_0^{T/2} f(t)\sin(n\omega_0 t)\,dt \tag{2.11}$$

EXAMPLE 2-6: Verify (2.10) and (2.11).

Solution: Since $f(t)$ is an odd function, the product $f(t)\cos n\omega_0 t$ is an odd function, and the product $f(t)\sin n\omega_0 t$ is an even function. Hence, according to (2.4) and (2.5),

$$a_n = 0, \qquad b_n = \frac{4}{T} \int_0^{T/2} f(t)\sin(n\omega_0 t)\,dt$$

C. Half-wave symmetric periodic functions

The Fourier series of any periodic function $f(t)$ that has half-wave symmetry contains *only odd harmonics*.

EXAMPLE 2-7: The coefficients a_n in the Fourier series expansion of a periodic function $f(t)$ are

$$a_n = \frac{2}{T} \int_{-T/2}^{T/2} f(t)\cos(n\omega_0 t)\,dt$$

$$= \frac{2}{T}\left[\int_{-T/2}^{0} f(t)\cos(n\omega_0 t)\,dt + \int_0^{T/2} f(t)\cos(n\omega_0 t)\,dt \right]$$

Changing the variable t to $(t - \tfrac{1}{2}T)$ in the first integral, we obtain

$$a_n = \frac{2}{T}\left\{ \int_0^{T/2} f\left(t - \frac{1}{2}T\right)\cos\left[n\omega_0\left(t - \frac{1}{2}T\right)\right]dt + \int_0^{T/2} f(t)\cos(n\omega_0 t)\,dt \right\}$$

Since $f(t)$ has half-wave symmetry, using the property $f(t) = -f(t - \tfrac{1}{2}T)$ from (2.7) and the fact that $\sin n\pi = 0$, we obtain

$$a_n = \frac{2}{T} \int_0^{T/2} [-f(t)\cos(n\omega_0 t)\cos n\pi + f(t)\cos(n\omega_0 t)]\,dt$$

$$= \frac{2}{T}[1 - (-1)^n] \int_0^{T/2} f(t)\cos(n\omega_0 t)\,dt$$

$$= \begin{cases} 0 & \text{for } n \text{ even} \\ \dfrac{4}{T} \displaystyle\int_0^{T/2} f(t)\cos(n\omega_0 t)\,dt & \text{for } n \text{ odd} \end{cases}$$

A similar investigation shows that

$$b_n = \begin{cases} 0 & \text{for } n \text{ even} \\ \dfrac{4}{T} \displaystyle\int_0^{T/2} f(t)\sin(n\omega_0 t)\,dt & \text{for } n \text{ odd} \end{cases}$$

D. Even quarter-wave symmetric functions

The Fourier series of any periodic function $f(t)$ that has even quarter-wave symmetry consists of *odd harmonics of cosine terms only*. That is,

$$f(t) = \sum_{n=1}^{\infty} a_{2n-1}\cos[(2n-1)\omega_0 t] \tag{2.12}$$

where $\omega_0 = 2\pi/T$, and

$$a_{2n-1} = \frac{8}{T} \int_0^{T/4} f(t)\cos[(2n-1)\omega_0 t]\,dt \tag{2.13}$$

EXAMPLE 2-8: Verify (2.12) and (2.13).

Solution: Since $f(t)$ has even quarter-wave symmetry,

$$f(t) = f(-t)$$

Even quarter-wave symmetry

$$f\left(t + \frac{1}{2}T\right) = -f(t)$$

Hence, from the results of Examples 2-5 and 2-7, we see that

$$\left.\begin{array}{l} b_n = 0 \\ a_{2n} = 0 \end{array}\right\} \quad \text{for all } n \text{ (including } a_0)$$

$$a_{2n-1} = \frac{4}{T}\left\{\int_0^{T/4} f(t)\cos[(2n-1)\omega_0 t]\,dt + \int_{T/4}^{T/2} f(t)\cos[(2n-1)\omega_0 t]\,dt\right\}$$

Changing the variable t to $(t + \frac{1}{2}T)$ in the second integral, we obtain

$$a_{2n-1} = \frac{4}{T}\left\{\int_0^{T/4} f(t)\cos[(2n-1)\omega_0 t]\,dt + \int_{-T/4}^0 f\left(t + \frac{1}{2}T\right)\cos\left[(2n-1)\omega_0\left(t + \frac{1}{2}T\right)\right]dt\right\}$$

Using the property $f(t) = -f(t + \frac{1}{2}T)$, we obtain

$$a_{2n-1} = \frac{4}{T}\left\{\int_0^{T/4} f(t)\cos[(2n-1)\omega_0 t]\,dt + \int_{-T/4}^0 f(t)\cos[(2n-1)\omega_0 t]\,dt\right\}$$

$$= \frac{4}{T}\int_{-T/4}^{T/4} f(t)\cos[(2n-1)\omega_0 t]\,dt$$

Since $f(-t) = f(t)$ and $f(t)\cos(2n-1)\omega_0 t$ is even, we obtain from (2.4),

$$a_{2n-1} = \frac{8}{T}\int_0^{T/4} f(t)\cos[(2n-1)\omega_0 t]\,dt$$

E. Odd quarter-wave symmetric functions

The Fourier series of any periodic function $f(t)$ that has odd quarter-wave symmetry consists of *odd harmonics of sine terms only*. That is,

$$f(t) = \sum_{n=1}^{\infty} b_{2n-1}\sin[(2n-1)\omega_0 t] \qquad (2.14)$$

where $\omega_0 = 2\pi/T$, and

$$b_{2n-1} = \frac{8}{T}\int_0^{T/4} f(t)\sin[(2n-1)\omega_0 t]\,dt \qquad (2.15)$$

EXAMPLE 2-9: Verify (2.14) and (2.15).

Solution: Since $f(t)$ has odd quarter-wave symmetry,

$$f(-t) = -f(t)$$

Odd quarter-wave symmetry

$$f\left(t + \frac{1}{2}T\right) = -f(t)$$

Hence, from the results of Examples 2-6 and 2-7,

$$\left.\begin{array}{l} a_n = 0 \\ b_{2n} = 0 \end{array}\right\} \quad \text{for all } n \text{ (including } a_0)$$

$$b_{2n-1} = \frac{4}{T}\int_0^{T/2} f(t)\sin[(2n-1)\omega_0 t]\,dt$$

Evaluating this integral as in Example 2-8, we get

$$b_{2n-1} = \frac{8}{T}\int_0^{T/4} f(t)\sin[(2n-1)\omega_0 t]\,dt$$

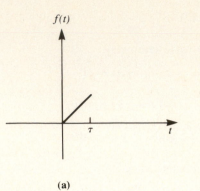

(a)

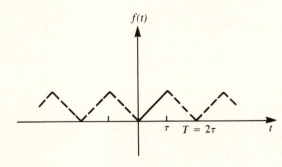

(b)

(c)

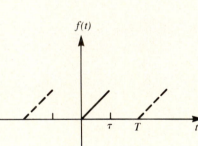

(d)

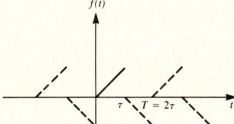

(e)

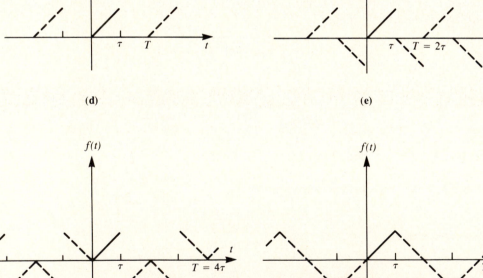

(f)

(g)

Figure 2-5   Fourier expansion.

(a) Given *f*(t). (b) Even symmetry: cosine terms, $\omega_0 = \pi/\tau$. (c) Odd symmetry: sine terms $\omega_0 = \pi/\tau$. (d) Sine and cosine terms, $\omega_0 = 2\pi/T$ (*T* is arbitrary). (e) Half-wave symmetry: sine and cosine terms; odd harmonics, $\omega_0 = \pi/\tau$. (f) Even quarter-wave symmetry: cosine terms and odd harmonics, $\omega_0 = \pi/(2\tau)$. (g) Odd quarter-wave symmetry: sine terms and odd harmonics, $\omega_0 = \pi/(2\tau)$.

2-3. Fourier Expansion of a Function over a Finite Interval

A. Fourier expansion technique

A nonperiodic function $f(t)$ defined over a certain finite interval $(0, \tau)$ can be expanded into a Fourier series defined only in the interval $(0, \tau)$. It is possible to expand $f(t)$ by a Fourier series with any desired fundamental frequency. Furthermore, $f(t)$ can be represented by sine or cosine terms alone. This can be done by constructing a proper periodic function identical to $f(t)$ over the interval $(0, \tau)$ and satisfying the symmetry conditions that yield the desired form of Fourier series. You can see an illustration of this procedure in Figure 2-5.

B. Half-range expansions

Let $f(t)$ have period $T = 2\tau$. If $f(t)$ is even, we have from (2.8) and (2.9) the Fourier cosine series:

Fourier cosine series
$$f(t) = \frac{1}{2}a_0 + \sum_{n=1}^{\infty} a_n \cos \frac{n\pi}{\tau} t \qquad (2.16)$$

with coefficients

$$a_n = \frac{2}{\tau} \int_0^\tau f(t)\cos\left(\frac{n\pi}{\tau} t\right) dt \qquad (2.17)$$

If $f(t)$ is odd, we have from (2.10) and (2.11) the Fourier sine series

Fourier sine series
$$f(t) = \sum_{n=1}^{\infty} b_n \sin \frac{n\pi}{\tau} t \qquad (2.18)$$

with coefficients

$$b_n = \frac{2}{\tau} \int_0^\tau f(t)\sin\left(\frac{n\pi}{\tau} t\right) dt \qquad (2.19)$$

Both the cosine series (2.16) and the sine series (2.18) represent the same given function $f(t)$ in the interval $(0, \tau)$. Outside this interval, the cosine series (2.16) will represent the even periodic extension of $f(t)$ having period $T = 2\tau$ (Figure 2-5b), and the sine series (2.18) will represent the odd periodic extension of $f(t)$ having period $T = 2\tau$ (Figure 2-5c). The cosine and sine series (2.16) and (2.18) with coefficients given, respectively, by (2.17) and (2.19) are called the **half-range expansions** of the given function $f(t)$.

SUMMARY

1. The effect of waveform symmetry in a Fourier series is summarized in the following:

Waveform Symmetry:	*Fourier Coefficients:*
Even symmetry	constant and cosine terms only
$f(-t) = f(t)$	a_n only, $\qquad b_n = 0$
Odd symmetry	sine terms only
$f(-t) = -f(t)$	b_n only, $\qquad a_n = 0$
Half-wave symmetry	odd harmonics only
$f(t) = -f(t + \frac{1}{2}T)$	a_{2n-1}, b_{2n-1} only, $\qquad a_{2n} = b_{2n} = 0$
Even quarter-wave symmetry	odd cosine harmonics only
$f(-t) = f(t), \qquad f(t) = -f(t + \frac{1}{2}T)$	a_{2n-1} only, $\qquad a_{2n} = b_n = 0$
Odd quarter-wave symmetry	odd sine harmonics only
$f(-t) = -f(t), \qquad f(t) = -f(t + \frac{1}{2}T)$	b_{2n-1} only, $\qquad a_n = b_{2n} = 0$

2. Half-range Fourier series expansion of a nonperiodic function $f(t)$ defined over $(0, \tau)$ is done by constructing a proper (even or odd) periodic function (with period $T = 2\tau$) identical to $f(t)$ over $(0, \tau)$.

RAISE YOUR GRADES

Can you explain ...?

☑ what waveform symmetry is
☑ what effect waveform symmetry has on Fourier coefficients
☑ what a half-range Fourier expansion is
☑ how to carry out half-range Fourier series expansion of a nonperiodic function

SOLVED PROBLEMS

Waveform Symmetry

PROBLEM 2-1   Show that any function $f(t)$ can be expressed as the sum of two component functions, one of which is even and the other odd.

Solution: Any function $f(t)$ can be expressed as

$$f(t) = \frac{1}{2}f(t) + \frac{1}{2}f(-t) + \frac{1}{2}f(t) - \frac{1}{2}f(-t)$$

$$= \frac{1}{2}[f(t) + f(-t)] + \frac{1}{2}[f(t) - f(-t)]$$

Let

$$\frac{1}{2}[f(t) + f(-t)] = f_e(t)$$

and

$$\frac{1}{2}[f(t) - f(-t)] = f_o(t)$$

where $f_e(t)$ is the even component of a given function $f(t)$ and $f_o(t)$ is the odd component of $f(t)$. Then

$$f_e(-t) = \frac{1}{2}[f(-t) + f(t)] = f_e(t)$$

and

$$f_o(-t) = \frac{1}{2}[f(-t) - f(t)] = -\frac{1}{2}[f(t) - f(-t)] = -f_o(t)$$

Hence,

$$f(t) = f_e(t) + f_o(t)$$

Alternate Solution: Let's assume that $f(t)$ can be expressed as

$$f(t) = f_e(t) + f_o(t) \tag{a}$$

where $f_e(t)$ and $f_o(t)$ denote the even and odd components of $f(t)$, respectively. According to the definition of even and odd functions $f_e(-t) = f_e(t)$ (2.1) and $f_o(-t) = -f_o(t)$ (2.2), it follows that

$$f(-t) = f_e(t) - f_o(t) \tag{b}$$

Addition and subtraction of eqs. (a) and (b) yield, respectively,

$$f_e(t) = \frac{1}{2}[f(t) + f(-t)]$$

and

$$f_o(t) = \frac{1}{2}[f(t) - f(-t)]$$

PROBLEM 2-2   Find the even and odd components of the function shown in Figure 2-6a and defined by

$$f(t) = \begin{cases} e^{-t} & t > 0 \\ 0 & t < 0 \end{cases}$$

Solution: For given $f(t)$,

$$f(-t) = \begin{cases} 0, & t > 0 \\ e^t, & t < 0 \end{cases}$$

Hence, by means of Problem 2-1,

$$f_e(t) = \frac{1}{2}[f(t) + f(-t)] = \begin{cases} \dfrac{1}{2}e^{-t}, & t > 0 \\ \dfrac{1}{2}e^t, & t < 0 \end{cases}$$

$$f_o(t) = \frac{1}{2}[f(t) - f(-t)] = \begin{cases} \dfrac{1}{2}e^{-t}, & t > 0 \\ -\dfrac{1}{2}e^t, & t < 0 \end{cases}$$

The even and odd components of $f(t)$ are shown, respectively, in Figure 2-6b and c.

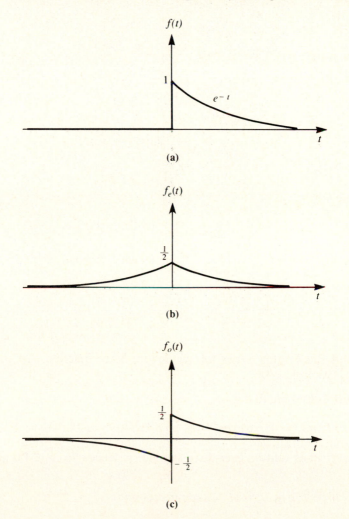

Figure 2-6

Fourier Coefficients of Symmetric Waveforms

PROBLEM 2-3   Find the Fourier series for the square wave function $f(t)$ shown in Figure 2-7.

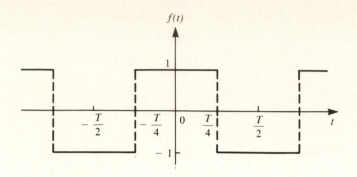

Figure 2-7

Solution: From Figure 2-7 we see that

$$f(-t) = f(t), \qquad f\left(t + \frac{1}{2}T\right) = -f(t)$$

That is, the function $f(t)$ has even quarter-wave symmetry. Hence, from (2.12) and (2.13) (proved in Example 2-8)

$$f(t) = \sum_{n=1}^{\infty} a_{2n-1}\cos[(2n-1)\omega_0 t], \qquad \omega_0 = \frac{2\pi}{T}$$

$$a_{2n-1} = \frac{8}{T}\int_0^{T/4} f(t)\cos[(2n-1)\omega_0 t]\, dt$$

Thus,

$$a_{2n-1} = \frac{8}{T}\int_0^{T/4} \cos[(2n-1)\omega_0 t]\, dt$$

$$= \frac{8}{(2n-1)\omega_0 T}\sin[(2n-1)\omega_0 t]\Big|_0^{T/4}$$

$$= \frac{4}{(2n-1)\pi}\sin\left[(2n-1)\frac{\pi}{2}\right]$$

$$= \begin{cases} \dfrac{4}{(2n-1)\pi} & \text{for } (2n-1) = 1, 5, \dots \\[2mm] \dfrac{-4}{(2n-1)\pi} & \text{for } (2n-1) = 3, 7, \dots \end{cases}$$

Hence,

$$f(t) = \frac{4}{\pi}\left(\cos\omega_0 t - \frac{1}{3}\cos 3\omega_0 t + \frac{1}{5}\cos 5\omega_0 t - \cdots\right)$$

PROBLEM 2-4   Find the Fourier series for the square wave function $f(t)$ shown in Figure 2-8. (*Hint:* See Chapter 1, Example 1-7.)

Solution: From Figure 2-8 we see that

$$f(-t) = -f(t), \qquad f\left(t + \frac{1}{2}T\right) = -f(t)$$

That is, the function $f(t)$ has odd quarter-wave symmetry. Hence, from (2.14) and (2.15) (proved in Example 2-9)

$$f(t) = \sum_{n=1}^{\infty} b_{2n-1}\sin[(2n-1)\omega_0 t], \qquad \omega_0 = \frac{2\pi}{T}$$

$$b_{2n-1} = \frac{8}{T}\int_0^{T/4} f(t)\sin[(2n-1)\omega_0 t]\, dt$$

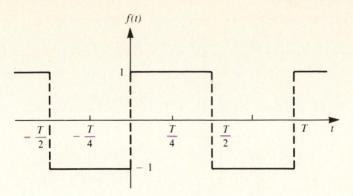

Figure 2-8

Thus,

$$b_{2n-1} = \frac{8}{T} \int_0^{T/4} \sin[(2n-1)\omega_0 t]\, dt$$

$$= \frac{-8}{(2n-1)\omega_0 T} \cos[(2n-1)\omega_0 t]\Big|_0^{T/4}$$

$$= \frac{4}{(2n-1)\pi}\left\{1 - \cos\left[(2n-1)\frac{\pi}{2}\right]\right\}$$

$$= \frac{4}{(2n-1)\pi}$$

Hence,

$$f(t) = \frac{4}{\pi}\left(\sin \omega_0 t + \frac{1}{3}\sin 3\omega_0 t + \frac{1}{5}\sin 5\omega_0 t + \cdots\right)$$

PROBLEM 2-5   Find the Fourier series for the function $f(t)$ shown in Figure 2-9a.

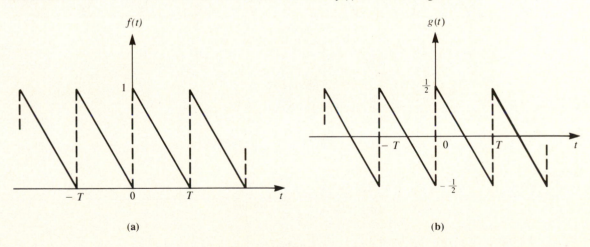

(a)   (b)

Figure 2-9

Solution: As shown in Figure 2-9b, the function $g(t) = [f(t) - \frac{1}{2}]$ is an odd function. Hence,

$$g(t) = \sum_{n=1}^{\infty} b_n \sin n\omega_0 t, \qquad \omega_0 = \frac{2\pi}{T}$$

$$b_n = \frac{2}{T} \int_{-T/2}^{T/2} g(t)\sin(n\omega_0 t)\, dt$$

Since $g(t)\sin n\omega_0 t$ is an even function, from (2.4) (proved in Example 2-2),

$$b_n = \frac{4}{T} \int_0^{T/2} g(t)\sin(n\omega_0 t)\, dt$$

Now, since

$$g(t) = \frac{1}{2} - \frac{1}{T} t \qquad \text{for} \quad 0 < t < T$$

then

$$b_n = \frac{4}{T} \int_0^{T/2} \left(\frac{1}{2} - \frac{1}{T} t \right) \sin(n\omega_0 t)\, dt$$

Integrating by parts,

$$b_n = \frac{4}{T} \left[-\left(\frac{1}{2} - \frac{1}{T} t \right) \frac{\cos n\omega_0 t}{n\omega_0} - \frac{\sin n\omega_0 t}{T(n\omega_0)^2} \right]\Bigg|_0^{T/2} = \frac{1}{n\pi}$$

Hence,

$$f(t) = \frac{1}{2} + g(t) = \frac{1}{2} + \frac{1}{\pi} \sum_{n=1}^{\infty} \frac{1}{n} \sin n\omega_0 t$$

$$= \frac{1}{2} + \frac{1}{\pi} \left(\sin \omega_0 t + \frac{1}{2} \sin 2\omega_0 t + \frac{1}{3} \sin 3\omega_0 t + \cdots \right)$$

PROBLEM 2-6 Using the result of Problem 2-5, find the Fourier series for the function $f(t)$ shown in Figure 2-10a.

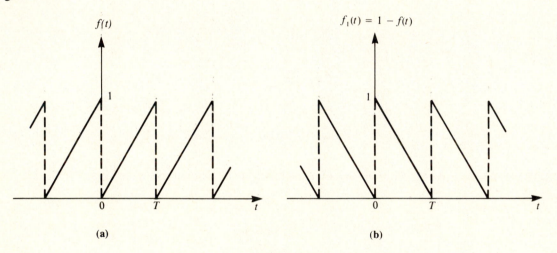

Figure 2-10

Solution: From Figure 2-10b and the result of Problem 2-5, we see that

$$f_1(t) = 1 - f(t) = \frac{1}{2} + \frac{1}{\pi} \sum_{n=1}^{\infty} \frac{1}{n} \sin n\omega_0 t$$

Hence,

$$f(t) = 1 - f_1(t) = 1 - \frac{1}{2} - \frac{1}{\pi} \sum_{n=1}^{\infty} \frac{1}{n} \sin n\omega_0 t$$

$$= \frac{1}{2} - \frac{1}{\pi} \left(\sin \omega_0 t + \frac{1}{2} \sin 2\omega_0 t + \frac{1}{3} \sin 3\omega_0 t + \cdots \right)$$

Fourier Expansion of a Function over a Finite Interval

PROBLEM 2-7 Given the function (Figure 2-11)

$$f(t) = \begin{cases} 0 & \text{for} \quad 0 < t < \frac{1}{2}\pi \\[2mm] 1 & \text{for} \quad \frac{1}{2}\pi < t < \pi \end{cases}$$

expand $f(t)$ in a Fourier cosine series and draw the corresponding periodic extension of $f(t)$.

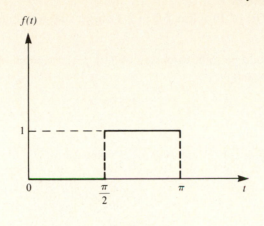

Figure 2-11

Solution: The graph of the even periodic extension of $f(t)$, $f_e(t)$, is shown in Figure 2-12. Since $f(t)$ is extended to be even,

$$b_n = 0, \qquad n = 1, 2, \ldots$$

From (2.17)

$$a_n = \frac{2}{\pi} \int_0^\pi f(t)\cos(nt)\, dt = \frac{2}{\pi} \int_{\pi/2}^\pi \cos(nt)\, dt$$

$$= \frac{2}{n\pi} \sin nt \Big|_{\pi/2}^\pi$$

$$= -\frac{2}{n\pi} \sin \frac{n\pi}{2}$$

That is,

$$a_n = \begin{cases} 0, & n \text{ even } (n \neq 0) \\[2mm] -\dfrac{2}{n\pi}, & n = 1, 5, \ldots \\[2mm] \dfrac{2}{n\pi}, & n = 3, 7, \ldots \end{cases}$$

For $n = 0$,

$$a_0 = \frac{2}{\pi} \int_{\pi/2}^\pi dt = 1$$

Thus,

$$f_e(t) = \frac{1}{2} - \frac{2}{\pi}\left(\cos t - \frac{1}{3} \cos 3t + \frac{1}{5} \cos 5t - \cdots \right)$$

for $0 < t < \pi$.

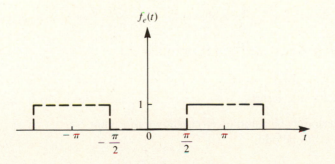

Figure 2-12

PROBLEM 2-8 Expand $f(t)$ of Problem 2-7 in a Fourier sine series and draw the corresponding periodic extension of $f(t)$.

Solution: The graph of the odd periodic extension of $f(t)$, $f_o(t)$, is shown in Figure 2-13.
Since $f(t)$ is extended to be odd,

$$a_n = 0, \qquad n = 0, 1, 2, \dots$$

From (2.19),

$$b_n = \frac{2}{\pi} \int_0^\pi f(t)\sin(nt)\, dt = \frac{2}{\pi} \int_{\pi/2}^\pi \sin(nt)\, dt$$

$$= -\frac{2}{n\pi} \cos nt \Big|_{\pi/2}^\pi$$

$$= -\frac{2}{n\pi} \left(\cos n\pi - \cos\frac{1}{2} n\pi \right)$$

That is,

$$b_n = \begin{cases} \dfrac{2}{n\pi}, & n = 1, 3, 5, \dots \\[2ex] -\dfrac{4}{n\pi}, & n = 2, 6, 10, \dots \\[2ex] 0, & n = 4, 8, 12, \dots \end{cases}$$

Therefore,

$$f_o(t) = \frac{2}{\pi}\left(\sin t + \frac{1}{3}\sin 3t + \frac{1}{5}\sin 5t + \cdots \right) - \frac{2}{\pi}\left(\sin 2t + \frac{1}{3}\sin 6t + \frac{1}{5}\sin 10t + \cdots \right)$$

for $0 < t < \pi$.

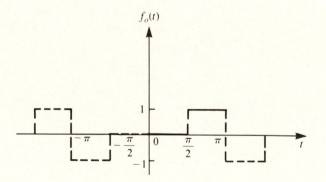

Figure 2-13

PROBLEM 2-9 Given the function (Figure 2-14)

$$f(t) = \begin{cases} \dfrac{2k}{l} t & \text{for } 0 < t < \frac{1}{2}l \\[2ex] \dfrac{2k}{l}(l - t) & \text{for } \frac{1}{2}l < t < l \end{cases}$$

expand $f(t)$ in a Fourier sine series.

Solution: The odd periodic extension of $f(t)$ is shown in Figure 2-15.
Since $f(t)$ is extended to be odd,

$$a_n = 0, \qquad n = 0, 1, 2, \dots$$

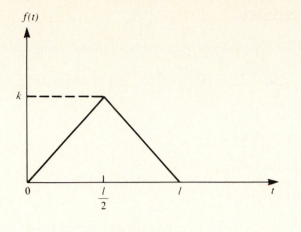

Figure 2-14

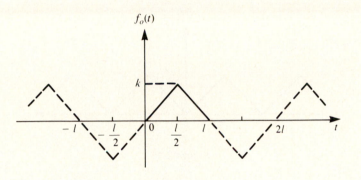

Figure 2-15

From (2.19),

$$b_n = \frac{2}{l} \int_0^l f(t)\sin\left(\frac{n\pi}{l}t\right)dt$$

$$= \frac{2}{l}\left[\frac{2k}{l}\int_0^{l/2} t\sin\left(\frac{n\pi}{l}t\right)dt + \frac{2k}{l}\int_{l/2}^l (l-t)\sin\left(\frac{n\pi}{l}t\right)dt\right]$$

Now integrating by parts,

$$\int_0^{l/2} t\sin\left(\frac{n\pi}{l}t\right)dt = -\frac{lt}{n\pi}\cos\frac{n\pi}{l}t\Big|_0^{l/2} + \frac{l}{n\pi}\int_0^{l/2}\cos\left(\frac{n\pi}{l}t\right)dt$$

$$= -\frac{l^2}{2n\pi}\cos\frac{1}{2}n\pi + \frac{l^2}{n^2\pi^2}\sin\frac{1}{2}n\pi$$

Similarly,

$$\int_{l/2}^l (l-t)\sin\left(\frac{n\pi}{l}t\right)dt = \frac{l^2}{2n\pi}\cos\frac{1}{2}n\pi + \frac{l^2}{n^2\pi^2}\sin\frac{1}{2}n\pi$$

Substituting these results into the first equation, we obtain

$$b_n = \frac{8k}{n^2\pi^2}\sin\frac{1}{2}n\pi$$

Thus,

$$f(t) = \frac{8k}{\pi^2}\left(\sin\frac{\pi}{l}t - \frac{1}{3^2}\sin 3\frac{\pi}{l}t + \frac{1}{5^2}\sin 5\frac{\pi}{l}t - \cdots\right)$$

Supplementary Exercises

PROBLEM 2-10   Find the even and odd components of the following functions:

$$\text{(a)} \quad e^t, \qquad \text{(b)} \quad \frac{t+1}{t-1}, \qquad \text{(c)} \quad t\sin t - \sin 2t$$

Answer:

(a) $f_e(t) = \cosh t, \qquad f_o(t) = \sinh t$

(b) $f_e(t) = \dfrac{t^2+1}{t^2-1}, \qquad f_o(t) = \dfrac{2t}{t^2-1}$

(c) $f_e(t) = t\sin t, \qquad f_o(t) = -\sin 2t$

PROBLEM 2-11   Find the Fourier series expansion of the function $f(t)$ defined by $f(t) = |t|$ for $(-\pi, \pi)$ and $f(t + 2\pi) = f(t)$. (See Figure 2-16.)

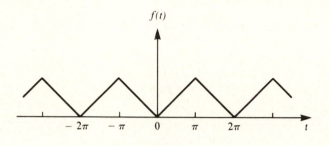

Figure 2-16

Answer: $\quad \dfrac{\pi}{2} - \dfrac{4}{\pi} \displaystyle\sum_{n=1}^{\infty} \dfrac{1}{(2n-1)^2} \cos(2n-1)t$

PROBLEM 2-12   Let $f(t)$ be a periodic function with period T defined over $(-T/2, T/2)$, whose Fourier series is

$$f(t) = \frac{a_0}{2} + \sum_{n=1}^{\infty} (a_n\cos n\omega_0 t + b_n\sin n\omega_0 t), \qquad \omega_0 = \frac{2\pi}{T}$$

If $f_e(t)$ and $f_o(t)$ are the even and odd components of $f(t)$, show that $f_e(t)$ and $f_o(t)$ have the Fourier series

$$f_e(t) = \frac{a_0}{2} + \sum_{n=1}^{\infty} a_n\cos n\omega_0 t \quad \text{and} \quad f_o(t) = \sum_{n=1}^{\infty} b_n\sin n\omega_0 t$$

PROBLEM 2-13   Show that the mean-square value of $f(t)$ is equal to the sum of the mean-square values of its even and odd components; that is,

$$\frac{1}{T} \int_{-T/2}^{T/2} [f(t)]^2 \, dt = \frac{1}{T} \int_{-T/2}^{T/2} [f_e(t)]^2 \, dt + \frac{1}{T} \int_{-T/2}^{T/2} [f_o(t)]^2 \, dt$$

PROBLEM 2-14   Let the function $f(t)$ be periodic with period T. If $f(\tfrac{1}{2}T - t) = f(t)$, determine the behavior of the Fourier coefficients a_n and b_n of $f(t)$. Illustrate $f(t)$ graphically.

Answer: $a_{2n+1} = 0, \qquad b_{2n} = 0$

PROBLEM 2-15   If the periodic function $f(t)$ with period T satisfies $f(\tfrac{1}{2}T - t) = -f(t)$, determine the behavior of the Fourier coefficients a_n and b_n of $f(t)$. Illustrate $f(t)$ graphically.

Answer: $a_{2n} = 0, \qquad b_{2n+1} = 0$

PROBLEM 2-16   Suppose $f(t) = 0$ for $-\tfrac{1}{2}T < t < 0$. If the Fourier series expansion of $f(t)$ on the interval $(-T/2, T/2)$ is

$$\frac{1}{2} a_0 + \sum_{n=1}^{\infty} (a_n\cos n\omega_0 t + b_n\sin n\omega_0 t), \qquad \omega_0 = 2\pi/T$$

show that the Fourier cosine series and the Fourier sine series of $f(t)$ on the interval $(0, T/2)$ are

$$a_0 + \sum_{n=1}^{\infty} 2a_n \cos n\omega_0 t \quad \text{and} \quad \sum_{n=1}^{\infty} 2b_n \sin n\omega_0 t$$

PROBLEM 2-17   Represent the following functions $f(t)$ by a Fourier cosine series and graph the corresponding periodic continuation of $f(t)$:

(a) $f(t) = t, \quad 0 < t < \pi;$ (b) $f(t) = \sin\dfrac{\pi}{l}\,t, \quad 0 < t < l$

Answer:

(a) $\dfrac{\pi}{2} - \dfrac{4}{\pi} \displaystyle\sum_{n=1}^{\infty} \dfrac{1}{(2n-1)^2} \cos(2n-1)t$

(b) $\dfrac{2}{\pi} - \dfrac{4}{\pi}\left[\dfrac{1}{1\cdot 3}\cos\left(\dfrac{2\pi}{l}\right)t + \dfrac{1}{3\cdot 5}\cos\left(\dfrac{4\pi}{l}\right)t + \dfrac{1}{5\cdot 7}\cos\left(\dfrac{6\pi}{l}\right)t + \cdots \right]$

PROBLEM 2-18   Represent the following functions $f(t)$ by a Fourier sine series and graph the corresponding periodic continuation of $f(t)$:

(a) $f(t) = \cos t, \quad 0 < t < \pi;$ (b) $f(t) = \pi - t, \quad 0 < t < \pi$

Answer:

(a) $\dfrac{8}{\pi} \displaystyle\sum_{n=1}^{\infty} \dfrac{n}{4n^2-1} \sin 2nt,$ (b) $2 \displaystyle\sum_{n=1}^{\infty} \dfrac{1}{n} \sin nt$

PROBLEM 2-19   Find the Fourier cosine and sine series of

$$f(t) = \begin{cases} \dfrac{1}{4}\pi t & \text{for} \ \ 0 < t < \dfrac{1}{2}\pi \\[2mm] \dfrac{1}{4}\pi t(\pi - t) & \text{for} \ \ \dfrac{1}{2}\pi < t < \pi \end{cases}$$

Answer: $\dfrac{\pi^2}{16} - \displaystyle\sum_{n=1}^{\infty} \dfrac{1}{n}\cos nt, \quad \displaystyle\sum_{n=1}^{\infty} \dfrac{(-1)^{n+1}}{(2n-1)^2}\sin(2n-1)t$

3 DISCRETE FREQUENCY SPECTRA

THIS CHAPTER IS ABOUT

☑ **The Complex Form of Fourier Series**
☑ **Orthogonality of Complex Exponential Functions**
☑ **Complex Frequency Spectra**
☑ **Power Content of a Periodic Function: Parseval's Theorem**

3-1. The Complex Form of Fourier Series

A. The complex Fourier series

Let $f(t)$ be a periodic function with period T. Then the complex Fourier series of $f(t)$ is given by

Complex Fourier series
$$f(t) = \sum_{n=-\infty}^{\infty} c_n e^{jn\omega_0 t} \tag{3.1}$$

where $\omega_0 = 2\pi/T$, and the coefficients c_n are the complex Fourier coefficients. Series (3.1) can be obtained from the trigonometric Fourier series (1.5), as Example 3-1 shows.

EXAMPLE 3-1: Derive expression (3.1).

Solution: The trigonometric Fourier series of a periodic function $f(t)$ given in Chapter 1 (1.5) is

$$f(t) = \frac{1}{2}a_0 + \sum_{n=1}^{\infty} (a_n \cos n\omega_0 t + b_n \sin n\omega_0 t)$$

where $\omega_0 = 2\pi/T$. The cosine and sine can be expressed in terms of the exponentials as

$$\cos n\omega_0 t = \frac{1}{2}(e^{jn\omega_0 t} + e^{-jn\omega_0 t}) \tag{3.2}$$

$$\sin n\omega_0 t = \frac{1}{2j}(e^{jn\omega_0 t} - e^{-jn\omega_0 t}) \tag{3.3}$$

Substituting the cosine (3.2) and sine (3.3) exponential expressions into the trigonometric Fourier series (1.5), we obtain

$$f(t) = \frac{1}{2}a_0 + \sum_{n=1}^{\infty} \left[a_n \frac{1}{2}(e^{jn\omega_0 t} + e^{-jn\omega_0 t}) + b_n \frac{1}{2j}(e^{jn\omega_0 t} - e^{-jn\omega_0 t}) \right] \tag{3.4}$$

Noting that $1/j = -j$, (3.4) can be rewritten as

$$f(t) = \frac{1}{2}a_0 + \sum_{n=1}^{\infty} \left[\frac{1}{2}(a_n - jb_n)e^{jn\omega_0 t} + \frac{1}{2}(a_n + jb_n)e^{-jn\omega_0 t} \right] \tag{3.5}$$

If we let

$$c_0 = \frac{1}{2}a_0, \qquad c_n = \frac{1}{2}(a_n - jb_n), \qquad c_{-n} = \frac{1}{2}(a_n + jb_n) \tag{3.6}$$

then (3.5) reduces to

$$f(t) = c_0 + \sum_{n=1}^{\infty} (c_n e^{jn\omega_0 t} + c_{-n} e^{-jn\omega_0 t})$$

Now, since

$$\sum_{n=1}^{\infty} c_{-n} e^{-jn\omega_0 t} = \sum_{n=-1}^{-\infty} c_n e^{jn\omega_0 t} = \sum_{n=-\infty}^{-1} c_n e^{jn\omega_0 t}$$

and

$$e^{j0\omega_0 t} = e^{j0} = 1$$

we obtain (3.1):

$$f(t) = \sum_{n=-\infty}^{-1} c_n e^{jn\omega_0 t} + c_0 e^{j0\omega_0 t} + \sum_{n=1}^{\infty} c_n e^{jn\omega_0 t}$$

$$= \sum_{n=-\infty}^{\infty} c_n e^{jn\omega_0 t}$$

B. Complex Fourier coefficients

The coefficients c_n can be easily evaluated in terms of a_n and b_n, which we already known. In fact, using the evaluation of Fourier coefficients (1.9)–(1.11) of Chapter 1, we have

$$c_0 = \frac{1}{2} a_0 = \frac{1}{T} \int_{-T/2}^{T/2} f(t)\, dt \tag{3.7}$$

With the use of the identity $e^{-j\theta} = \cos\theta - j\sin\theta$,

$$c_n = \frac{1}{2}(a_n - jb_n)$$

$$= \frac{1}{T}\left[\int_{-T/2}^{T/2} f(t)\cos(n\omega_0 t)\, dt - j\int_{-T/2}^{T/2} f(t)\sin(n\omega_0 t)\, dt \right]$$

$$= \frac{1}{T}\left\{ \int_{-T/2}^{T/2} f(t)[\cos(n\omega_0 t) - j\sin(n\omega_0 t)]\, dt \right\}$$

$$= \frac{1}{T}\int_{-T/2}^{T/2} f(t) e^{-jn\omega_0 t}\, dt \tag{3.8}$$

In a similar fashion, we can obtain

$$c_{-n} = \frac{1}{2}(a_n + jb_n) = \frac{1}{T}\int_{-T/2}^{T/2} f(t) e^{jn\omega_0 t}\, dt \tag{3.9}$$

Equations (3.7), (3.8), and (3.9) may be combined into a single formula

Complex Fourier coefficients
$$c_n = \frac{1}{T}\int_{-T/2}^{T/2} f(t) e^{-jn\omega_0 t}\, dt, \qquad n = 0, \pm 1, \pm 2, \dots \tag{3.10}$$

Since $f(t)e^{-jn\omega_0 t}$ is periodic with period T and since $\int_0^T f(t)\, dt = \int_{-T/2}^{T/2} f(t)\, dt$, which is obtained by setting $a = -T/2$ in (1.4), c_n can also be found from the formula

$$c_n = \frac{1}{T}\int_0^T f(t) e^{-jn\omega_0 t}\, dt \tag{3.11}$$

Next, if $f(t)$ is real, then

Complex coefficients for real f(t)
$$c_n = |c_n| e^{j\phi_n}$$

$$c_{-n} = c_n^* = |c_n| e^{-j\phi_n} \tag{3.12}$$

where the asterisk (*) indicates the complex conjugate, and

Relation of c_n to trigonometric Fourier series coefficients
$$|c_n| = \frac{1}{2}\sqrt{a_n^2 + b_n^2} \tag{3.13}$$

$$\phi_n = \tan^{-1}\left(-\frac{b_n}{a_n}\right) \tag{3.14}$$

for all n except $n = 0$. In that case, c_0 is real and $c_0 = \frac{1}{2} a_0$.

3-2. Orthogonality of Complex Exponential Functions

A. Definition of orthogonality for complex functions

The orthogonality of the sine and cosine functions has been demonstrated in Section 1-3. However, for functions that assume complex values, the concept of orthogonality has to be modified. We call the set of complex functions $\{f_n(t)\}$ orthogonal over the interval $a < t < b$ if

Orthogonality relation
$$\int_a^b f_n(t) f_m^*(t)\, dt = \begin{cases} 0 & \text{for } n \neq m \\ r_n & \text{for } n = m \end{cases} \tag{3.15}$$

where $f_m^*(t)$ is the complex conjugate of $f_m(t)$. For example, if

$$f_m(t) = e^{jm\omega_0 t} = \cos m\omega_0 t + j \sin m\omega_0 t$$

then its complex conjugate is

$$f_m^*(t) = e^{-jm\omega_0 t} = \cos m\omega_0 t - j \sin m\omega_0 t$$

B. Orthogonality relation for complex exponential functions

Let us consider a set of complex exponential functions $\{e^{jn\omega_0 t}\}$, $n = 0, \pm 1, \pm 2, \ldots$, where $\omega_0 = 2\pi/T$. Using elementary calculus, we can show that (see Example 3-2)

Orthogonality relations for complex exponential functions
$$\int_{-T/2}^{T/2} e^{jn\omega_0 t} \cdot 1\, dt = 0 \qquad \text{for } n \neq 0 \tag{3.16}$$

$$\int_{-T/2}^{T/2} e^{jn\omega_0 t} (e^{jm\omega_0 t})^*\, dt = \begin{cases} 0 & \text{for } n \neq m \\ T & \text{for } n = m \end{cases} \tag{3.17}$$

The relations (3.16) and (3.17) show that the functions $\{e^{jn\omega_0 t}\}$, $n = 0, \pm 1, \pm 2, \ldots$, form an orthogonal set of functions over the interval $-T/2 < t < T/2$.

EXAMPLE 3-2: Verify the orthogonality relations (3.16) and (3.17).

Solution: First, we know that

$$\int_{-T/2}^{T/2} e^{jn\omega_0 t} \cdot 1\, dt = \frac{1}{jn\omega_0} e^{jn\omega_0 t} \Big|_{-T/2}^{T/2} = \frac{1}{jn\omega_0} (e^{jn\pi} - e^{-jn\pi}) = 0 \qquad \text{for } n \neq 0$$

Since $e^{jn\pi} = e^{-jn\pi} = (-1)^n$.

Next,

$$\int_{-T/2}^{T/2} e^{jn\omega_0 t} (e^{jm\omega_0 t})^*\, dt = \int_{-T/2}^{T/2} e^{jn\omega_0 t} e^{-jm\omega_0 t}\, dt$$

$$= \int_{-T/2}^{T/2} e^{j(n-m)\omega_0 t}\, dt$$

$$= \frac{1}{j(n-m)\omega_0} e^{j(n-m)\omega_0 t} \Big|_{-T/2}^{T/2}$$

$$= \frac{1}{j(n-m)\omega_0} (e^{j(n-m)\pi} - e^{-j(n-m)\pi})$$

$$= 0 \qquad \text{for } n \neq m$$

since $e^{j(n-m)\pi} = e^{-j(n-m)\pi} = (-1)^{n-m}$.

When $n = m$,

$$\int_{-T/2}^{T/2} e^{jn\omega_0 t} (e^{jn\omega_0 t})^*\, dt = \int_{-T/2}^{T/2} e^{jn\omega_0 t} e^{-jn\omega_0 t}\, dt$$

$$= \int_{-T/2}^{T/2} e^0\, dt = \int_{-T/2}^{T/2} dt = T$$

since $e^0 = 1$.

C. Determination of complex Fourier coefficients

Using the orthogonality of the complex exponential functions set $\{e^{jn\omega_0 t}\}$, we can easily obtain the coefficients of the complex Fourier series (3.1).

EXAMPLE 3-3: Verify expression (3.10).

Solution: Let $f(t)$ be a periodic function with period T, and let the complex exponential Fourier series of $f(t)$ be given by

$$f(t) = \sum_{n=-\infty}^{\infty} c_n e^{jn\omega_0 t}, \qquad \omega_0 = 2\pi/T$$

Now, multiplying both sides by $e^{-jm\omega_0 t}$ and integrating over $-T/2$ to $T/2$, we obtain

$$\int_{-T/2}^{T/2} f(t) e^{-jm\omega_0 t}\, dt = \int_{-T/2}^{T/2} \left(\sum_{n=-\infty}^{\infty} c_n e^{jn\omega_0 t} \right) e^{-jm\omega_0 t}\, dt$$

$$= \sum_{n=-\infty}^{\infty} c_n \left[\int_{-T/2}^{T/2} e^{j(n-m)\omega_0 t}\, dt \right]$$

In view of the orthogonality relations (3.16) and (3.17), the quantity in square brackets is zero except when $n = m$; therefore,

$$\int_{-T/2}^{T/2} f(t) e^{-jm\omega_0 t}\, dt = c_m \int_{-T/2}^{T/2} e^{j0}\, dt = c_m \int_{-T/2}^{T/2} dt = c_m T$$

Hence, by changing m to n, we obtain the combined formula for evaluating complex Fourier series coefficients (3.10):

$$c_n = \frac{1}{T} \int_{-T/2}^{T/2} f(t) e^{-jn\omega_0 t}\, dt$$

3-3. Complex Frequency Spectra

A. Line spectra

A plot of the magnitude of the complex Fourier coefficients c_n in the series (3.1) versus the angular frequency ω is called the **amplitude spectrum** of the periodic function $f(t)$. A plot of the phase angle ϕ_n of c_n versus ω is called the **phase spectrum** of $f(t)$. Since the index n assumes only integers, the amplitude and phase spectra are not continuous curves but appear only at the discrete frequency $n\omega_0$. These are therefore referred to as **discrete frequency spectra** or **line spectra**.

EXAMPLE 3-4: Find the line spectra for the periodic function $f(t)$ shown in Figure 3-1, which consists of a train of identical rectangular pulses of magnitude A and duration d, for **(a)** $d = 1/20$, $T = 1/4$ seconds; and **(b)** $d = 1/20$, $T = 1/2$ seconds.

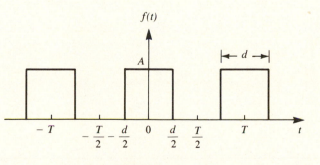

Figure 3-1

Solution: The functions $f(t)$ can be expressed over one period as follows:

$$f(t) = \begin{cases} A & \text{for } -d/2 < t < d/2 \\ 0 & \text{for } -T/2 < t < -d/2, \quad d/2 < t < T/2 \end{cases}$$

Then from (3.10), with $\omega_0 = 2\pi/T$,

$$c_n = \frac{1}{T} \int_{-T/2}^{T/2} f(t)e^{-jn\omega_0 t}\, dt = \frac{A}{T} \int_{-d/2}^{d/2} e^{-jn\omega_0 t}\, dt$$

$$= \frac{A}{T} \frac{1}{-jn\omega_0} e^{-jn\omega_0 t} \Big|_{-d/2}^{d/2}$$

$$= \frac{A}{T} \frac{1}{jn\omega_0} (e^{jn\omega_0 d/2} - e^{-jn\omega_0 d/2})$$

$$= \frac{Ad}{T} \frac{1}{(n\omega_0 d/2)} \frac{1}{2j} (e^{jn\omega_0 d/2} - e^{-jn\omega_0 d/2})$$

$$= \frac{Ad}{T} \frac{\sin(n\omega_0 d/2)}{(n\omega_0 d/2)} \tag{3.18}$$

But $n\omega_0 d/2 = n\pi d/T$; hence,

$$c_n = \frac{Ad}{T} \frac{\sin(n\pi d/T)}{(n\pi d/T)} \tag{3.19}$$

It is seen from (3.18) or (3.19) that c_n is real. The amplitude spectrum is obtained by plotting $|c_n|$ versus the discrete frequency $n\omega_0$. Equation (3.18) has values only for the discrete frequency $n\omega_0$; that is, the magnitude spectrum exists only at

$$\omega = 0, \pm 2\pi/T, \pm 4\pi/T, \ldots, \text{etc.}$$

(a) For $d = 1/20$ and $T = 1/4$ second,

$$\omega_0 = 2\pi/T = 8\pi, \qquad d/T = 1/5$$

Hence, the amplitude spectrum exists at

$$\omega = 0, \pm 8\pi, \pm 16\pi, \ldots, \text{etc.}$$

and is shown in Figure 3-2a.

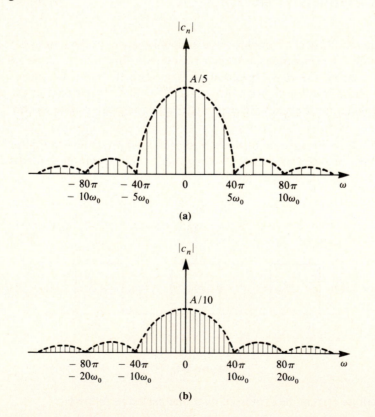

(a)

(b)

Figure 3-2 *The amplitude spectra: (a) $d = 1/20$, $T = 1/4$, $d/T = 1/5$; $\omega_0 = 2\pi/T = 8\pi$. (b) $d = 1/20$, $T = 1/2$, $d/T = 1/10$; $\omega_0 = 2\pi/T = 4\pi$.*

Since $d/T = 1/5$, the amplitude spectrum becomes zero at the value of $n\omega_0$, for which

$$n\omega_0(d/2) = m\pi \quad \text{or} \quad n\pi(d/T) = n\pi(1/5) = m\pi \qquad (m = \pm 1, \pm 2, \ldots)$$

That is, at $\omega = \pm 5\omega_0 = \pm 40\pi$, $\pm 10\omega_0 = \pm 80\pi$, $\pm 15\omega_0 = \pm 120\pi, \ldots$.
   (b) For $d = 1/20$ and $T = 1/2$ second,

$$\omega_0 = 2\pi/T = 4\pi, \qquad d/T = 1/10$$

Hence, the amplitude spectrum exists at $\omega = 0, \pm 4\pi, \pm 8\pi, \ldots$ and becomes zero at the value of $n\omega_0$, for which

$$n\omega_0 d/2 = m\pi \quad \text{or} \quad n\pi d/T = n\pi(1/10) = m\pi \qquad (m = \pm 1, \pm 2, \ldots)$$

That is, at $\omega = \pm 10\omega_0 = \pm 40\pi$, $\pm 20\omega_0 = \pm 80\pi$, $\pm 30\omega_0 = \pm 120\pi, \ldots$. The amplitude spectrum for this case is shown in Figure 3-2b.

B. Sampling function

In (3.18) or (3.19), if we set $n\omega_0 d/2 = n\pi d/T = x_n$, then

$$c_n = \frac{Ad}{T} \frac{\sin x_n}{x_n}$$

The envelope of c_n is a continuous function found by replacing $n\omega_0$ with ω or replacing x_n by x. The function

Sampling function $$Sa(x) = \frac{\sin x}{x} \qquad (3.20)$$

is known as the **sampling function.** The sampling function is shown in Figure 3-3. Note that the function has zeros at $x = \pm n\pi$, $n = 1, 2, \ldots$, etc.

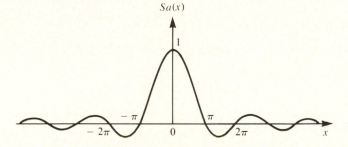

Figure 3-3   The sampling function.

3-4. Power Content of a Periodic Function: Parseval's Theorem

A. Power content

The **power content** of a periodic function $f(t)$ with period T is defined as the mean-square value

Power content $$\frac{1}{T} \int_{-T/2}^{T/2} [f(t)]^2 \, dt \qquad (3.21)$$

If we assume that the function $f(t)$ is a voltage or a current waveform, then (3.21) represents the average power delivered by $f(t)$ to a 1-Ω resistor.

B. Parseval's theorem

Parseval's theorem states that if $f(t)$ is a real periodic function with period T, then

Parseval's theorem $$\frac{1}{T} \int_{-T/2}^{T/2} [f(t)]^2 \, dt = \sum_{n=-\infty}^{\infty} |c_n|^2 \qquad (3.22)$$

where the c_n's are the complex Fourier coefficients of $f(t)$.
   Let $f_1(t)$ and $f_2(t)$ be two periodic functions having the same period T. Then

$$\frac{1}{T} \int_{-T/2}^{T/2} f_1(t) f_2(t) \, dt = \sum_{n=-\infty}^{\infty} (c_1)_n (c_2)_{-n} \qquad (3.23)$$

where $(c_1)_n$ and $(c_2)_n$ are the complex Fourier coefficients of $f_1(t)$ and $f_2(t)$, respectively.

EXAMPLE 3-5: Prove (3.23).

Proof: Let

$$f_1(t) = \sum_{n=-\infty}^{\infty} (c_1)_n e^{jn\omega_0 t}, \qquad \omega_0 = 2\pi/T \tag{3.24}$$

where

$$(c_1)_n = \frac{1}{T} \int_{-T/2}^{T/2} f_1(t) e^{-jn\omega_0 t} dt \tag{3.25}$$

Let

$$f_2(t) = \sum_{n=-\infty}^{\infty} (c_2)_n e^{jn\omega_0 t} \tag{3.26}$$

where

$$(c_2)_n = \frac{1}{T} \int_{-T/2}^{T/2} f_2(t) e^{-jn\omega_0 t} dt \tag{3.27}$$

Then,

$$\int_{-T/2}^{T/2} f_1(t) f_2(t) dt = \frac{1}{T} \int_{-T/2}^{T/2} \left[\sum_{n=-\infty}^{\infty} (c_1)_n e^{jn\omega_0 t} f_2(t) dt \right]$$

$$= \sum_{n=-\infty}^{\infty} (c_1)_n \left[\frac{1}{T} \int_{-T/2}^{T/2} f_2(t) e^{jn\omega_0 t} dt \right] \tag{3.28}$$

In view of (3.27), we have

$$\frac{1}{T} \int_{-T/2}^{T/2} f_2(t) e^{jn\omega_0 t} dt = \frac{1}{T} \int_{-T/2}^{T/2} f_2(t) e^{-j(-n)\omega_0 t} dt = (c_2)_{-n}$$

Hence, (3.28) reduces to

$$\frac{1}{T} \int_{-T/2}^{T/2} f_1(t) f_2(t) dt = \sum_{n=-\infty}^{\infty} (c_1)_n (c_2)_{-n}$$

EXAMPLE 3-6: Prove Parseval's theorem (3.22).

Proof: By putting $f_1(t) = f_2(t) = f(t)$ in (3.23), we obtain

$$\frac{1}{T} \int_{-T/2}^{T/2} [f(t)]^2 dt = \sum_{n=-\infty}^{\infty} c_n c_{-n} \tag{3.29}$$

If $f(t)$ is real, then, from (3.12),

$$c_{-n} = c_n^*$$

Hence, Parseval's theorem (3.22) is proved:

$$\frac{1}{T} \int_{-T/2}^{T/2} [f(t)]^2 dt = \sum_{n=-\infty}^{\infty} c_n c_n^* = \sum_{n=-\infty}^{\infty} |c_n|^2$$

SUMMARY

1. A periodic function $f(t)$ with period T can be represented by the complex Fourier series

$$f(t) = \sum_{n=-\infty}^{\infty} c_n e^{jn\omega_0 t}, \qquad \omega_0 = 2\pi/T$$

2. The coefficients of the complex Fourier series, c_n, are given by

$$c_n = \frac{1}{T} \int_{-T/2}^{T/2} f(t) e^{-jn\omega_0 t} dt, \qquad (n = 0, \pm 1, \pm 2, \ldots)$$

3. If $f(t)$ is real, then

$$c_n = |c_n| e^{j\phi_n}, \qquad c_{-n} = c_n^* = |c_n| e^{-j\phi_n}$$

and the c_n's are related to the trigonometric Fourier series coefficients by

$$|c_n| = \frac{1}{2}\sqrt{a_n^2 + b_n^2}, \qquad \phi_n = \tan^{-1}\left(-\frac{b_n}{a_n}\right)$$

4. Plots of $|c_n|$ versus ω are called the amplitude spectrum of $f(t)$, and plots of ϕ_n versus ω are called the phase spectrum of $f(t)$.
5. The power content of a periodic function $f(t)$ is defined as

$$\frac{1}{T}\int_{-T/2}^{T/2} [f(t)]^2\, dt$$

6. Parseval's theorem states that

$$\frac{1}{T}\int_{-T/2}^{T/2} [f(t)]^2\, dt = \sum_{n=-\infty}^{\infty} |c_n|^2$$

RAISE YOUR GRADES

Can you explain...?

☑ the relationship between the complex Fourier series representation and the trigonometric Fourier series representation of a periodic function
☑ what the orthogonality properties of exponential complex functions over a period are
☑ what line spectra of a periodic function are

SOLVED PROBLEMS

Complex Form of Fourier Series

PROBLEM 3-1 Find the complex Fourier series of the sawtooth function $f(t)$ shown in Figure 3-4 and defined by

$$f(t) = \frac{A}{T}t, \qquad 0 < t < T; \qquad f(t + T) = f(t)$$

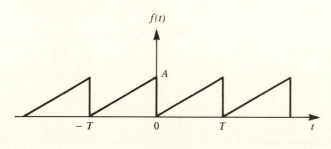

Figure 3-4

Solution: The complex Fourier series of $f(t)$ is given by (3.1):

$$f(t) = \sum_{n=-\infty}^{\infty} c_n e^{jn\omega_0 t}, \qquad \omega_0 = 2\pi/T$$

The coefficients c_n can be found from (3.11):

$$c_n = \frac{1}{T}\int_0^T f(t)e^{-jn\omega_0 t}\, dt$$

Thus

$$c_n = \frac{A}{T^2} \int_0^T t e^{-jn\omega_0 t} \, dt$$

$$= \frac{A}{T^2} \left(\frac{t e^{-jn\omega_0 t}}{-jn\omega_0} \bigg|_0^T + \frac{1}{jn\omega_0} \int_0^T e^{-jn\omega_0 t} \, dt \right)$$

$$= \frac{A}{T^2} \left[\frac{T e^{-jn2\pi}}{-jn\omega_0} - \frac{1}{(jn\omega_0)^2} (e^{-jn2\pi} - 1) \right]$$

Since $e^{-jn2\pi} = 1$ and $e^{j\pi/2} = j$,

$$c_n = j \frac{A}{n\omega_0 T} = j \frac{A}{2\pi n} = \frac{A}{2\pi n} e^{j\pi/2}$$

Certainly, this has no meaning for $n = 0$. Therefore, for $n = 0$, from (3.7),

$$c_0 = \frac{1}{T} \int_0^T f(t) \, dt = \frac{A}{T^2} \int_0^T t \, dt = \frac{1}{2} A$$

Hence,

$$f(t) = \frac{A}{2} + j \frac{A}{2\pi} \sum_{n=-\infty}^{\infty}{}' \frac{1}{n} e^{jn\omega_0 t} = \frac{A}{2} + \frac{A}{2\pi} \sum_{n=-\infty}^{\infty}{}' \frac{1}{n} e^{j(n\omega_0 t + \pi/2)}$$

where $\displaystyle\sum_{n=-\infty}^{\infty}{}'$ means that the summation is over nonzero integers.

PROBLEM 3-2   Find the complex Fourier series of a rectified sine wave periodic function $f(t)$ shown in Figure 3-5 and defined by

$$f(t) = A \sin \pi t, \qquad 0 < t < 1; \qquad f(t + T) = f(t), \qquad T = 1$$

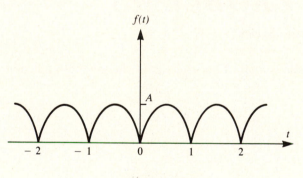

Figure 3-5

Solution: Since the period $T = 1$, $\omega_0 = 2\pi/T = 2\pi$; hence, the complex Fourier series is given by

$$f(t) = \sum_{n=-\infty}^{\infty} c_n e^{j2\pi nt}$$

From (3.11), the coefficients c_n are

$$c_n = \frac{1}{T} \int_0^T f(t) e^{-j2\pi nt} \, dt = \int_0^1 A \sin \pi t \, e^{-j2\pi nt} \, dt$$

$$= A \int_0^1 \frac{1}{2j} (e^{j\pi t} - e^{-j\pi t}) e^{-j2\pi nt} \, dt$$

$$= \frac{A}{2j} \int_0^1 \left[e^{-j\pi(2n-1)t} - e^{-j\pi(2n+1)t} \right] dt$$

$$= \frac{A}{2j} \left[\frac{e^{-j\pi(2n-1)t}}{-j\pi(2n-1)} - \frac{e^{-j\pi(2n+1)t}}{-j\pi(2n+1)} \right] \Bigg|_0^1$$

Since $e^{\pm j2\pi n} = 1$ and $e^{+j\pi} = e^{-j\pi}$,

$$c_n = \frac{-2A}{\pi(4n^2 - 1)}$$

We can use (3.7) to check this result for $n = 0$; thus,

$$c_0 = \frac{1}{T} \int_0^T f(t)\, dt = \frac{2A}{\pi}$$

Hence,

$$f(t) = -\frac{2A}{\pi} \sum_{n=-\infty}^{\infty} \frac{1}{4n^2 - 1} e^{j2\pi nt}$$

PROBLEM 3-3 Find the complex Fourier series for the function $f(t)$ defined by $f(t) = \sin^4 t$ for $(0, \pi)$ and $f(t + \pi) = f(t)$.

Solution: As in Problem 1-10, we shall make use of the identity $\sin t = (e^{jt} - e^{-jt})/2j$. Now we can write

$$\sin^4 t = \left(\frac{e^{jt} - e^{-jt}}{2j}\right)^4 = \frac{1}{16}(e^{j4t} - 4e^{j2t} + 6 - 4e^{-j2t} + e^{-j4t})$$

which is the required complex Fourier series. Here the complex Fourier series has only five terms.

Orthogonality of Complex Functions

PROBLEM 3-4 A set of infinite complex functions $\{\psi_n(t)\}$, where $n = 1, 2, \ldots$, is said to be an orthonormal set on the interval (a, b) if

$$\int_a^b \psi_n(t)\psi_m^*(t)\, dt = \delta_{mn}$$

where $*$ denotes the complex conjugate and δ_{mn} is Kronecker's delta, defined by

$$\delta_{mn} = \begin{cases} 1, & m = n \\ 0, & m \neq n \end{cases}$$

Let $f(t)$ be a function defined on the interval (a, b) and suppose that $f(t)$ can be represented as

$$f(t) = c_1\psi_1(t) + c_2\psi_2(t) + \cdots = \sum_{n=1}^{\infty} c_n\psi_n(t)$$

on (a, b) everywhere, where c_n are constants. Show that

$$c_n = \int_a^b f(t)\psi_n^*(t)\, dt, \qquad n = 1, 2, \ldots$$

Solution: Multiplying $f(t)$ by $\psi_m^*(t)$ and integrating over the interval (a, b), we obtain

$$\int_a^b f(t)\psi_m^*(t)\, dt = \int_a^b \left[\sum_{n=1}^{\infty} c_n\psi_n(t)\right]\psi_m^*(t)\, dt$$

$$= \sum_{n=1}^{\infty} c_n \int_a^b \psi_n(t)\psi_m^*(t)\, dt$$

$$= \sum_{n=1}^{\infty} c_n\delta_{nm} = c_m$$

Changing m to n, we obtain

$$c_n = \int_a^b f(t)\psi_n^*(t)\, dt$$

Complex Frequency Spectra

PROBLEM 3-5 Find the frequency spectra for the periodic function $f(t)$ shown in Figure 3-6.

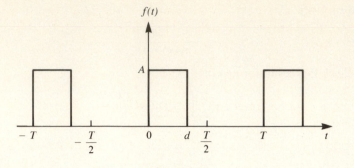

Figure 3-6

Solution: From (3.11), with $\omega_0 = 2\pi/T$, we obtain

$$c_n = \frac{1}{T}\int_0^T f(t)e^{-jn\omega_0 t}\,dt = \frac{A}{T}\int_0^d e^{-jn\omega_0 t}\,dt$$

$$= \frac{A}{T}\frac{1}{-jn\omega_0}e^{-jn\omega_0 t}\Bigg|_0^d$$

$$= \frac{A}{T}\frac{1}{jn\omega_0}(1 - e^{-jn\omega_0 d})$$

$$= \frac{A}{T}\frac{1}{jn\omega_0}e^{-jn\omega_0 d/2}(e^{jn\omega_0 d/2} - e^{-jn\omega_0 d/2})$$

$$= \frac{Ad}{T}\frac{\sin(n\omega_0 d/2)}{(n\omega_0 d/2)}e^{-jn\omega_0 d/2}$$

Hence,

$$|c_n| = \frac{Ad}{T}\left|\frac{\sin(n\omega_0 d/2)}{(n\omega_0 d/2)}\right|$$

since $|e^{-jn\omega_0 d/2}| = 1$.

The amplitude spectrum is exactly the same as that of Example 3-4 and is not affected by the shift of the origin, but the phase spectrum is changed by an amount of $-n\omega_0 d/2 = -n\pi d/T$ radians at frequency $n\omega_0$.

PROBLEM 3-6   Show that the time displacement in a periodic function has no effect on the magnitude spectrum, but changes the phase spectrum by an amount of $-n\omega_0\tau$ radians for the component at the frequency $n\omega_0$ if the time displacement is τ.

Solution: Let $f(t)$ be a periodic function with period T, and let its complex Fourier series be given by

$$f(t) = \sum_{n=-\infty}^{\infty} c_n e^{jn\omega_0 t}, \qquad \omega_0 = 2\pi/T$$

Now, changing t to $t - \tau$, we have

$$f(t-\tau) = \sum_{n=-\infty}^{\infty} c_n e^{jn\omega_0(t-\tau)} = \sum_{n=-\infty}^{\infty} c_n e^{-jn\omega_0\tau}e^{jn\omega_0 t} = \sum_{n=-\infty}^{\infty} c_n' e^{jn\omega_0 t}$$

where

$$c_n' = c_n e^{-jn\omega_0\tau}$$

Hence, if

$$c_n = |c_n|e^{j\phi_n}$$

then

$$c_n' = |c_n|e^{j(\phi_n - n\omega_0\tau)}$$

Thus it is seen that the magnitude spectra of $f(t)$ and $f(t - \tau)$ are the same; however, the phase spectra are different. The time displacement τ causes a lag of phase by an amount of $n\omega_0\tau$ radians in the frequency component $n\omega_0$.

Power Content of a Periodic Function

PROBLEM 3-7 Using Parseval's theorem (3.22), derive Parseval's identity (1.27)

$$\frac{1}{T}\int_{-T/2}^{T/2}[f(t)]^2\,dt = \frac{1}{4}a_0^2 + \frac{1}{2}\sum_{n=1}^{\infty}(a_n^2 + b_n^2)$$

where a_n and b_n are the trigonometric Fourier coefficients of $f(t)$.

Solution: From (3.6), we have

$$c_0 = \frac{1}{2}a_0, \qquad c_n = \frac{1}{2}(a_n - jb_n), \qquad c_{-n} = \frac{1}{2}(a_n + jb_n)$$

Hence,

$$c_0^2 = \frac{1}{4}a_0^2, \qquad |c_n|^2 = \frac{1}{4}(a_n^2 + b_n^2) = |c_{-n}|^2$$

Substituting these results into (3.22), we obtain

$$\frac{1}{T}\int_{-T/2}^{T/2}[f(t)]^2\,dt = \sum_{n=-\infty}^{\infty}|c_n|^2$$

$$= |c_0|^2 + 2\sum_{n=1}^{\infty}|c_n|^2 = \frac{1}{4}a_0^2 + \frac{1}{2}\sum_{n=1}^{\infty}(a_n^2 + b_n^2)$$

PROBLEM 3-8 Show that the mean-square value of a periodic function $f(t)$ is the sum of the mean-square values of its harmonics.

Solution: From eq. (1.6) of Chapter 1, we have

$$f(t) = C_0 + \sum_{n=1}^{\infty}C_n\cos(n\omega_0 t - \theta_n)$$

The mean-square value of the nth harmonic of $f(t)$, $C_n\cos(n\omega_0 t - \theta_n)$, can be evaluated as follows:

$$\frac{1}{T}\int_{-T/2}^{T/2}[C_n\cos(n\omega_0 t - \theta_n)]^2\,dt = \frac{1}{T}\int_{-T/2}^{T/2}C_n^2\cos^2(n\omega_0 t - \theta_n)\,dt$$

$$= \frac{1}{T}\int_{-T/2}^{T/2}C_n^2\frac{1}{2}[1 + \cos 2(n\omega_0 t - \theta_n)]\,dt$$

$$= \frac{1}{2}C_n^2$$

with the use of identity $\cos^2\theta = \frac{1}{2}(1 + \cos 2\theta)$ and

$$\int_{-T/2}^{T/2}\cos 2(n\omega_0 t - \theta_n)\,dt = 0$$

Thus, the mean-square value of the nth harmonic is $C_n^2/2$.

From Example 1-4 of Chapter 1, we have

$$C_n = \sqrt{a_n^2 + b_n^2} = 2|c_n|, \qquad C_0 = \frac{1}{2}a_0 = |c_0|$$

Hence,

$$|c_n|^2 = \frac{1}{4}C_n^2, \qquad |c_0|^2 = C_0^2$$

Then, from the equation obtained in Problem 3-7, we get

$$\frac{1}{T}\int_{-T/2}^{T/2}[f(t)]^2\,dt = |c_0|^2 + 2\sum_{n=1}^{\infty}|c_n|^2 = C_0^2 + \sum_{n=1}^{\infty}\left|\frac{C_n}{\sqrt{2}}\right|^2$$

which indicates that the mean-square value of a periodic function $f(t)$ is the sum of the mean-square values of its harmonics.

Supplementary Exercises

PROBLEM 3-9   Show that the complex Fourier coefficients of an even periodic function are real and those of an odd periodic function are purely imaginary.

PROBLEM 3-10   If $f(t)$ and $g(t)$ are periodic functions with period T and their Fourier expansions are

$$f(t) = \sum_{n=-\infty}^{\infty} c_n e^{jn\omega_0 t}, \qquad g(t) = \sum_{n=-\infty}^{\infty} d_n e^{jn\omega_0 t} \qquad \text{for } \omega_0 = \frac{2\pi}{T}$$

show that the function

$$h(t) = \frac{1}{T} \int_{-T/2}^{T/2} f(t - \tau) g(\tau) \, d\tau$$

is a periodic function with the same period T and can be expressed as

$$h(t) = \sum_{n=-\infty}^{\infty} c_n d_n e^{jn\omega_0 t}$$

PROBLEM 3-11   If $f(t)$ and $g(t)$ are periodic functions with period T and their Fourier expansions are

$$f(t) = \sum_{n=-\infty}^{\infty} c_n e^{jn\omega_0 t}, \qquad g(t) = \sum_{n=-\infty}^{\infty} d_n e^{jn\omega_0 t} \qquad \text{for } \omega_0 = \frac{2\pi}{T}$$

show that the function $h(t) = f(t)g(t)$ is a periodic function with the same period T and can be expressed as

$$h(t) = \sum_{n=-\infty}^{\infty} \alpha_n e^{jn\omega_0 t}$$

where $\alpha_n = \sum_{k=-\infty}^{\infty} c_{n-k} d_k$.
[*Hint:* Show that $\alpha_n = \sum_{k=-\infty}^{\infty} c_{n-k} d_k$ are the Fourier coefficients of $h(t)$.]

PROBLEM 3-12   Find the complex Fourier series for the function $f(t)$ defined by $f(t) = e^t$ for $(0, 2\pi)$ and $f(t + 2\pi) = f(t)$ by direct integration.

Answer: $\dfrac{e^{2\pi} - 1}{2\pi} \displaystyle\sum_{n=-\infty}^{\infty} \dfrac{1}{1 - jn} e^{jnt}$

PROBLEM 3-13   Reduce the result of Problem 3-12 to the trigonometric form of the Fourier series.

Answer: $\dfrac{e^{2\pi} - 1}{\pi} \left[\dfrac{1}{2} + \displaystyle\sum_{n=1}^{\infty} \dfrac{1}{1 + n^2} (\cos nt - n \sin nt) \right]$

PROBLEM 3-14   Find the complex Fourier coefficients and sketch the frequency spectra for a half-rectified sine wave $f(t)$ defined by

$$f(t) = \begin{cases} A \sin \omega_0 t & \text{for} \quad 0 < t < T/2 \\ 0 & \text{for} \quad T/2 < t < T \end{cases}$$

and $f(t + T) = f(t)$, where $\omega_0 = 2\pi/T$.

Answer: $c_n = \dfrac{1}{2\pi(1 - n^2)} (1 + e^{-jn\pi})$, and note that $c_1 = c_{-1} = -\dfrac{j}{4}$ and $c_{2m+1} = 0$, where $m = 1, 2, \ldots$.

PROBLEM 3-15   Find the complex Fourier coefficients and sketch the frequency spectra for the sawtooth wave function $f(t)$ defined by $f(t) = -t/T + \frac{1}{2}$ for $0 < t < T$ and $f(t + T) = f(t)$.

Answer: $c_n = \dfrac{1}{j2\pi n}$, $c_0 = 0$

PROBLEM 3-16   Apply Parseval's theorem (3.22) to the result of Problem 3-15 to prove that

$$\sum_{n=1}^{\infty} \frac{1}{n^2} = \frac{\pi^2}{6}$$

PROBLEM 3-17   Let $f_1(t)$ and $f_2(t)$ be two periodic functions having the same period T. Show that

$$\frac{1}{T} \int_{-T/2}^{T/2} f_1(t + \tau) f_2(t)\, dt = \sum_{n=-\infty}^{\infty} (c_1)_n (c_2)_{-n} e^{jn\omega_0 \tau}$$

where $(c_1)_n$ and $(c_2)_n$ are the complex Fourier coefficients of $f_1(t)$ and $f_2(t)$, respectively, and $\omega_0 = 2\pi/T$.

PROBLEM 3-18   Show that if $f(t)$ is a real periodic function with period T, then

$$\frac{1}{T} \int_{-T/2}^{T/2} f(t + \tau) f(t)\, dt = \sum_{n=-\infty}^{\infty} |c_n|^2 e^{jn\omega_0 \tau}$$

where the c_n's are the complex Fourier coefficients of $f(t)$, and $\omega_0 = 2\pi/T$.

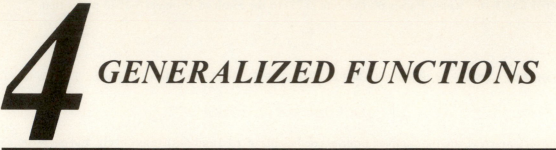

4 GENERALIZED FUNCTIONS

THIS CHAPTER IS ABOUT

☑ **The Unit Impulse Function $\delta(t)$**
☑ **The Unit Step Function $u(t)$**
☑ **Generalized Derivatives**
☑ **Generalized Derivative of a Discontinuous Function**
☑ **Fourier Series of Derivatives of Discontinuous Periodic Functions**
☑ **Evaluation of Fourier Coefficients by Differentiation**

4-1. The Unit Impulse Function $\delta(t)$

The **unit impulse function $\delta(t)$**, also known as the **delta function** (δ-function), may be defined in many ways. Usually it is expressed by the relation

Unit impulse
function
(δ-function)

$$\delta(t) = \begin{cases} 0 & \text{if} \quad t \neq 0 \\ \infty & \text{if} \quad t = 0 \end{cases}$$

$$\int_{-\infty}^{\infty} \delta(t)\,dt = \int_{-\varepsilon}^{\varepsilon} \delta(t)\,dt = 1, \qquad \varepsilon > 0$$

If we use this kind of definition, we may encounter some mathematical difficulties. The δ-function can also be defined in terms of its integral properties alone.

A. Definition of $\delta(t)$ as a generalized function

A **testing function $\phi(t)$** is a function that is continuous, has continuous derivatives of all orders, and vanishes identically outside some finite interval. Using the concept of testing functions, the δ-function $\delta(t)$ can be defined in the sense of a so-called **generalized** (or **symbolic**) **function** by the relation

δ-Function

$$\int_{-\infty}^{\infty} \delta(t)\phi(t)\,dt = \phi(0) \tag{4.1}$$

Expression (4.1) has no meaning as an ordinary integral. Instead, the integral and the function $\delta(t)$ are merely defined by the number $\phi(0)$ assigned to the function $\phi(t)$.

With this understanding, we can handle $\delta(t)$ as if it were an ordinary function, except that we'll never talk about the *value* of $\delta(t)$. Instead, we talk about the *values of integrals* involving $\delta(t)$.

B. The equivalence property

The equivalence property is one of the basic properties used in studying a generalized function, such as $\delta(t)$, by means of the testing function $\phi(t)$. The **equivalence property** states that two generalized functions possessing the same evaluation on each testing function are identical. Let $g_1(t)$ and $g_2(t)$ be generalized functions. Then the equivalence property states that $g_1(t) = g_2(t)$ if and only if

Equivalence property

$$\int_{-\infty}^{\infty} g_1(t)\phi(t)\,dt = \int_{-\infty}^{\infty} g_2(t)\phi(t)\,dt \tag{4.2}$$

for all testing functions $\phi(t)$.

C. Properties of $\delta(t)$

1. $\qquad \displaystyle\int_{-\infty}^{\infty} \delta(t - t_0)\phi(t)\,dt = \phi(t_0)$ $\hfill$ (4.3)

2. $\qquad \delta(at) = \dfrac{1}{|a|}\,\delta(t)$ $\hfill$ (4.4)

3. $\qquad \delta(-t) = \delta(t)$ $\hfill$ (4.5)

4. $\qquad f(t)\delta(t) = f(0)\delta(t),$ $\qquad$ where $f(t)$ is continuous at $t = 0$ $\hfill$ (4.6)

EXAMPLE 4-1: Prove property (4.3): $\int_{-\infty}^{\infty} \delta(t - t_0)\phi(t)\,dt = \phi(t_0)$.

Proof: With a formal change in the independent variable; $t - t_0 = \tau$, and hence $t = \tau + t_0, dt = d\tau$, we have

$$\int_{-\infty}^{\infty} \delta(t - t_0)\phi(t)\,dt = \int_{-\infty}^{\infty} \delta(\tau)\phi(\tau + t_0)\,d\tau = \int_{-\infty}^{\infty} \delta(t)\phi(t + t_0)\,dt \qquad (4.7)$$

Then, by definition (4.1), we get

$$\int_{-\infty}^{\infty} \delta(t)\phi(t + t_0)\,dt = \phi(t + t_0)\Big|_{t=0} = \phi(t_0) \qquad (4.8)$$

Note that property (4.3) can also be used as a definition of $\delta(t - t_0)$.

EXAMPLE 4-2: Prove property (4.4): $\delta(at) = (1/|a|)\,\delta(t)$.

Proof: With a change of variable, $at = \tau$, and hence $t = \tau/a, dt = (1/a)\,d\tau$, we obtain the following equations:

If $a > 0$,

$$\int_{-\infty}^{\infty} \delta(at)\phi(t)\,dt = \frac{1}{a}\int_{-\infty}^{\infty} \delta(\tau)\phi\left(\frac{\tau}{a}\right)d\tau$$

$$= \frac{1}{a}\int_{-\infty}^{\infty} \delta(t)\phi\left(\frac{t}{a}\right)dt = \frac{1}{a}\,\phi\left(\frac{t}{a}\right)\Big|_{t=0} = \frac{1}{|a|}\,\phi(0)$$

If $a < 0$,

$$\int_{-\infty}^{\infty} \delta(at)\phi(t)\,dt = \frac{1}{a}\int_{\infty}^{-\infty} \delta(\tau)\phi\left(\frac{\tau}{a}\right)d\tau$$

$$= \frac{1}{-a}\int_{-\infty}^{\infty} \delta(t)\phi\left(\frac{t}{a}\right)dt = \frac{1}{-a}\,\phi(0) = \frac{1}{|a|}\,\phi(0)$$

Thus, for any a

$$\int_{-\infty}^{\infty} \delta(at)\phi(t)\,dt = \frac{1}{|a|}\,\phi(0) \qquad (4.9)$$

Now, substituting (4.1) into (4.9), we obtain

$$\int_{-\infty}^{\infty} \delta(at)\phi(t)\,dt = \frac{1}{|a|}\,\phi(0)$$

$$= \frac{1}{|a|}\int_{-\infty}^{\infty} \delta(t)\phi(t)\,dt = \int_{-\infty}^{\infty} \frac{1}{|a|}\,\delta(t)\phi(t)\,dt \qquad (4.10)$$

for any $\phi(t)$. Then, applying the equivalence property (4.2), we prove (4.4):

$$\delta(at) = \frac{1}{|a|}\,\delta(t)$$

EXAMPLE 4-3: Prove property (4.5): $\delta(-t) = \delta(t)$.

Proof: Setting $a = -1$ in (4.4), we obtain

$$\delta(-t) = \frac{1}{|-1|}\delta(t)$$

which shows that $\delta(-t) = \delta(t)$ (4.4); that is, $\delta(t)$ is an even function.

EXAMPLE 4-4: Prove property (4.6): $f(t)\delta(t) = f(0)\delta(t)$.

Proof: If $f(t)$ is a continuous function, then

$$\int_{-\infty}^{\infty} [f(t)\delta(t)]\phi(t)\,dt = \int_{-\infty}^{\infty} \delta(t)[f(t)\phi(t)]\,dt = f(0)\phi(0)$$

$$= f(0)\int_{-\infty}^{\infty} \delta(t)\phi(t)\,dt = \int_{-\infty}^{\infty} [f(0)\delta(t)]\phi(t)\,dt \qquad (4.11)$$

for all $\phi(t)$. Hence, by the equivalence property (4.2), we conclude that (4.6) holds:

$$f(t)\delta(t) = f(0)\delta(t)$$

For convenience, $\delta(t)$ and $\delta(t - t_0)$ are shown schematically in Figure 4-1.

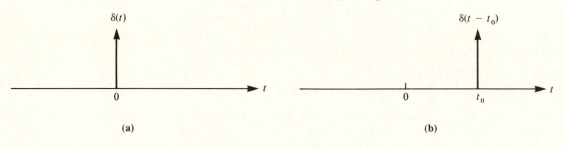

(a)   (b)

Figure 4-1 The unit impulse function.

4-2. The Unit Step Function $u(t)$

A. Conventional definition of $u(t)$

The ***unit step function $u(t)$***, also known as the **Heaviside unit function**, is customarily defined as (see Figure 4-2)

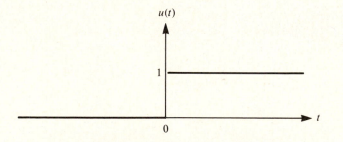

Figure 4-2 The unit step function.

Unit step function (conventional)	$$u(t) = \begin{cases} 1 & \text{for } t > 0 \\ 0 & \text{for } t < 0 \end{cases}$$	(4.12)

B. Definition of $u(t)$ as a generalized function

Using the concept of the testing functions $\phi(t)$, the unit step function can also be defined by the following relation:

Unit step function (generalized)	$$\int_{-\infty}^{\infty} u(t)\phi(t)\,dt = \int_{0}^{\infty} \phi(t)\,dt$$	(4.13)

Note that $\int_{0}^{\infty} \phi(t)\,dt$ is the area under $\phi(t)$ over $0 < t < \infty$.

C. Equivalence of definitions (4.12) and (4.13)

Using the definition of $u(t)$ as a generalized function (4.13), the conventional definition (4.12) can be derived as follows.

EXAMPLE 4-5: Derive expression (4.12) by definition (4.13).

Solution: Rewriting definition (4.13) as

$$\int_{-\infty}^{\infty} u(t)\phi(t)dt = \int_{-\infty}^{0} u(t)\phi(t)dt + \int_{0}^{\infty} u(t)\phi(t)dt = \int_{0}^{\infty} \phi(t)dt$$

we obtain

$$\int_{-\infty}^{0} u(t)\phi(t)\,dt = \int_{0}^{\infty} [1 - u(t)]\phi(t)\,dt$$

This can be true only if

$$\int_{-\infty}^{0} u(t)\phi(t)\,dt = 0 \quad \text{and} \quad \int_{0}^{\infty} [1 - u(t)]\phi(t)\,dt = 0$$

These conditions imply that

$$u(t)\phi(t) = 0 \qquad \text{for } t < 0$$

and

$$[1 - u(t)]\phi(t) = 0 \qquad \text{for } t > 0$$

Since $\phi(t)$ is arbitrary, we have

$$u(t) = 0 \quad \text{for } t < 0 \qquad \text{and} \qquad 1 - u(t) = 0 \quad \text{for } t > 0$$

that is,

$$u(t) = \begin{cases} 1 & \text{for } t > 0 \\ 0 & \text{for } t < 0 \end{cases}$$

Similarly, we can define (see Figure 4-3)

$$u(t - t_0) = \begin{cases} 1 & \text{for } t - t_0 > 0 \quad \text{or} \quad t > t_0 \\ 0 & \text{for } t - t_0 < 0 \quad \text{or} \quad t < t_0 \end{cases} \tag{4.14}$$

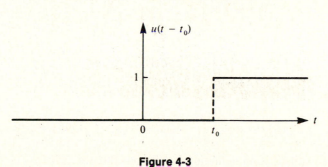

Figure 4-3

4-3. Generalized Derivatives

A. Definition of a generalized derivative

If $g(t)$ is a generalized function, its generalized derivative $g'(t)$ is defined by the expression

Generalized derivative
$$\int_{-\infty}^{\infty} g'(t)\phi(t)\,dt = -\int_{-\infty}^{\infty} g(t)\phi'(t)\,dt \tag{4.15}$$

where $\phi(t)$ is the testing function and $\phi'(t)$ is the derivative of $\phi(t)$.

B. High-order derivatives

Denote the nth derivative of $g(t)$ as $g^{(n)}(t)$. Then by a repeated application of definition (4.15), $g^{(n)}(t)$ is defined by the expression

Generalized nth derivative
$$\int_{-\infty}^{\infty} g^{(n)}(t)\phi(t)\,dt = (-1)^n \int_{-\infty}^{\infty} g(t)\phi^{(n)}(t)\,dt \tag{4.16}$$

where $\phi^{(n)}(t)$ is the nth derivative of $\phi(t)$.

The definition of a generalized derivative (4.15) is consistent with an ordinary definition of a derivative of $f(t)$ if $f(t)$ is an ordinary function with a continuous first derivative.

EXAMPLE 4-6: Verify (4.15) for an ordinary function $f(t)$.

Solution: Consider the integral given by

$$\int_{-\infty}^{\infty} f'(t)\phi(t)\,dt$$

Integrating by parts, we obtain

$$\int_{-\infty}^{\infty} f'(t)\phi(t)\,dt = f(t)\phi(t)\Big|_{-\infty}^{\infty} - \int_{-\infty}^{\infty} f(t)\phi'(t)\,dt$$

Recalling that the testing function $\phi(t)$ is such that it vanishes outside some interval, it is zero at $t = \pm\infty$, and thus

$$\int_{-\infty}^{\infty} f'(t)\phi(t)\,dt = -\int_{-\infty}^{\infty} f(t)\phi'(t)\,dt \tag{4.17}$$

C. Derivative of the unit step function

Using definition (4.15), we obtain an important relation—the derivative of the unit step function is $\delta(t)$:

Derivative of the unit step function
$$u'(t) = \frac{du(t)}{dt} = \delta(t) \tag{4.18}$$

EXAMPLE 4-7: Prove that the derivative of the unit step function is $\delta(t)$ (4.18).

Proof: From definition (4.15), we have

$$\int_{-\infty}^{\infty} u'(t)\phi(t)\,dt = -\int_{-\infty}^{\infty} u(t)\phi'(t)\,dt \tag{4.19}$$

But from (4.13),

$$\int_{-\infty}^{\infty} u(t)\phi'(t)\,dt = \int_{0}^{\infty} \phi'(t)\,dt$$

Thus, we obtain

$$\int_{-\infty}^{\infty} u'(t)\phi(t)\,dt = -\int_{0}^{\infty} \phi'(t)\,dt = -[\phi(\infty) - \phi(0)] = \phi(0) \tag{4.20}$$

because $\phi(\infty) = 0$. Then substituting definition (4.1) into (4.20), we have

$$\int_{-\infty}^{\infty} u'(t)\phi(t)\,dt = \int_{-\infty}^{\infty} \delta(t)\phi(t)\,dt \tag{4.21}$$

for any $\phi(t)$. Consequently, by the equivalence property (4.2), we conclude that (4.18) holds:

$$u'(t) = \frac{du(t)}{dt} = \delta(t)$$

Note that by changing t to $t - t_0$ in (4.18), we obtain

$$u'(t - t_0) = \delta(t - t_0) \tag{4.22}$$

Equations (4.18) and (4.22) are illustrated in Figure 4-4a and b, respectively.

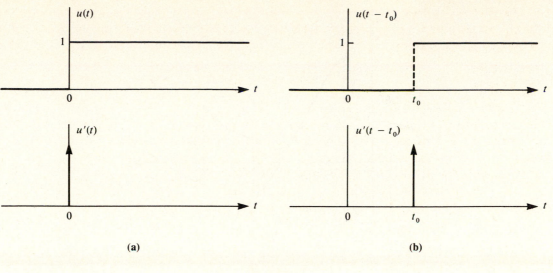

Figure 4-4

D. Derivatives of the δ-function

Using the definition of a generalized derivative (4.15), the derivative of $\delta(t)$, denoted by $\delta'(t)$, is defined by the expression

Derivative of the δ-function

$$\int_{-\infty}^{\infty} \delta'(t)\phi(t)\,dt = -\int_{-\infty}^{\infty} \delta(t)\phi'(t)\,dt = -\phi'(0) \tag{4.23}$$

where

$$\delta'(t) = \frac{d\delta(t)}{dt}, \qquad \phi'(0) = \frac{d\phi}{dt}\bigg|_{t=0}$$

Equation (4.23) shows that $\delta'(t)$ is a generalized function that assigns the value $-\phi'(0)$ to a testing function $\phi(t)$.

The nth derivative of $\delta(t)$

$$\delta^{(n)}(t) = \frac{d^n\delta(t)}{dt^n}$$

can be similarly defined by (4.16); that is,

High-order derivative of the δ-function

$$\int_{-\infty}^{\infty} \delta^{(n)}(t)\phi(t)\,dt = (-1)^n \phi^{(n)}(0) \tag{4.24}$$

where

$$\phi^{(n)}(0) = \frac{d^n\phi(t)}{dt^n}\bigg|_{t=0}$$

4-4. Generalized Derivative of a Discontinuous Function

A. Generalized derivative of a piecewise continuous function having jump discontinuities

Using (4.18) and (4.22), we can define a generalized derivative of a piecewise continuous function having jump discontinuities. Let $f(t)$ be a piecewise continuous function having jump discontinuities $a_1, a_2, \ldots,$ at $t_1, t_2, \ldots$ (see Figure 4-5a), where $a_i = f(t_i+) - f(t_i-)$ and $f(t_i+) = \lim_{\varepsilon \to 0} f(t_i + \varepsilon)$ and $f(t_i-) = \lim_{\varepsilon \to 0} f(t_i - \varepsilon)$. Then the generalized derivative of $f(t)$, denoted by $f'(t)$, is given by

Generalized derivative of a piecewise continuous function having jump discontinuities

$$f'(t) = g'(t) + \sum_k a_k \delta(t - t_k) \tag{4.25}$$

where $g'(t)$ is the ordinary derivative of $f(t)$ where it exists.

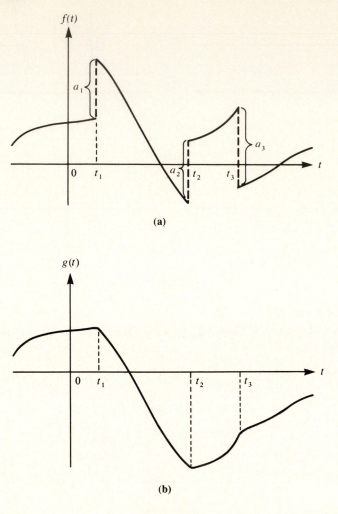

(a)

(b)

Figure 4-5

Equation (4.25) shows that the generalized derivative of a piecewise differentiable function with jumps is the ordinary derivative, where it exists, plus the sum of the δ-functions at the discontinuities multiplied by the magnitude of the jumps.

EXAMPLE 4-8: Derive (4.25).

Solution: Consider the function $g(t)$ (see Figure 4-5b)

$$g(t) = f(t) - \sum_k a_k u(t - t_k) \tag{4.26}$$

where

$$u(t - t_k) = \begin{cases} 1 & \text{for } t > t_k \\ 0 & \text{for } t < t_k \end{cases}$$

The function $g(t)$ is everywhere continuous, and it has—except at a finite number of points—a derivative equal to $f'(t)$. Hence, differentiation of (4.26) gives

$$g'(t) = f'(t) - \sum_k a_k \delta(t - t_k) \tag{4.27}$$

with the use of eq. (4.22). From eq. (4.27), we obtain the generalized derivative (4.25):

$$f'(t) = g'(t) + \sum_k a_k \delta(t - t_k)$$

EXAMPLE 4-9: Find the generalized derivative of the waveform of Figure 4-6.

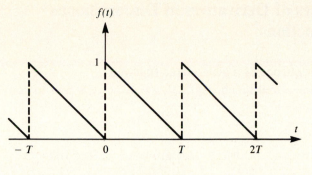

Figure 4-6

Solution: From Figure 4-6 it is seen that $f(t)$ has discontinuities at $t = nT$ ($n = 0, \pm 1, \pm 2, \ldots$) with jumps of $+1$. The derivative of $f(t)$ is $-1/T$ between $nT < t < (n + 1)T$ ($n = 0, \pm 1, \pm 2, \ldots$). Thus, from (4.25), $f'(t)$ is given by (see Figure 4-7)

$$f'(t) = -\frac{1}{T} + \sum_{n=-\infty}^{\infty} \delta(t - nT) \qquad \textbf{(4.28)}$$

Equation (4.28) indicates that the generalized derivative of $f(t)$ consists of a constant term $-1/T$ and a periodic train of unit impulses.

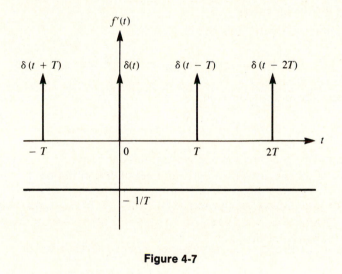

Figure 4-7

B. A periodic train of unit impulses

The periodic train of unit impulses (Figure 4-8) is a very useful generalized function, and hence it is convenient to denote this function by a special symbol, $\delta_T(t)$. Thus,

Periodic train of unit impulses

$$\delta_T(t) = \sum_{n=-\infty}^{\infty} \delta(t - nT) \qquad \textbf{(4.29)}$$

Figure 4-8   The periodic train of unit impulses.

4-5. Fourier Series of Derivatives of Discontinuous Periodic Functions

A. Convergence of a sequence of a generalized function

A sequence of a generalized function $\{f_n(t)\}$, $n = 1, 2, \ldots$, is said to converge to the generalized function $f(t)$ if and only if

Convergence of a sequence of a generalized function

$$\lim_{n \to \infty} \int_{-\infty}^{\infty} f_n(t)\phi(t)\,dt = \int_{-\infty}^{\infty} f(t)\phi(t)\,dt \tag{4.30}$$

for every testing function $\phi(t)$.

Similarly, a series $\sum_{n=1}^{k} f_n(t)$ of generalized functions that converge to the generalized function $f(t)$ can be differentiated term-by-term. In other words,

$$f'(t) = \sum_{n=1}^{k} f'_n(t) \tag{4.31}$$

In this case, we say that the series converges in the sense of generalized functions. But in the ordinary sense, the derivative of a convergent series of differentiable functions may not, in general, converge. (See Example 4-10.)

B. Differentiation of a Fourier series

Now, if $f(t)$ is periodic and continuous and given by

$$f(t) = \frac{1}{2} a_0 + \sum_{n=1}^{\infty} (a_n \cos n\omega_0 t + b_n \sin n\omega_0 t) \tag{4.32}$$

then $f'(t)$ is also periodic and can be obtained by term-by-term differentiation; that is,

$$f'(t) = \sum_{n=1}^{\infty} (-n\omega_0 a_n \sin n\omega_0 t + n\omega_0 b_n \cos n\omega_0 t) \tag{4.33}$$

Using the concepts of the δ-function and generalized derivatives, we can now study the Fourier series of derivatives of waveforms with a finite number of discontinuities in one period.

EXAMPLE 4-10: Find the Fourier series of the derivative of the waveform of Figure 4-6 in Example 4-9.

Solution: From the result of Problem 2-5, the Fourier series of $f(t)$ is given by

$$f(t) = \frac{1}{2} + \frac{1}{\pi} \sum_{n=1}^{\infty} \frac{1}{n} \sin n\omega_0 t = \frac{1}{2} + \frac{1}{\pi} \sum_{n=1}^{\infty} \frac{1}{n} \sin \frac{n2\pi}{T} t$$

By term-by-term differentiation, we obtain

$$f'(t) = \frac{2}{T} \sum_{n=1}^{\infty} \cos \frac{n2\pi}{T} t \tag{4.34}$$

On the other hand, from (4.28), we have

$$f'(t) = -\frac{1}{T} + \sum_{n=-\infty}^{\infty} \delta(t - nT)$$

We see that the Fourier series (4.34) is not a converging series in the ordinary sense, but we can say that the series (4.34) converges to the generalized function (4.28) in the sense of a generalized function.

C. Fourier series expression of the periodic train of unit impulses $\delta_T(t)$

Equating (4.28) and (4.34), we have

$$-\frac{1}{T} + \sum_{n=-\infty}^{\infty} \delta(t - nT) = \frac{2}{T} \sum_{n=1}^{\infty} \cos \frac{n2\pi}{T} t$$

Hence,

Fourier series of $\delta_T(t)$

$$\delta_T(t) = \sum_{n=-\infty}^{\infty} \delta(t - nT) = \frac{1}{T} + \frac{2}{T}\sum_{n=1}^{\infty} \cos n\omega_0 t \tag{4.35}$$

where $\omega_0 = 2\pi/T$.

Equation (4.35) is the Fourier series expression of $\delta_T(t)$, which consists of a constant term $1/T$ and a sum of harmonics, each having exactly the same amplitude of $2/T$.

EXAMPLE 4-11: Derive the complex Fourier series of $\delta_T(t)$.

Solution: Substituting the identity $\cos n\omega_0 t = \frac{1}{2}(e^{jn\omega_0 t} + e^{-jn\omega_0 t})$ into the Fourier series expression (4.35) and rearranging, we obtain

$$\sum_{n=-\infty}^{\infty} \delta(t - nT) = \frac{1}{T} + \frac{2}{T}\sum_{n=1}^{\infty} \cos n\omega_0 t = \frac{1}{T} + \frac{2}{T}\sum_{n=1}^{\infty} \frac{1}{2}(e^{jn\omega_0 t} + e^{-jn\omega_0 t})$$

$$= \frac{1}{T}\left[1 + \sum_{n=1}^{\infty}(e^{jn\omega_0 t} + e^{-jn\omega_0 t})\right]$$

Since $e^{j0} = 1$, and

$$\sum_{n=1}^{\infty} e^{-jn\omega_0 t} = \sum_{n=-1}^{-\infty} e^{jn\omega_0 t} = \sum_{n=-\infty}^{-1} e^{jn\omega_0 t}$$

we obtain

$$\delta_T(t) = \sum_{n=-\infty}^{\infty} \delta(t - nT)$$

Complex Fourier series of $\delta_T(t)$

$$= \frac{1}{T}\left(\sum_{n=-\infty}^{-1} e^{jn\omega_0 t} + e^{j0} + \sum_{n=1}^{\infty} e^{jn\omega_0 t}\right)$$

$$= \frac{1}{T}\sum_{n=-\infty}^{\infty} e^{jn\omega_0 t} \tag{4.36}$$

4-6. Evaluation of Fourier Coefficients by Differentiation

The use of the Fourier series expressions of $\delta_T(t)$, (4.35) and (4.36) in conjunction with differentiation techniques facilitates the computation of the Fourier coefficients of certain periodic functions.

EXAMPLE 4-12: Find the Fourier series for the waveform of Figure 4-9a by first differentiating $f(t)$.

Solution: Let

$$f(t) = \frac{1}{2}a_0 + \sum_{n=1}^{\infty}(a_n\cos n\omega_0 t + b_n\sin n\omega_0 t) \tag{4.37}$$

$$f'(t) = \frac{1}{2}\alpha_0 + \sum_{n=1}^{\infty}(\alpha_n\cos n\omega_0 t + \beta_n\sin n\omega_0 t) \tag{4.38}$$

with $\omega_0 = 2\pi/T$.

Differentiating (4.37) term-by-term and equating to (4.38), we obtain

$$\alpha_n = n\omega_0 b_n, \qquad \beta_n = -n\omega_0 a_n \qquad (n \neq 0)$$

Hence,

$$a_n = -\frac{\beta_n}{n\omega_0}, \qquad b_n = \frac{\alpha_n}{n\omega_0} \qquad (n \neq 0) \tag{4.39}$$

The derivative $f'(t)$ of Figure 4-9b can be expressed as

$$f'(t) = A\left[\sum_{n=-\infty}^{\infty} \delta\left(t + \frac{1}{2}d - nT\right)\right] - A\left[\sum_{n=-\infty}^{\infty} \delta\left(t - \frac{1}{2}d - nT\right)\right] \tag{4.40}$$

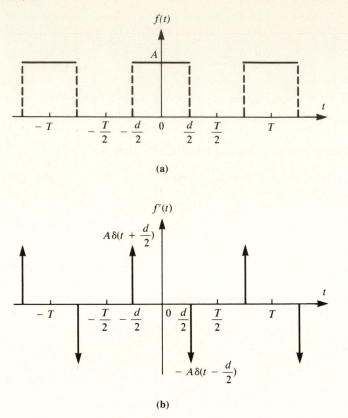

(a)

(b)

Figure 4-9

From (4.35), we obtain

$$\sum_{n=-\infty}^{\infty} \delta\left(t + \frac{1}{2}d - nT\right) = \frac{1}{T} + \frac{2}{T}\sum_{n=1}^{\infty} \cos\left[n\omega_0\left(t + \frac{1}{2}d\right)\right] \qquad (4.41)$$

$$\sum_{n=-\infty}^{\infty} \delta\left(t - \frac{1}{2}d - nT\right) = \frac{1}{T} + \frac{2}{T}\sum_{n=1}^{\infty} \cos\left[n\omega_0\left(t - \frac{1}{2}d\right)\right] \qquad (4.42)$$

Substituting equations (4.41) and (4.42) into equation (4.40) and using the trigonometric identity $\cos(A + B) - \cos(A - B) = -2\sin A \sin B$, we have

$$f'(t) = \frac{2A}{T}\sum_{n=1}^{\infty}\left[\cos\left(n\omega_0 t + \frac{n\pi d}{T}\right) - \cos\left(n\omega_0 t - \frac{n\pi d}{T}\right)\right]$$

$$= -\frac{4A}{T}\sum_{n=1}^{\infty}\sin\left(\frac{n\pi d}{T}\right)\sin(n\omega_0 t)$$

Hence,

$$\beta_n = -\frac{4A}{T}\sin\left(\frac{n\pi d}{T}\right), \qquad \alpha_n = 0, \qquad (n \neq 0)$$

Thus, by (4.39)

$$a_n = -\frac{\beta_n}{n\omega_0} = \frac{4A}{n\omega_0 T}\sin\left(\frac{n\pi d}{T}\right) = \frac{2A}{n\pi}\sin\left(\frac{n\pi d}{T}\right) = \frac{2Ad}{T}\frac{\sin(n\pi d/T)}{(n\pi d/T)}$$

$$b_n = \frac{\alpha_n}{n\omega_0} = 0$$

Since the constant term $\frac{1}{2}a_0$ vanishes with the differentiation process, using (1.11), we have

$$\frac{1}{2}a_0 = \frac{1}{T}\int_{-T/2}^{T/2} f(t)\,dt = \frac{Ad}{T} \qquad (4.43)$$

Hence,

$$f(t) = \frac{Ad}{T} + \frac{2Ad}{T} \sum_{n=1}^{\infty} \frac{\sin(n\pi d/T)}{(n\pi d/T)} \cos\left(n\frac{2\pi}{T}t\right)$$

EXAMPLE 4-13: Find the complex Fourier coefficients for the waveform of Figure 4-9a by the differentiation technique.

Solution: Assume that

$$f(t) = \sum_{n=-\infty}^{\infty} c_n e^{jn\omega_0 t}, \qquad \omega_0 = 2\pi/T \tag{4.44}$$

By term-by-term differentiation,

$$f'(t) = \sum_{n=-\infty}^{\infty} (jn\omega_0)c_n e^{jn\omega_0 t} \tag{4.45}$$

From (4.40),

$$f'(t) = A\left[\sum_{n=-\infty}^{\infty} \delta\left(t + \frac{1}{2}d - nT\right)\right] - A\left[\sum_{n=-\infty}^{\infty} \delta\left(t - \frac{1}{2}d - nT\right)\right]$$

Then from (4.36), we have

$$\sum_{n=-\infty}^{\infty} \delta\left(t + \frac{1}{2}d - nT\right) = \frac{1}{T} \sum_{n=-\infty}^{\infty} e^{jn\omega_0(t+d/2)} \tag{4.46}$$

$$\sum_{n=-\infty}^{\infty} \delta\left(t - \frac{1}{2}d - nT\right) = \frac{1}{T} \sum_{n=-\infty}^{\infty} e^{jn\omega_0(t-d/2)} \tag{4.47}$$

Substituting (4.46) and (4.47) into (4.40) and using the identity

$$\sin\theta = \frac{1}{2j}(e^{j\theta} - e^{-j\theta})$$

we obtain

$$f'(t) = \frac{A}{T} \sum_{n=-\infty}^{\infty} e^{jn\omega_0 t}(e^{jn\omega_0 d/2} - e^{-jn\omega_0 d/2})$$

$$= j\frac{2A}{T} \sum_{n=-\infty}^{\infty} \sin(n\omega_0 d/2)e^{jn\omega_0 t} \tag{4.48}$$

Equating (4.45) and (4.48), we obtain

$$jn\omega_0 c_n = j\frac{2A}{T} \sin(n\omega_0 d/2), \qquad (n \neq 0)$$

Hence,

$$c_n = \frac{2A}{n\omega_0 T} \sin(n\omega_0 d/2) = A\frac{d}{T}\frac{\sin(n\omega_0 d/2)}{(n\omega_0 d/2)}$$

which is exactly the same as (3.18).

From (3.7) and (4.43), we have

$$c_0 = \frac{1}{2}a_0 = \frac{Ad}{T}$$

SUMMARY

1. The unit impulse function $\delta(t)$ is defined as a generalized function by the relation

$$\int_{-\infty}^{\infty} \delta(t)\phi(t)\,dt = \phi(0)$$

2. The generalized derivative $g'(t)$ of a generalized function $g(t)$ is defined by the relation

$$\int_{-\infty}^{\infty} g'(t)\phi(t)\,dt = -\int_{-\infty}^{\infty} g(t)\phi'(t)\,dt$$

3. The generalized derivative of the unit step function $u(t)$ is $\delta(t)$; that is, $u'(t) = \delta(t)$.
4. The generalized derivative of a piecewise continuous function $f(t)$ having jump discontinuities a_1, $a_2, \ldots$ at $t_1, t_2, \ldots$ is given by

$$f'(t) = g'(t) + \sum_k a_k \delta(t - t_k)$$

where $g'(t)$ is the ordinary derivative of $f(t)$ where it exists.
5. The Fourier series representations of the periodic train of unit impulses $\delta_T(t)$ are given by

$$\delta_T(t) = \sum_{n=-\infty}^{\infty} \delta(t - nT) = \frac{1}{T} + \frac{2}{T} \sum_{n=1}^{\infty} \cos n\omega_0 t$$

and

$$\delta_T(t) = \sum_{n=-\infty}^{\infty} \delta(t - nT) = \frac{1}{T} \sum_{n=-\infty}^{\infty} e^{jn\omega_0 t}$$

6. The computation of the Fourier coefficients of certain periodic functions may be facilitated by the use of the Fourier series representations of $\delta_T(t)$ in conjunction with the differentiation technique.

RAISE YOUR GRADES

Can you explain . . .?

☑ how the unit impulse function $\delta(t)$ is defined and how the various relationships involving $\delta(t)$ are obtained
☑ how to obtain the derivative of a function having discontinuities
☑ how to use the Fourier series representations of the periodic train of unit impulses and the differentiation technique to evaluate Fourier coefficients

SOLVED PROBLEMS

The Unit Impulse Function $\delta(t)$

PROBLEM 4-1   Show that

$$t\delta(t) = 0 \tag{a}$$
$$e^t \delta(t) = \delta(t) \tag{b}$$
$$\sin t\, \delta(t) = 0 \tag{c}$$
$$\cos t\, \delta(t) = \delta(t) \tag{d}$$

Solution: Using property (4.6), $f(t)\delta(t) = f(0)\delta(t)$, we obtain

(a) $t\delta(t) = 0\delta(t) = 0$
(b) $e^t \delta(t) = e^0 \delta(t) = 1\delta(t) = \delta(t)$
(c) $\sin t\, \delta(t) = \sin 0\, \delta(t) = 0\delta(t) = 0$
(d) $\cos t\, \delta(t) = \cos 0\, \delta(t) = 1\delta(t) = \delta(t)$

PROBLEM 4-2   Show that $f(t)\delta(t - t_0) = f(t_0)\delta(t - t_0)$.

Solution: From the equivalence property (4.2), we have

$$\int_{-\infty}^{\infty} f(t)\delta(t - t_0)\phi(t)\, dt = \int_{-\infty}^{\infty} \delta(t - t_0)[f(t)\phi(t)]\, dt$$

$$= f(t_0)\phi(t_0)$$

$$\int_{-\infty}^{\infty} f(t)\delta(t-t_0)\phi(t)\,dt = f(t_0)\int_{-\infty}^{\infty} \delta(t-t_0)\phi(t)\,dt$$

$$= \int_{-\infty}^{\infty} f(t_0)\delta(t-t_0)\phi(t)\,dt$$

Since $\phi(t)$ is arbitrary, we conclude that $f(t)\delta(t-t_0) = f(t_0)\delta(t-t_0)$.

The Unit Step Function $u(t)$

PROBLEM 4-3   Show that

$$u(-t) = \begin{cases} 1, & t < 0 \\ 0, & t > 0 \end{cases}$$

Solution: Let $\tau = -t$. Then by property (4.12)

$$u(-t) = u(\tau) = \begin{cases} 1, & \tau > 0 \\ 0, & \tau < 0 \end{cases}$$

Since $\tau > 0$ and $\tau < 0$ imply, respectively, that $t < 0$ and $t > 0$, we get

$$u(-t) = \begin{cases} 1, & t < 0 \\ 0, & t > 0 \end{cases}$$

which is shown in Figure 4-10.

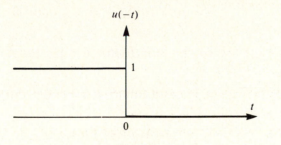

Figure 4-10

PROBLEM 4-4   If sgn t (pronounced "signum t") is defined as (see Figure 4-11)

$$\text{sgn}\,t = \begin{cases} 1, & t > 0 \\ -1, & t < 0 \end{cases}$$

show that

$$\text{sgn}\,t = u(t) - u(-t)$$

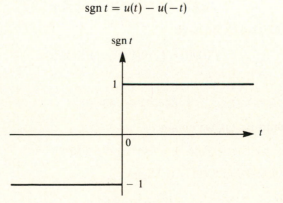

Figure 4-11

Solution: Since

$$u(t) = \begin{cases} 1, & t > 0 \\ 0, & t < 0 \end{cases}$$

and

$$u(-t) = \begin{cases} 1, & t < 0 \\ 0, & t > 0 \end{cases}$$

we have

$$u(t) - u(-t) = \begin{cases} 1, & t > 0 \\ -1, & t < 0 \end{cases}$$

Hence,

$$\text{sgn } t = u(t) - u(-t)$$

Generalized Derivatives

PROBLEM 4-5 If $f(t)$ is a continuous differentiable function, show that the product rule of differentiation, $[f(t)\delta(t)]' = f(t)\delta'(t) + f'(t)\delta(t)$, remains valid.

Solution: Using the definition of a generalized derivative (4.15), we have

$$\int_{-\infty}^{\infty} [f(t)\delta(t)]'\phi(t)\,dt = -\int_{-\infty}^{\infty} [f(t)\delta(t)]\phi'(t)\,dt$$

$$= -\int_{-\infty}^{\infty} \delta(t)[f(t)\phi'(t)]\,dt$$

$$= -\int_{-\infty}^{\infty} \delta(t)\{[f(t)\phi(t)]' - f'(t)\phi(t)\}\,dt$$

$$= -\int_{-\infty}^{\infty} \delta(t)[f(t)\phi(t)]'\,dt + \int_{-\infty}^{\infty} \delta(t)[f'(t)\phi(t)]\,dt$$

$$= \int_{-\infty}^{\infty} \delta'(t)[f(t)\phi(t)]\,dt + \int_{-\infty}^{\infty} [\delta(t)f'(t)]\phi(t)\,dt$$

$$= \int_{-\infty}^{\infty} [\delta'(t)f(t) + \delta(t)f'(t)]\phi(t)\,dt$$

Hence,

$$[f(t)\delta(t)]' = f(t)\delta'(t) + f'(t)\delta(t)$$

PROBLEM 4-6 Prove that $f(t)\delta'(t) = f(0)\delta'(t) - f'(0)\delta(t)$ and $t\delta'(t) = -\delta(t)$.

Proof: From the product rule of differentiation (see Problem 4-5), we have

$$f'(t)\delta(t) = [f(t)\delta(t)]' - f'(t)\delta(t) \tag{a}$$

Since, from $f(t)\delta(t) = f(0)\delta(t)$ (4.6), we have

$$f'(t)\delta(t) = [f(0)\delta(t)]' - f'(0)\delta(t) \tag{b}$$

Now,

$$[f(0)\delta(t)]' = f(0)\delta'(t) \tag{c}$$

Thus, substituting eq. (c) into eq. (a), we obtain

$$f(t)\delta'(t) = f(0)\delta'(t) - f'(0)\delta(t)$$

Next, let $f(t) = t$; then $f(0) = 0$ and $f'(0) = 1$. Then,

$$t\delta'(t) = -\delta(t)$$

PROBLEM 4-7 Find the generalized derivative of the square pulse function $p(t)$ shown in Figure 4-12.

Solution: Since $p(t)$ is constant except for a jump of magnitude A at the origin and a jump of magnitude $-A$ at $t = t_0$, by (4.25) we obtain

$$p'(t) = A\delta(t) - A\delta(t - t_0)$$

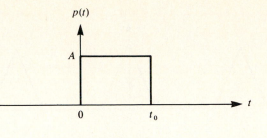

Figure 4-12

Alternate Solution: Using the unit step function, $p(t)$ can be expressed as

$$p(t) = Au(t) - Au(t - t_0)$$

Thus, using the derivatives of the unit step function (4.18) and (4.22), we obtain

$$p'(t) = Au'(t) - Au'(t - t_0)$$
$$= A\delta(t) - A\delta(t - t_0)$$

PROBLEM 4-8   Show that

$$\frac{d}{dt}\operatorname{sgn} t = 2\delta(t)$$

Solution: From Figure 4-11, shown in Problem 4-4, it is seen that sgn t is constant except for a jump of magnitude 2 at the origin. Thus, by (4.25), and noting that the derivative of a constant is zero, we have

$$\frac{d}{dt}\operatorname{sgn} t = 2\delta(t)$$

Evaluation of Fourier Coefficients by Differentiation

PROBLEM 4-9   Find the Fourier coefficients for the waveform of Figure 4-13a by differentiation.

Solution: Assume that

$$f(t) = \frac{1}{2}a_0 + \sum_{n=1}^{\infty}(a_n\cos n\omega_0 t + b_n\sin n\omega_0 t) \tag{a}$$

where $\omega_0 = 2\pi/T$. Then, term-by-term differentiation gives

$$f'(t) = \sum_{n=1}^{\infty}(-n\omega_0 a_n\sin n\omega_0 t + n\omega_0 b_n\cos n\omega_0 t) \tag{b}$$

$$f''(t) = \sum_{n=1}^{\infty}[-(n\omega_0)^2 a_n\cos n\omega_0 t - (n\omega_0)^2 b_n\sin n\omega_0 t] \tag{c}$$

From Figure 4-13c, $f''(t)$ can be expressed as

$$f''(t) = \frac{A}{t_1}\sum_{n=-\infty}^{\infty}\delta(t + t_1 - nT) - \frac{2A}{t_1}\sum_{n=-\infty}^{\infty}\delta(t - nT) + \frac{A}{t_1}\sum_{n=-\infty}^{\infty}\delta(t - t_1 - nT) \tag{d}$$

We know from (4.35) that

$$\sum_{n=-\infty}^{\infty}\delta(t - nT) = \frac{1}{T} + \frac{2}{T}\sum_{n=1}^{\infty}\cos n\omega_0 t \tag{e}$$

Changing t to $t + t_1$ and $t - t_1$, in expression (e), we have, respectively,

$$\sum_{n=-\infty}^{\infty}\delta(t + t_1 - nT) = \frac{1}{T} + \frac{2}{T}\sum_{n=1}^{\infty}\cos n\omega_0(t + t_1) \tag{f}$$

$$\sum_{n=-\infty}^{\infty}\delta(t - t_1 - nT) = \frac{1}{T} + \frac{2}{T}\sum_{n=1}^{\infty}\cos n\omega_0(t - t_1) \tag{g}$$

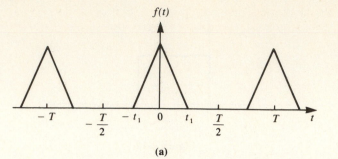

(a)

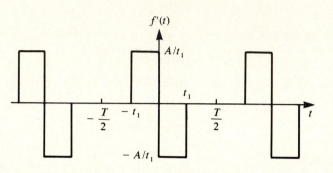

(b)

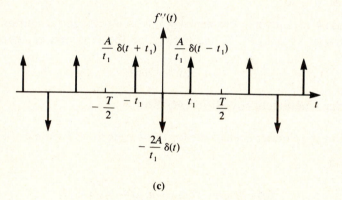

(c)

Figure 4-13

Substituting eqs. (e), (f), and (g) into eq. (d), we obtain

$$f''(t) = \frac{A}{t_1}\left[\frac{1}{T} + \frac{2}{T}\sum_{n=1}^{\infty}\cos n\omega_0(t + t_1)\right]$$

$$-\frac{2A}{t_1}\left[\frac{1}{T} + \frac{2}{T}\sum_{n=1}^{\infty}\cos n\omega_0 t\right]$$

$$+\frac{A}{t_1}\left[\frac{1}{T} + \frac{2}{T}\sum_{n=1}^{\infty}\cos n\omega_0(t - t_1)\right]$$

$$= \frac{A}{t_1}\frac{2}{T}\sum_{n=1}^{\infty}\left[-2\cos n\omega_0 t + \cos n\omega_0(t + t_1) + \cos n\omega_0(t - t_1)\right]$$

$$= \frac{2A}{t_1 T}\sum_{n=1}^{\infty}(-2\cos n\omega_0 t + 2\cos n\omega_0 t\cos n\omega_0 t_1)$$

$$= \frac{-4A}{t_1 T}\sum_{n=1}^{\infty}(1 - \cos n\omega_0 t_1)\cos n\omega_0 t$$

$$= -\frac{8A}{t_1 T}\sum_{n=1}^{\infty}\sin^2(n\omega_0 t_1/2)\cos n\omega_0 t$$

(h)

Thus, equating eqs. (h) and (c), we have

$$-(n\omega_0)^2 a_n = -\frac{8A}{t_1 T}\sin^2(n\omega_0 t_1/2)$$

$$-(n\omega_0)^2 b_n = 0$$

Hence,

$$a_n = \frac{8A}{t_1 T(n\omega_0)^2}\sin^2(n\omega_0 t_1/2) = 2A\left(\frac{t_1}{T}\right)\frac{\sin^2(n\omega_0 t_1/2)}{(n\omega_0 t_1/2)^2}$$

$$b_n = 0$$

By (1.11),

$$\frac{1}{2}a_0 = \frac{1}{T}\int_{-T/2}^{T/2} f(t)\,dt = At_1/T$$

PROBLEM 4-10 Find the complex Fourier coefficients for the function $f(t)$ of Problem 4-9 by differentiation.

Solution: Assume that

$$f(t) = \sum_{n=-\infty}^{\infty} c_n e^{jn\omega_0 t}, \qquad \omega_0 = 2\pi/T \tag{a}$$

By term-by-term differentiation,

$$f'(t) = \sum_{n=-\infty}^{\infty} (jn\omega_0)c_n e^{jn\omega_0 t} \tag{b}$$

$$f''(t) = \sum_{n=-\infty}^{\infty} (jn\omega_0)^2 c_n e^{jn\omega_0 t} = -\sum_{n=-\infty}^{\infty} (n\omega_0)^2 c_n e^{jn\omega_0 t} \tag{c}$$

From Figure 4-13c of Problem 4-9, $f''(t)$ can be expressed as

$$f''(t) = \frac{A}{t_1}\sum_{n=-\infty}^{\infty}\delta(t + t_1 - nT) - \frac{2A}{t_1}\sum_{n=-\infty}^{\infty}\delta(t - nT) + \frac{A}{t_1}\sum_{n=-\infty}^{\infty}\delta(t - t_1 - nT) \tag{d}$$

From (4.36), we have

$$\sum_{n=-\infty}^{\infty}\delta(t - nT) = \frac{1}{T}\sum_{n=-\infty}^{\infty}e^{jn\omega_0 t} \tag{e}$$

Changing t to $t + t_1$ and $t - t_1$, in eq. (e), we have, respectively,

$$\sum_{n=-\infty}^{\infty}\delta(t + t_1 - nT) = \frac{1}{T}\sum_{n=-\infty}^{\infty}e^{jn\omega_0(t+t_1)} = \frac{1}{T}\sum_{n=-\infty}^{\infty}e^{jn\omega_0 t_1}e^{jn\omega_0 t} \tag{f}$$

$$\sum_{n=-\infty}^{\infty}\delta(t - t_1 - nT) = \frac{1}{T}\sum_{n=-\infty}^{\infty}e^{jn\omega_0(t-t_1)} = \frac{1}{T}\sum_{n=-\infty}^{\infty}e^{-jn\omega_0 t_1}e^{jn\omega_0 t} \tag{g}$$

Substituting eqs. (e), (f), and (g) into eq. (d), we obtain

$$f''(t) = \frac{A}{Tt_1}\sum_{n=-\infty}^{\infty}(e^{jn\omega_0 t_1} + e^{-jn\omega_0 t_1} - 2)e^{jn\omega_0 t}$$

$$= \frac{2A}{Tt_1}\sum_{n=-\infty}^{\infty}(\cos n\omega_0 t_1 - 1)e^{jn\omega_0 t} \tag{h}$$

Therefore, equating eqs. (h) and (c), we have

$$-(n\omega_0)^2 c_n = \frac{2A}{Tt_1}(\cos n\omega_0 t_1 - 1)$$

Hence,

$$c_n = \frac{2A}{Tt_1}\left[\frac{(1 - \cos n\omega_0 t_1)}{(n\omega_0)^2}\right] = A\frac{t_1}{T}\left[\frac{\sin(n\omega_0 t_1/2)}{(n\omega_0 t_1/2)}\right]^2$$

PROBLEM 4-11   Find the Fourier series for the function $f(t) = |A \sin \omega_0 t|$, where $\omega_0 = 2\pi/T$, by the differentiation technique.

Solution: The function $f(t)$ is shown in Figure 4-14a.

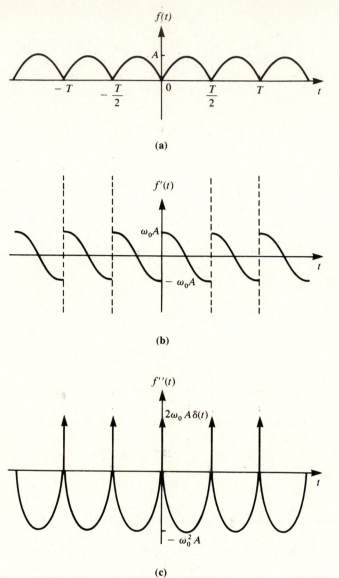

(a)

(b)

(c)

Figure 4-14

Since the period of $f(t)$ is $T/2 = T_1$, we assume that

$$f(t) = \frac{1}{2} a_0 + \sum_{n=1}^{\infty} (a_n \cos n\omega_1 t + b_n \sin n\omega_1 t) \tag{a}$$

where $\omega_1 = 2\pi/T_1 = 4\pi/T = 2\omega_0$. Term-by-term differentiation gives

$$f'(t) = \sum_{n=1}^{\infty} (-n\omega_1 a_n \sin n\omega_1 t + n\omega_1 b_n \cos n\omega_1 t) \tag{b}$$

$$f''(t) = \sum_{n=1}^{\infty} [-(n\omega_1)^2 a_n \cos n\omega_1 t - (n\omega_1)^2 b_n \sin n\omega_1 t] \tag{c}$$

From Figure 4-14c, $f''(t)$ can be expressed as

$$f''(t) = 2\omega_0 A \sum_{n=-\infty}^{\infty} \delta(t - nT_1) - \omega_0^2 f(t) \tag{d}$$

From (4.35), we have

$$\sum_{n=-\infty}^{\infty} \delta(t - nT_1) = \frac{1}{T_1} + \frac{2}{T_1} \sum_{n=1}^{\infty} \cos n\omega_1 t \tag{e}$$

Substituting eq. (e) into eq. (d), we obtain

$$f''(t) = 2\omega_0 A\left[\frac{1}{T_1} + \frac{2}{T_1} \sum_{n=1}^{\infty} \cos n\omega_1 t\right] - \omega_0^2 f(t) \tag{f}$$

or

$$\omega_0^2 f(t) + f''(t) = \frac{2\omega_0 A}{T_1} + \frac{4\omega_0 A}{T_1} \sum_{n=1}^{\infty} \cos n\omega_1 t \tag{g}$$

Substituting eqs. (a) and (c) into eq. (g) and equating the coefficients of both sides, we obtain

$$\omega_0^2 \frac{1}{2} a_0 = 2\omega_0 A/T_1$$

$$\omega_0^2 a_n - (n\omega_1)^2 a_n = 4\omega_0 A/T_1$$

$$\omega_0^2 b_n - (n\omega_1)^2 b_n = 0$$

Hence,

$$\frac{1}{2} a_0 = 2A/(\omega_0 T_1) = 4A/(\omega_0 T) = 4A/(2\pi) = 2A/\pi$$

$$a_n = \frac{1}{\omega_0^2 - (n\omega_1)^2}\left(\frac{4\omega_0 A}{T_1}\right) = \frac{1}{(1 - 4n^2)}\left(\frac{8A}{\omega_0 T}\right) = \frac{4A}{\pi(1 - 4n^2)}$$

$$b_n = 0$$

Thus, from eq. (a), we obtain

$$f(t) = \frac{2A}{\pi} + \frac{4A}{\pi} \sum_{n=1}^{\infty} \frac{1}{(1 - 4n^2)} \cos n\omega_1 t$$

$$= \frac{2A}{\pi} + \frac{4A}{\pi} \sum_{n=1}^{\infty} \frac{1}{(1 - 4n^2)} \cos 2n\omega_0 t$$

Supplementary Exercises

PROBLEM 4-12   Show that **(a)** $\delta'(-t) = -\delta'(t)$ and **(b)** $\delta^n(-t) = (-1)^n \delta^n(t)$.

PROBLEM 4-13   Show that if $f(t)$ is a continuous differentiable function, then $f(t)\delta'(t - t_0) = f(t_0)\delta'(t - t_0) - f'(t_0)\delta(t - t_0)$.

PROBLEM 4-14   Use differentiation to find the Fourier coefficients for the function $f(t)$ defined by $f(t) = t$ for $(-\pi, \pi)$ and $f(t + 2\pi) = f(t)$.

Answer: See Problem 1-21.

PROBLEM 4-15   Use differentiation to find the Fourier coefficients for the fully rectified sine wave $f(t) = |A \sin \omega_0 t|$.

Answer: See Problem 1-23.

PROBLEM 4-16   Use differentiation to find the Fourier coefficients for the function whose waveform is shown in Problem 1-8.

Answer: See Problem 1-8.

PROBLEM 4-17   Use differentiation to find the Fourier coefficients for the half-rectified sine wave of Problem 1-9.

Answer: See Problem 1-9.

PROBLEM 4-18   Use the result of Problem 4-15 to deduce the Fourier series for the half-rectified sine wave of Problem 1-9. [*Hint:* Note that $f(t)$ can be expressed as $f(t) = \frac{1}{2}A \sin \omega_0 t + \frac{1}{2}|A \sin \omega_0 t|$.]

PROBLEM 4-19   Use differentiation to find the complex Fourier series for the function of Problem 3-12. [*Hint:* Note that $f'(t) = f(t) - (e^{2\pi} - 1)\delta_{2\pi}(t)$, where $\delta_{2\pi}(t) = \sum_{n=-\infty}^{\infty} \delta(t - 2\pi n)$.]

PROBLEM 4-20   Use differentiation to find the complex Fourier series of the sawtooth wave function of Problem 3-1.

Answer: See Problem 3-1.

PROBLEM 4-21   Use differentiation to find the complex Fourier series of the fully rectified sine wave of Problem 3-2.

Answer: See Problem 3-2.

5 FOURIER TRANSFORMS

THIS CHAPTER IS ABOUT

☑ **Fourier Transforms**
☑ **Properties of Fourier Transforms**
☑ **Differentiation and Integration Theorems**
☑ **Parseval's Formula**
☑ **Multidimensional Fourier Transforms**

5-1. Fourier Transforms

We have seen that Fourier series are powerful tools in treating various problems involving periodic functions. Since many practical problems do not involve periodic functions, we need to develop a method of Fourier analysis that includes nonperiodic functions. In this chapter, we shall discuss a frequency representation of nonperiodic functions by means of the Fourier transforms.

A. Definition of Fourier transforms

The **Fourier transform of** $f(t)$ (symbolized by $\mathscr{F}$) is defined by

Fourier transform
$$F(\omega) = \mathscr{F}[f(t)] = \int_{-\infty}^{\infty} f(t)e^{-j\omega t}\, dt \tag{5.1}$$

The **inverse Fourier transform of** $F(\omega)$ (symbolized by $\mathscr{F}^{-1}$) is defined by

Inverse Fourier transform
$$f(t) = \mathscr{F}^{-1}[F(\omega)] = \frac{1}{2\pi}\int_{-\infty}^{\infty} F(\omega)e^{j\omega t}\, d\omega \tag{5.2}$$

Equations (5.1) and (5.2) are often called the **Fourier transform pair**. The condition for the existence of $F(\omega)$ is usually given by

$$\int_{-\infty}^{\infty} |f(t)|\, dt < \infty \tag{5.3}$$

In other words, the function $f(t)$ is absolutely integrable.

Note that condition (5.3) is a sufficient but not a necessary condition for the existence of $\mathscr{F}[f(t)]$. Functions that do not satisfy (5.3) may have Fourier transforms. (We'll discuss these functions in Chapter 6.)

B. Fourier spectra

The function $F(\omega) = \mathscr{F}[f(t)]$ is, in general, complex, and

$$F(\omega) = R(\omega) + jX(\omega) = |F(\omega)|e^{j\phi(\omega)} \tag{5.4}$$

where $|F(\omega)|$ is called the **magnitude spectrum** of $f(t)$ and $\phi(\omega)$ is the **phase spectrum** of $f(t)$.

EXAMPLE 5-1: Find the Fourier transform of $f(t)$ defined by

$$f(t) = \begin{cases} e^{-\alpha t}, & t > 0 \\ 0, & t < 0 \end{cases}$$

where $\alpha > 0$ (Figure 5-1a). Also plot the magnitude spectrum and phase spectrum of $f(t)$.

75

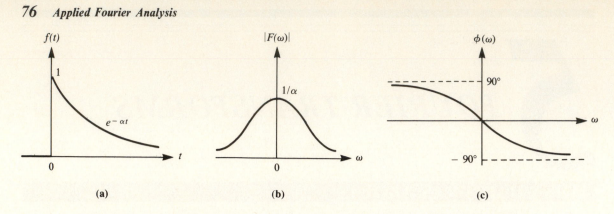

Figure 5-1   (a) Exponential function; (b) magnitude spectrum; (c) phase spectrum.

Solution: By means of definition (5.1), we have

$$F(\omega) = \int_{-\infty}^{\infty} f(t)e^{-j\omega t}\, dt$$

$$= \int_{0}^{\infty} e^{-\alpha t}e^{-j\omega t}\, dt = \int_{0}^{\infty} e^{-(\alpha + j\omega)t}\, dt$$

$$= \frac{1}{-(\alpha + j\omega)}\, e^{-(\alpha + j\omega)t}\bigg|_{0}^{\infty} = \frac{1}{\alpha + j\omega}$$

$$F(\omega) = \frac{1}{\alpha + j\omega} = \frac{1}{\sqrt{\alpha^2 + \omega^2}}\, e^{-j\tan^{-1}(\omega/\alpha)} = |F(\omega)|e^{j\phi(\omega)}$$

where

$$|F(\omega)| = \frac{1}{\sqrt{\alpha^2 + \omega^2}} \quad \text{and} \quad \phi(\omega) = -\tan^{-1}(\omega/\alpha)$$

The magnitude spectrum $|F(\omega)|$ and the phase spectrum $\phi(\omega)$ of $f(t)$ are plotted in Figure 5-1b and 5-1c, respectively.

5-2. Properties of Fourier Transforms

We use the notation

$$f(t) \leftrightarrow F(\omega)$$

to denote a transform pair (repeated here):

$$F(\omega) = \int_{-\infty}^{\infty} f(t)e^{-j\omega t}\, dt$$

$$f(t) = \frac{1}{2\pi} \int_{-\infty}^{\infty} F(\omega)e^{j\omega t}\, d\omega$$

A. Elementary properties of Fourier transforms

1. *Linearity:* If

$$f_1(t) \leftrightarrow F_1(\omega) \quad \text{and} \quad f_2(t) \leftrightarrow F_2(\omega)$$

then

Linearity $\qquad\qquad a_1 f_1(t) + a_2 f_2(t) \leftrightarrow a_1 F_1(\omega) + a_2 F_2(\omega)$ (5.5)

where a_1 and a_2 are constants.

2. *Time Shifting:* If

$$f(t) \leftrightarrow F(\omega)$$

then

Time shifting $\qquad\qquad f(t - t_0) \leftrightarrow F(\omega)e^{-j\omega t_0}$ (5.6)

3. Frequency Shifting: If

$$f(t) \leftrightarrow F(\omega)$$

then

Frequency shifting

$$f(t)e^{j\omega_0 t} \leftrightarrow F(\omega - \omega_0) \tag{5.7}$$

4. Scaling: If

$$f(t) \leftrightarrow F(\omega)$$

then for a real constant a

Scaling

$$f(at) \leftrightarrow \frac{1}{|a|} F\left(\frac{\omega}{a}\right) \tag{5.8}$$

5. Time-reversal: If

$$f(t) \leftrightarrow F(\omega)$$

then

Time-reversal

$$f(-t) \leftrightarrow F(-\omega) \tag{5.9}$$

6. Symmetry: If

$$f(t) \leftrightarrow F(\omega)$$

then

Symmetry

$$F(t) \leftrightarrow 2\pi f(-\omega) \tag{5.10}$$

EXAMPLE 5-2: Prove the linearity property (5.5).

Proof:

$$\mathcal{F}[a_1 f_1(t) + a_2 f_2(t)] = \int_{-\infty}^{\infty} [a_1 f_1(t) + a_2 f_2(t)]e^{-j\omega t}\, dt$$

$$= a_1 \int_{-\infty}^{\infty} f_1(t)e^{-j\omega t}\, dt + a_2 \int_{-\infty}^{\infty} f_2(t)e^{-j\omega t}\, dt$$

$$= a_1 F_1(\omega) + a_2 F_2(\omega)$$

EXAMPLE 5-3: Prove the time-shifting property (5.6).

Proof:

$$\mathcal{F}[f(t - t_0)] = \int_{-\infty}^{\infty} f(t - t_0)e^{-j\omega t}\, dt$$

Letting $t - t_0 = x$, then $dt = dx$; hence,

$$\mathcal{F}[f(t - t_0)] = \int_{-\infty}^{\infty} f(x)e^{-j\omega(t_0 + x)}\, dx$$

$$= e^{-j\omega t_0} \int_{-\infty}^{\infty} f(x)e^{-j\omega x}\, dx$$

$$= e^{-j\omega t_0} F(\omega)$$

EXAMPLE 5-4: Prove the frequency-shifting property (5.7).

Proof:

$$\mathcal{F}[f(t)e^{j\omega_0 t}] = \int_{-\infty}^{\infty} [f(t)e^{j\omega_0 t}]e^{-j\omega t}\, dt = \int_{-\infty}^{\infty} f(t)e^{-j(\omega - \omega_0)t}\, dt$$

$$= F(\omega - \omega_0)$$

EXAMPLE 5-5: Prove the scaling property (5.8).

Proof: For $a > 0$

$$\mathscr{F}[f(at)] = \int_{-\infty}^{\infty} f(at)e^{-j\omega t}\,dt$$

Let $at = x$; then

$$\mathscr{F}[f(at)] = \frac{1}{a}\int_{-\infty}^{\infty} f(x)e^{-j(\omega/a)x}\,dx$$

Since the dummy variable can be represented by any symbol,

$$\mathscr{F}[f(at)] = \frac{1}{a}\int_{-\infty}^{\infty} f(t)e^{-j(\omega/a)t}\,dt = \frac{1}{|a|}F\left(\frac{\omega}{a}\right)$$

For $a < 0$,

$$\mathscr{F}[f(at)] = \int_{-\infty}^{\infty} f(at)e^{-j\omega t}\,dt$$

Again, let $at = x$; then

$$\mathscr{F}[f(at)] = \frac{1}{a}\int_{\infty}^{-\infty} f(x)e^{-j(\omega/a)x}\,dx$$

$$= -\frac{1}{a}\int_{-\infty}^{\infty} f(t)e^{-j(\omega/a)t}\,dt$$

$$= \frac{1}{|a|}F\left(\frac{\omega}{a}\right)$$

Consequently, combining these two results, we obtain

$$\mathscr{F}[f(at)] = \frac{1}{|a|}F\left(\frac{\omega}{a}\right)$$

EXAMPLE 5-6: Prove the time-reversal property (5.9).

Proof: Letting $a = -1$ in the scaling property (5.8), we have

$$\mathscr{F}[f(-t)] = F(-\omega)$$

EXAMPLE 5-7: Prove the symmetry property (5.10).

Proof: From definition (5.2), we have

$$2\pi f(t) = \int_{-\infty}^{\infty} F(\omega)e^{j\omega t}\,d\omega$$

Changing t to $-t$, we obtain

$$2\pi f(-t) = \int_{-\infty}^{\infty} F(\omega)e^{-j\omega t}\,d\omega$$

Now interchanging t and ω, we obtain

$$2\pi f(-\omega) = \int_{-\infty}^{\infty} F(t)e^{-j\omega t}\,dt = \mathscr{F}[F(t)]$$

B. The modulation theorem

The **modulation theorem** states that if

$$f(t) \leftrightarrow F(\omega)$$

then

Modulation theorem $$f(t)\cos \omega_0 t \leftrightarrow \frac{1}{2}F(\omega - \omega_0) + \frac{1}{2}F(\omega + \omega_0)$$ (5.11)

EXAMPLE 5-8: Prove the modulation theorem (5.11).

Proof: With the identity $\cos \omega_0 t = \frac{1}{2}(e^{j\omega_0 t} + e^{-j\omega_0 t})$, the frequency-shifting property (5.7), and the linearity property (5.5), we obtain

$$\mathscr{F}[f(t)\cos \omega_0 t] = \mathscr{F}\left[\frac{1}{2}f(t)e^{j\omega_0 t} + \frac{1}{2}f(t)e^{-j\omega_0 t}\right]$$

$$= \frac{1}{2}\mathscr{F}[f(t)e^{j\omega_0 t}] + \frac{1}{2}\mathscr{F}[f(t)e^{-j\omega_0 t}]$$

$$= \frac{1}{2}F(\omega - \omega_0) + \frac{1}{2}F(\omega + \omega_0)$$

C. Additional properties when $f(t)$ is real

Let

$$f(t) \leftrightarrow F(\omega)$$

If $f(t)$ is real, then let

$$f(t) = f_e(t) + f_o(t)$$

where $f_e(t)$ and $f_o(t)$ are the even and odd components of $f(t)$, respectively.

Let

$$F(\omega) = R(\omega) + jX(\omega)$$

where $R(\omega)$ and $jX(\omega)$ are the real and imaginary parts of $F(\omega)$, respectively. Then

$$R(\omega) = \int_{-\infty}^{\infty} f(t)\cos \omega t\, dt \qquad (5.12)$$

$$X(\omega) = -\int_{-\infty}^{\infty} f(t)\sin \omega t\, dt \qquad (5.13)$$

$$R(\omega) = R(-\omega) \qquad (5.14)$$

$$X(\omega) = -X(-\omega) \qquad (5.15)$$

Condition for a real function $\qquad F(-\omega) = F^*(\omega) \qquad (5.16)$

even component: $\qquad f_e(t) \leftrightarrow R(\omega) \qquad (5.17)$

odd component: $\qquad f_o(t) \leftrightarrow jX(\omega) \qquad (5.18)$

where $F^*(\omega)$ is the complex conjugate of $F(\omega)$.

EXAMPLE 5-9: Verify properties (5.12)–(5.16).

Solution: If $f(t)$ is real, then using the identity

$$e^{-j\omega t} = \cos \omega t - j \sin \omega t.$$

we can rewrite the Fourier transform (5.1) as

$$F(\omega) = \int_{-\infty}^{\infty} f(t)e^{-j\omega t}\, dt$$

$$= \int_{-\infty}^{\infty} f(t)\cos \omega t\, dt - j \int_{-\infty}^{\infty} f(t)\sin \omega t\, dt$$

$$= R(\omega) + jX(\omega)$$

Thus, equating the real and imaginary parts, we have (5.12) and (5.13):

$$R(\omega) = \int_{-\infty}^{\infty} f(t)\cos \omega t\, dt$$

$$X(\omega) = -\int_{-\infty}^{\infty} f(t)\sin \omega t\, dt$$

Next, since $f(t)$ is real,

$$R(-\omega) = \int_{-\infty}^{\infty} f(t)\cos(-\omega t)\,dt = \int_{-\infty}^{\infty} f(t)\cos \omega t\,dt = R(\omega)$$

$$X(-\omega) = -\int_{-\infty}^{\infty} f(t)\sin(-\omega t)\,dt = \int_{-\infty}^{\infty} f(t)\sin \omega t\,dt = -X(\omega)$$

Hence, $R(\omega)$ is an even function of ω and $X(\omega)$ is an odd function of ω ((5.14) and (5.15)). From (5.14) and (5.15), we see that (5.16) holds:

$$F(-\omega) = R(-\omega) + jX(-\omega) = R(\omega) - jX(\omega) = F^*(\omega)$$

Thus, (5.16) is a necessary condition for $f(t)$ to be real.

Equation (5.16) is a necessary and sufficient condition for $f(t)$ to be real. We have already shown that $F(-\omega) = F^*(\omega)$ (5.16) is a necessary condition for $f(t)$ to be real (Example 5-9). Now we can prove that (5.16) is also a sufficient condition for $f(t)$ to be real.

EXAMPLE 5-10: Prove that (5.16) is a sufficient condition for $f(t)$ to be real.

Proof: Let

$$f(t) = f_1(t) + j f_2(t)$$

where $f_1(t)$ and $f_2(t)$ are real functions. Then from the inverse Fourier transform (5.2),

$$f(t) = f_1(t) + j f_2(t)$$

$$= \frac{1}{2\pi} \int_{-\infty}^{\infty} F(\omega) e^{j\omega t}\,d\omega$$

$$= \frac{1}{2\pi} \int_{-\infty}^{\infty} [R(\omega) + jX(\omega)](\cos \omega t + j\sin \omega t)\,d\omega$$

$$= \frac{1}{2\pi} \int_{-\infty}^{\infty} [R(\omega)\cos \omega t - X(\omega)\sin \omega t]\,d\omega$$

$$+ j\frac{1}{2\pi} \int_{-\infty}^{\infty} [R(\omega)\sin \omega t + X(\omega)\cos \omega t]\,d\omega$$

Hence,

$$f_1(t) = \frac{1}{2\pi} \int_{-\infty}^{\infty} [R(\omega)\cos \omega t - X(\omega)\sin \omega t]\,d\omega \qquad (5.19)$$

$$f_2(t) = \frac{1}{2\pi} \int_{-\infty}^{\infty} [R(\omega)\sin \omega t + X(\omega)\cos \omega t]\,d\omega \qquad (5.20)$$

Now if $F(-\omega) = F^*(\omega)$, then

$$R(-\omega) = R(\omega) \quad \text{and} \quad X(-\omega) = -X(\omega)$$

Consequently, from the results of Example 2-1 (the product of an even and an odd function is an odd function), $R(\omega)\sin \omega t$ and $X(\omega)\cos \omega t$ are odd functions of ω, and the integrand in (5.20) is an odd function of ω. Hence, from (2.5), we conclude that

$$f_2(t) = 0$$

That is, $f(t)$ is real.

EXAMPLE 5-11: Verify (5.17) and (5.18).

Solution: Let

$$f(t) = f_e(t) + f_o(t)$$

Then from the result of Problem 2-1 (any function can be expressed as the sum of an even and an odd component), we have

$$f_e(t) = \frac{1}{2}[f(t) + f(-t)]$$

$$f_o(t) = \frac{1}{2}[f(t) - f(-t)]$$

Now, if $f(t)$ is real, then from (5.9) and (5.16) we have

$$\mathscr{F}[f(t)] = F(\omega) = R(\omega) + jX(\omega)$$
$$\mathscr{F}[f(-t)] = F(-\omega) = F^*(\omega) = R(\omega) - jX(\omega)$$

Thus, by the linearity property (5.5) we obtain (5.17):

$$\mathscr{F}[f_e(t)] = \frac{1}{2}F(\omega) + \frac{1}{2}F^*(\omega)$$

$$= \frac{1}{2}[R(\omega + jX(\omega)] + \frac{1}{2}[R(\omega) - jX(\omega)]$$

$$= R(\omega)$$

and (5.18):

$$\mathscr{F}[f_o(t)] = \frac{1}{2}F(\omega) - \frac{1}{2}F^*(\omega)$$

$$= \frac{1}{2}[R(\omega) + jX(\omega)] - \frac{1}{2}[R(\omega) - jX(\omega)]$$

$$= jX(\omega)$$

5-3. Differentiation and Integration Theorems

A. Differentiation theorems

If

$$f(t) \leftrightarrow F(\omega)$$

then

Time domain differentiation theorem
$$f'(t) \leftrightarrow j\omega F(\omega) \tag{5.21}$$

provided that $f(t) \to 0$ as $t \to \pm\infty$.

Frequency domain differentiation theorem
$$(-jt)f(t) \leftrightarrow F'(\omega) \tag{5.22}$$

EXAMPLE 5-12: Prove the time domain differentiation theorem (5.21).

Proof: On integration by parts, we have

$$\mathscr{F}[f'(t)] = \int_{-\infty}^{\infty} f'(t)e^{-j\omega t}\,dt = f(t)e^{-j\omega t}\Big|_{-\infty}^{\infty} + j\omega\int_{-\infty}^{\infty} f(t)e^{-j\omega t}\,dt \tag{5.23}$$

Since $f(t) \to 0$ as $t \to \pm\infty$, we obtain

$$\mathscr{F}[f'(t)] = j\omega\int_{-\infty}^{\infty} f(t)e^{-j\omega t}\,dt = j\omega F(\omega)$$

Equation (5.21) shows that differentiation in the time domain corresponds to multiplication of the Fourier transform by $j\omega$, provided that $f(t) \to 0$ as $t \to \pm\infty$.

Note that if $f(t)$ has a finite number of jump discontinuities, then $f'(t)$ contains impulses (see eq. (4.25) of Chapter 4). Then the Fourier transform of $f'(t)$ for this case must contain the Fourier transform of the impulses in $f'(t)$, which we'll discuss in Chapter 6.

By repeated application of (5.21), we have

$$\mathscr{F}[f^{(n)}(t)] = (j\omega)^n F(\omega) = (j\omega)^n \mathscr{F}[f(t)], \qquad n = 1, 2, \dots \tag{5.24}$$

Note that (5.24) does not guarantee the *existence* of the Fourier transform of $f^{(n)}(t)$—it only indicates that *if the transform exists, then it is given by* $(j\omega)^n F(\omega)$.

EXAMPLE 5-13: Prove the frequency domain differentiation theorem (5.22).

Proof: Since

$$F(\omega) = \int_{-\infty}^{\infty} f(t) e^{-j\omega t} dt$$

we have

$$\frac{dF(\omega)}{d\omega} = \frac{d}{d\omega} \int_{-\infty}^{\infty} f(t) e^{-j\omega t} dt$$

Changing the order of differentiation and integration, we obtain

$$\frac{dF(\omega)}{d\omega} = \int_{-\infty}^{\infty} f(t) \frac{\partial}{\partial \omega} (e^{-j\omega t}) dt = \int_{-\infty}^{\infty} [-jtf(t)] e^{-j\omega t} dt = \mathscr{F}[-jtf(t)]$$

B. Integration theorem

If

$$f(t) \leftrightarrow F(\omega)$$

and

$$\int_{-\infty}^{\infty} f(t) dt = F(0) = 0 \tag{5.25}$$

then

Integration theorem
$$\left[\int_{-\infty}^{t} f(x) dx \right] \leftrightarrow \frac{1}{j\omega} F(\omega) \tag{5.26}$$

EXAMPLE 5-14: Prove the integration theorem (5.26).

Proof: Consider the function

$$\phi(t) = \int_{-\infty}^{t} f(x) dx$$

Then $\phi'(t) = f(t)$. Hence, if $\mathscr{F}[\phi(t)] = \Phi(\omega)$, then from (5.21) we have

$$\mathscr{F}[\phi'(t)] = \mathscr{F}[f(t)] = j\omega\Phi(\omega)$$

provided that

$$\lim_{t \to \infty} \phi(t) = \int_{-\infty}^{\infty} f(x) dx = \int_{-\infty}^{\infty} f(t) dt = F(0) = 0$$

Therefore,

$$\Phi(\omega) = \frac{1}{j\omega} \mathscr{F}[f(t)] = \frac{1}{j\omega} F(\omega)$$

that is,

$$\mathscr{F}\left[\int_{-\infty}^{t} f(x) dx \right] = \frac{1}{j\omega} F(\omega) = \frac{1}{j\omega} \mathscr{F}[f(t)]$$

5-4. Parseval's Formula

Let

$$f(t) \leftrightarrow F(\omega)$$

and

$$g(t) \leftrightarrow G(\omega)$$

Then **Parseval's formula** is given by

$$\int_{-\infty}^{\infty} f(x)G(x)\,dx = \int_{-\infty}^{\infty} F(x)g(x)\,dx \qquad (5.27)$$

Equation (5.27) can be written as

$$\int_{-\infty}^{\infty} f(\omega)\mathscr{F}[g(t)]\,d\omega = \int_{-\infty}^{\infty} \mathscr{F}[f(t)]g(\omega)\,d\omega \qquad (5.28)$$

Since $f(t) = \mathscr{F}^{-1}[F(\omega)]$ and $g(t) = \mathscr{F}^{-1}[G(\omega)]$, eq. (5.27) can also be written as

$$\int_{-\infty}^{\infty} \mathscr{F}^{-1}[F(\omega)]G(t)\,dt = \int_{-\infty}^{\infty} F(t)\mathscr{F}^{-1}[G(\omega)]\,dt \qquad (5.29)$$

EXAMPLE 5-15: Derive Parseval's formula (5.27).

Solution: From the definition of the Fourier transform (5.1), we have

$$F(y) = \int_{-\infty}^{\infty} f(x)e^{-jxy}\,dx$$

$$G(x) = \int_{-\infty}^{\infty} g(y)e^{-jxy}\,dy$$

Then

$$\int_{-\infty}^{\infty} f(x)G(x)\,dx = \int_{-\infty}^{\infty} f(x)\left[\int_{-\infty}^{\infty} g(y)e^{-jxy}\,dy\right]dx$$

Interchanging the order of integration,

$$\int_{-\infty}^{\infty} f(x)G(x)\,dx = \int_{-\infty}^{\infty} g(y)\left[\int_{-\infty}^{\infty} f(x)e^{-jxy}\,dx\right]dy$$

$$= \int_{-\infty}^{\infty} g(y)F(y)\,dy$$

and because we can change the dummy variable's symbol, we obtain (5.27):

$$\int_{-\infty}^{\infty} f(x)G(x)\,dx = \int_{-\infty}^{\infty} F(x)g(x)\,dx$$

5-5. Multidimensional Fourier Transforms

The theory of Fourier transforms of functions of a single variable can be extended to functions of several variables.

A. The two-dimensional Fourier transform

The **two-dimensional Fourier transform** $F(u, v)$ of a two-dimensional function $f(x, y)$ can be defined as a double integral:

Two-dimensional Fourier transform
$$F(u, v) = \int_{-\infty}^{\infty}\int_{-\infty}^{\infty} f(x, y)e^{-j(ux + vy)}\,dx\,dy \qquad (5.30)$$

Then $f(x, y)$ can be found from the inversion formula

Two-dimensional inverse Fourier transform
$$f(x, y) = \frac{1}{(2\pi)^2}\int_{-\infty}^{\infty}\int_{-\infty}^{\infty} F(u, v)e^{j(ux + vy)}\,du\,dv \qquad (5.31)$$

EXAMPLE 5-16: Using the one-dimensional Fourier transform technique, derive the inversion formula (5.31).

Solution: We denote by $G(u, y)$ the Fourier transform of the function $f(x, y)$, where the transform is taken with respect to x; that is,

$$G(u, y) = \int_{-\infty}^{\infty} f(x, y)e^{-jux}\,dx \qquad (5.32)$$

Then from the one-dimensional inversion formula (5.2), we have

$$f(x, y) = \frac{1}{2\pi} \int_{-\infty}^{\infty} G(u, y)e^{jux} \, du \tag{5.33}$$

We now take the Fourier transform $F(u, v)$ of $G(u, y)$ with respect to y, considering x as a parameter:

$$F(u, v) = \int_{-\infty}^{\infty} G(u, y)e^{-jvy} \, dy \tag{5.34}$$

The inversion formula (5.2) gives

$$G(u, y) = \frac{1}{2\pi} \int_{-\infty}^{\infty} F(u, v)e^{jvy} \, dv \tag{5.35}$$

Substituting (5.35) into (5.33), we obtain (5.31):

$$f(x, y) = \frac{1}{(2\pi)^2} \int_{-\infty}^{\infty} \int_{-\infty}^{\infty} F(u, v)e^{j(ux + vy)} \, du \, dv$$

Combining (5.34) and (5.32), we have (5.30):

$$F(u, v) = \int_{-\infty}^{\infty} \int_{-\infty}^{\infty} f(x, y)e^{-j(ux + vy)} \, dx \, dy$$

B. The three-dimensional Fourier transform

In a similar way, the three-dimensional Fourier transform pair is defined as a triple integral:

Three-dimensional Fourier transform pair

$$F(u, v, w) = \int_{-\infty}^{\infty} \int_{-\infty}^{\infty} \int_{-\infty}^{\infty} f(x, y, z)e^{-j(ux + vy + wz)} \, dx \, dy \, dz \tag{5.36}$$

$$f(x, y, z) = \frac{1}{(2\pi)^3} \int_{-\infty}^{\infty} \int_{-\infty}^{\infty} \int_{-\infty}^{\infty} F(u, v, w)e^{j(ux + vy + wz)} \, du \, dv \, dw \tag{5.37}$$

From the above, the generalization to a greater number of variables is obvious.

SUMMARY

1. The Fourier transform pair denoted by

$$f(t) \leftrightarrow F(\omega)$$

is defined by

$$F(\omega) = \mathscr{F}[f(t)] = \int_{-\infty}^{\infty} f(t)e^{-j\omega t} \, dt$$

$$f(t) = \mathscr{F}^{-1}[F(\omega)] = \frac{1}{2\pi} \int_{-\infty}^{\infty} F(\omega)e^{j\omega t} \, d\omega$$

2. Properties of the Fourier transform pair:

Linearity $\qquad a_1 f_1(t) + a_2 f_2(t) \leftrightarrow a_1 F_1(\omega) + a_2 F_2(\omega)$

Time shifting $\qquad f(t - t_0) \leftrightarrow F(\omega)e^{-j\omega t_0}$

Frequency shifting $\qquad f(t)e^{j\omega_0 t} \leftrightarrow F(\omega - \omega_0)$

Scaling $\qquad f(at) \leftrightarrow \frac{1}{|a|} F\left(\frac{\omega}{a}\right) \qquad$ (where a is real)

Time reversal $\qquad f(-t) \leftrightarrow F(-\omega)$

Symmetry $\qquad F(t) \leftrightarrow 2\pi f(-\omega)$

Modulation $\qquad f(t)\cos \omega_0 t \leftrightarrow \frac{1}{2} F(\omega - \omega_0) + \frac{1}{2} F(\omega + \omega_0)$

Differentiation $\qquad f'(t) \leftrightarrow j\omega F(\omega) \qquad$ (provided $f(\pm \infty) = 0$)

$\qquad\qquad\qquad (-jt)f(t) \leftrightarrow F'(\omega)$

Integration $\qquad \int_{-\infty}^{t} f(x) \, dx \leftrightarrow \frac{1}{j\omega} F(\omega) \qquad$ ((provided $F(0) = 0$)

3. Additional properties when $f(t)$ is real:

$$f(t) = f_e(t) + f_o(t) \leftrightarrow F(\omega) = R(\omega) + jX(\omega)$$

where $f_e(t)$ and $f_o(t)$ are the even and odd components of $f(t)$, respectively.

$$f_e(t) \leftrightarrow R(\omega)$$
$$f_o(t) \leftrightarrow jX(\omega)$$
$$R(-\omega) = R(\omega), \qquad X(-\omega) = -X(\omega), \qquad F(-\omega) = F^*(\omega)$$

4. Parseval's formula is given by

$$\int_{-\infty}^{\infty} f(x)G(x)\,dx = \int_{-\infty}^{\infty} F(x)g(x)\,dx$$

where $f(t) \leftrightarrow F(\omega)$ and $g(t) \leftrightarrow G(\omega)$.

5. Fourier transforms of functions of several variables can be obtained by simple extension of Fourier transforms of functions of a single variable.

RAISE YOUR GRADES

Can you explain ...?

☑ how to define the Fourier transform pair
☑ how to utilize the properties of the Fourier transforms to derive new transform pairs

SOLVED PROBLEMS

Fourier Transforms

PROBLEM 5-1 Show that $\int_{-\infty}^{\infty} |f(t)|\,dt < \infty$ (5.3) is a sufficient condition for the existence of the Fourier transform of $f(t)$.

Solution: Since

$$e^{-j\omega t} = \cos \omega t - j \sin \omega t$$

and hence

$$|e^{-j\omega t}| = \sqrt{\cos^2 \omega t + \sin^2 \omega t} = 1$$

and

$$|f(t)e^{-j\omega t}| = |f(t)|$$

Thus, it follows that if

$$\int_{-\infty}^{\infty} |f(t)|\,dt = \int_{-\infty}^{\infty} |f(t)e^{-j\omega t}|\,dt$$

is finite, then

$$\int_{-\infty}^{\infty} f(t)e^{-j\omega t}\,dt$$

is finite; that is, $\mathscr{F}[f(t)]$ exists.

PROBLEM 5-2 Find the Fourier transform of the rectangular pulse $p_d(t)$ (shown in Figure 5-2a) defined by

$$p_d(t) = \begin{cases} 1, & |t| < \dfrac{1}{2}d \\[2mm] 0, & |t| > \dfrac{1}{2}d \end{cases}$$

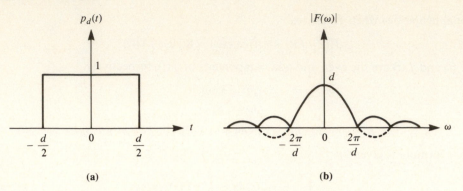

Figure 5-2 **(a) Rectangular pulse; (b) magnitude spectrum.**

Solution: From the definition of the Fourier transform (5.1), we have

$$F(\omega) = \mathscr{F}[p_d(t)] = \int_{-\infty}^{\infty} p_d(t)e^{-j\omega t}\, dt = \int_{-d/2}^{d/2} e^{-j\omega t}\, dt$$

$$= \frac{1}{-j\omega} e^{-j\omega t}\Big|_{-d/2}^{d/2} = \frac{1}{j\omega}[e^{j\omega d/2} - e^{-j\omega d/2}]$$

$$= \frac{2}{\omega}\sin(\omega d/2) = d\,\frac{\sin(\omega d/2)}{(\omega d/2)}$$

In Figure 5-2b, the solid line is the magnitude spectrum $|F(\omega)|$, and the dotted line shows $F(\omega)$.

PROBLEM 5-3   Find the Fourier transform of $f(t) = e^{-a|t|}$, where $a > 0$ (see Figure 5-3a).

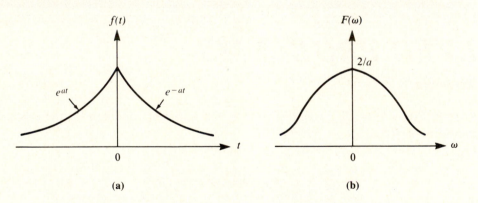

Figure 5-3 **(a) Exponential function $e^{-a|t|}$; (b) Fourier transform.**

Solution:

$$f(t) = e^{-a|t|} = \begin{cases} e^{-at}, & t > 0 \\ e^{at}, & t < 0 \end{cases}$$

Thus, from definition (5.1),

$$F(\omega) = \int_{-\infty}^{\infty} f(t)e^{-j\omega t}\, dt$$

$$= \int_{-\infty}^{0} e^{at}e^{-j\omega t}\, dt + \int_{0}^{\infty} e^{-at}e^{-j\omega t}\, dt$$

$$= \int_{-\infty}^{0} e^{-(-a+j\omega)t}\, dt + \int_{0}^{\infty} e^{-(a+j\omega)t}\, dt$$

$$= \frac{-1}{-a+j\omega} + \frac{1}{a+j\omega} = \frac{2a}{a^2+\omega^2}$$

which is plotted in the Figure 5-3b.

Properties of Fourier Transforms

PROBLEM 5-4   Give an interpretation of the scaling property of the Fourier transform

$$f(at) \leftrightarrow \frac{1}{|a|} F\left(\frac{\omega}{a}\right)$$

Solution: The function $f(at)$ represents the function $f(t)$ compressed in the time scale by a factor of a. Similarly, the function $F(\omega/a)$ represents the function $F(\omega)$ expanded in the frequency scale by the same factor a. The scaling property therefore states that compression in the time domain is equivalent to expansion in the frequency domain, and vice versa.

PROBLEM 5-5   If $F(\omega) = \mathscr{F}[f(t)]$, find the Fourier transform of $f(t)\sin \omega_0 t$.

Solution: With the identity

$$\sin \omega_0 t = \frac{1}{2j} (e^{j\omega_0 t} - e^{-j\omega_0 t})$$

and using the frequency-shifting property (5.7) and the linearity property (5.5), we obtain

$$\mathscr{F}[f(t)\sin \omega_0 t] = \mathscr{F}\left[\frac{1}{2j} f(t)e^{j\omega_0 t} - \frac{1}{2j} f(t)e^{-j\omega_0 t}\right]$$

$$= \frac{1}{2j} F(\omega - \omega_0) - \frac{1}{2j} F(\omega + \omega_0)$$

$$= \frac{1}{2j} [F(\omega - \omega_0) - F(\omega + \omega_0)]$$

PROBLEM 5-6   Find the Fourier transform of the function

$$f(t) = \frac{\sin at}{\pi t}$$

Solution: From the result of Problem 5-2, we have

$$\mathscr{F}[p_d(t)] = \frac{2}{\omega} \sin\left(\frac{\omega d}{2}\right) \tag{a}$$

Now from the symmetry property of the Fourier transform (5.10), we have

$$\mathscr{F}\left[\frac{2}{t} \sin\left(\frac{dt}{2}\right)\right] = 2\pi p_d(-\omega) \tag{b}$$

or

$$\mathscr{F}\left[\frac{\sin\left(\frac{1}{2} dt\right)}{\pi t}\right] = p_d(-\omega) \tag{c}$$

Since $p_d(\omega)$ is defined by (see Problem 5-2)

$$p_d(\omega) = \begin{cases} 1 & \text{for } |\omega| < \frac{1}{2} d \\ 0 & \text{for } |\omega| > \frac{1}{2} d \end{cases}$$

it is an even function of ω. Hence,

$$p_d(-\omega) = p_d(\omega)$$

Letting $\frac{1}{2} d = a$ in eq. (c), we obtain

$$\mathscr{F}\left(\frac{\sin at}{\pi t}\right) = p_{2a}(\omega) \tag{d}$$

where

$$p_{2a}(\omega) = \begin{cases} 1 & \text{for} \quad |\omega| < a \\ 0 & \text{for} \quad |\omega| > a \end{cases}$$

Plots of $f(t) = \sin at/\pi t$ and its Fourier transform $F(\omega)$ are shown in Figure 5-4a and 5-4b, respectively.

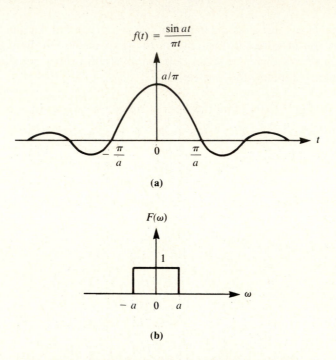

(a)

(b)

Figure 5-4   **(a) Function $\sin at/\pi t$; (b) Fourier transform.**

PROBLEM 5-7   Find the Fourier transform of $f(t) = 1/(a^2 + t^2)$.

Solution: From the result of Problem 5-3, we have

$$e^{-a|t|} \leftrightarrow \frac{2a}{a^2 + \omega^2}$$

Thus, applying the symmetry property (5.10), we obtain

$$\frac{2a}{a^2 + t^2} \leftrightarrow 2\pi e^{-a|-\omega|} = 2\pi e^{-a|\omega|}$$

Dividing both sides by $2a$ (linearity property), we have

$$\frac{1}{a^2 + t^2} \leftrightarrow \frac{\pi}{a} e^{-a|\omega|}$$

Plots of $f(t)$ and $F(\omega)$ are shown in Figure 5-5a and 5-5b, respectively

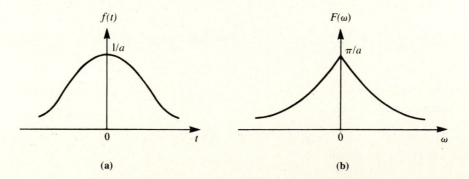

(a)   (b)

Figure 5-5   **(a) Function $1/(a^2 + t^2)$; (b) Fourier transform.**

PROBLEM 5-8 Find the Fourier transform of the cosine function of finite duration d.

Solution: The cosine function of finite duration d (see Figure 5-6a) can be expressed as a pulse-modulated function, i.e.,

$$f(t) = p_d(t)\cos \omega_0 t$$

where

$$p_d(t) = \begin{cases} 1 & \text{for } |t| < \frac{1}{2}d \\ 0 & \text{for } |t| > \frac{1}{2}d \end{cases}$$

Now from the result of Problem 5-2, we have

$$\mathscr{F}[p_d(t)] = \frac{2}{\omega}\sin\left(\frac{\omega d}{2}\right)$$

Then applying the modulation theorem (5.11), we obtain

$$F(\omega) = \mathscr{F}[p_d(t)\cos \omega_0 t]$$

$$= \frac{\sin \frac{1}{2}d(\omega - \omega_0)}{\omega - \omega_0} + \frac{\sin \frac{1}{2}d(\omega + \omega_0)}{\omega + \omega_0}$$

The Fourier transform $F(\omega)$ is plotted in Figure 5-6b.

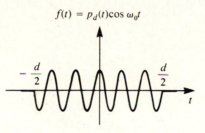

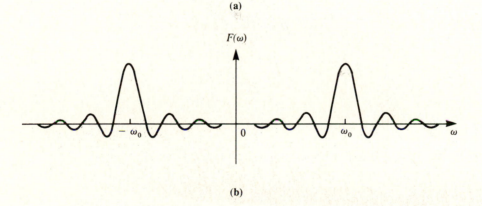

Figure 5-6   (a) Cosine function of finite duration; (b) Fourier transform.

PROBLEM 5-9 Show that if the Fourier transform of a real function $f(t)$ is real, then $f(t)$ is an even function of t, and if the Fourier transform of a real function $f(t)$ is purely imaginary, then $f(t)$ is an odd function of t.

Solution: Let

$$\mathscr{F}[f(t)] = F(\omega) = R(\omega) + jX(\omega)$$

Then we use (5.12) and (5.13):

$$R(\omega) = \int_{-\infty}^{\infty} f(t)\cos \omega t \, dt \tag{a}$$

$$X(\omega) = -\int_{-\infty}^{\infty} f(t)\sin \omega t \, dt \tag{b}$$

If $F(\omega) = R(\omega)$ and $X(\omega) = 0$, then by eq. (2.5) of Chapter 2 the integrand of eq. (b) must be odd with respect to t. Since $\sin \omega t$ is an odd function of t, $f(t)$ must be an even function of t.

Similarly, if $F(\omega) = jX(\omega)$, that is, $R(\omega) = 0$, then the integrand of eq. (a) must be odd with respect to t. Since $\cos \omega t$ is an even function of t, $f(t)$ must be an odd function of t.

Alternate Solution: If

$$f(t) \leftrightarrow F(\omega) = R(\omega) + jX(\omega)$$

and

$$f(t) = f_e(t) + f_o(t)$$

where $f_e(t)$ and $f_o(t)$ are the even and odd components of $f(t)$, respectively, then by (5.17) and (5.18), we have

$$f_e(t) \leftrightarrow R(\omega)$$

$$f_o(t) \leftrightarrow jX(\omega)$$

Thus, if $F(\omega) = R(\omega)$, then $X(\omega) = 0$, which implies that $f_o(t) = 0$; that is, $f(t) = f_e(t)$. Next, if $F(\omega) = jX(\omega)$, then $R(\omega) = 0$, which implies that $f_e(t) = 0$; that is, $f(t) = f_o(t)$.

PROBLEM 5-10 If $f(t)$ is real, show that its magnitude spectrum $|F(\omega)|$ is an even function of ω and its phase spectrum $\phi(\omega)$ is an odd function of ω.

Solution: If $f(t)$ is real, then, from the condition for a real function (5.16),

$$F(-\omega) = F^*(\omega)$$

Now, from (5.4),

$$F^*(\omega) = |F(\omega)|e^{-j\phi(\omega)}$$

Thus,

$$F(-\omega) = |F(-\omega)|e^{j\phi(-\omega)}$$

Hence,

$$|F(-\omega)|e^{j\phi(-\omega)} = |F(\omega)|e^{-j\phi(\omega)}$$

and therefore,

$$|F(-\omega)| = |F(\omega)|$$

$$\phi(-\omega) = -\phi(\omega)$$

indicating that $|F(\omega)|$ is an even function of ω and $\phi(\omega)$ is an odd function of ω.

PROBLEM 5-11 If $f(t)$ is real and even, then show that

$$F(\omega) = R(\omega) = 2\int_0^\infty f(t)\cos \omega t \, dt$$

and

$$f(t) = \frac{1}{\pi}\int_0^\infty R(\omega)\cos \omega t \, d\omega$$

Solution: From the result of Problem 5-9, if $f(t)$ is real and even, then we have $X(\omega) = 0$ and

$$F(\omega) = R(\omega) = \int_{-\infty}^\infty f(t)\cos \omega t \, dt$$

Now if $f(t)$ is even, $\cos \omega t$ is even; thus, $f(t)\cos \omega t$ is also even and by eq. (2.4), we have

$$\int_{-\infty}^\infty f(t)\cos \omega t \, dt = 2\int_0^\infty f(t)\cos \omega t \, dt$$

Thus,

$$F(\omega) = R(\omega) = 2\int_0^\infty f(t)\cos \omega t \, dt$$

Next, from (5.19) with $X(\omega) = 0$, we have

$$f(t) = \frac{1}{2\pi} \int_{-\infty}^{\infty} R(\omega)\cos \omega t\, d\omega$$

Now, by (5.14), $R(\omega)$ is even; thus, $R(\omega)\cos \omega t$ is even and—again by (2.4)—we have

$$\int_{-\infty}^{\infty} R(\omega)\cos \omega t\, d\omega = 2 \int_{0}^{\infty} R(\omega)\cos \omega t\, d\omega$$

Thus, we obtain

$$f(t) = \frac{1}{\pi} \int_{0}^{\infty} R(\omega)\cos \omega t\, d\omega$$

PROBLEM 5-12   If $f(t)$ is real and odd, then show that

$$F(\omega) = jX(\omega) = -j2 \int_{0}^{\infty} f(t)\sin \omega t\, dt$$

and

$$f(t) = -\frac{1}{\pi} \int_{0}^{\infty} X(\omega)\sin \omega t\, d\omega$$

Solution: From the result of Problem 5-9, if $f(t)$ is real and odd, then we have $R(\omega) = 0$, and

$$F(\omega) = jX(\omega) = -j \int_{-\infty}^{\infty} f(t)\sin \omega t\, dt$$

Now $f(t)$ is odd and $\sin \omega t$ is also odd; thus, $f(t)\sin \omega t$ is even, and by (2.4), we have

$$\int_{-\infty}^{\infty} f(t)\sin \omega t\, dt = 2 \int_{0}^{\infty} f(t)\sin \omega t\, dt$$

Thus,

$$F(\omega) = jX(\omega) = -j2 \int_{0}^{\infty} f(t)\sin \omega t\, dt$$

Next, from (5.19) with $R(\omega) = 0$, we have

$$f(t) = -\frac{1}{2\pi} \int_{-\infty}^{\infty} X(\omega)\sin \omega t\, d\omega$$

Now, by (5.15), $X(\omega)$ is odd; thus, $X(\omega)\sin \omega t$ is even and we have

$$\int_{-\infty}^{\infty} X(\omega)\sin \omega t\, d\omega = 2 \int_{0}^{\infty} X(\omega)\sin \omega t\, d\omega$$

Thus, we obtain

$$f(t) = -\frac{1}{\pi} \int_{0}^{\infty} X(\omega)\sin \omega t\, d\omega$$

Differentiation and Integration Theorems

PROBLEM 5-13
(a) Find the Fourier transform of the pulse $f_1(t)$ shown in Figure 5-7a.
(b) The pulse $f_2(t)$ shown in Figure 5-7b is the integral of $f_1(t)$. Use the result of part (a) to obtain the Fourier transform of $f_2(t)$.
(c) Check the result by direct integration.

Solution:

(a)

$$f_1(t) = \begin{cases} A/T, & -T < t < 0 \\ -A/T, & 0 < t < T \\ 0, & \text{otherwise} \end{cases}$$

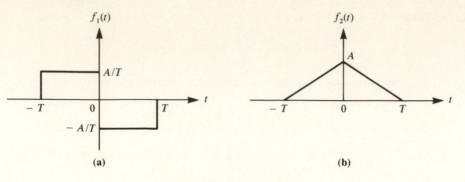

Figure 5-7

By definition (5.1), we have

$$F_1(\omega) = \int_{-\infty}^{\infty} f_1(t)e^{-j\omega t}\,dt$$

$$= \int_{-T}^{0} (A/T)e^{-j\omega t}\,dt + \int_{0}^{T} (-A/T)e^{-j\omega t}\,dt$$

$$= \frac{A}{T}\left(\frac{1}{-j\omega}\right)(1 - e^{j\omega T}) - \frac{A}{T}\left(\frac{1}{-j\omega}\right)(e^{-j\omega T} - 1)$$

$$= \frac{A}{j\omega T}(e^{j\omega T} + e^{-j\omega T} - 2)$$

$$= \frac{2A}{j\omega T}\left[\frac{1}{2}(e^{j\omega T} + e^{-j\omega T}) - 1\right]$$

$$= \frac{2A}{j\omega T}(\cos \omega T - 1) = -\frac{4A}{j\omega T}\sin^2(\omega T/2) \tag{a}$$

(b) Now

$$f_2(t) = \int_{-\infty}^{t} f_1(x)\,dx \quad \text{and} \quad \int_{-\infty}^{\infty} f_1(t)\,dt = 0$$

Thus, by the integration theorem (5.26), we have

$$F_2(\omega) = \frac{1}{j\omega}F_1(\omega) = \frac{1}{j\omega}\left(-\frac{4A}{j\omega T}\right)\sin^2(\omega T/2)$$

$$= \frac{4A}{\omega^2 T}\sin^2(\omega T/2) = AT\frac{\sin^2(\omega T/2)}{(\omega T/2)^2} \tag{b}$$

(c)

$$f_2(t) = \begin{cases} (A/T)t + A, & -T < t < 0 \\ (-A/T)t + A, & 0 < t < T \\ 0, & \text{otherwise} \end{cases}$$

Thus, by definition (5.1), we have

$$F_2(\omega) = \int_{-\infty}^{\infty} f_2(t)e^{-j\omega t}\,dt$$

$$= \int_{-T}^{0}\left(\frac{A}{T}t + A\right)e^{-j\omega t}\,dt + \int_{0}^{T}\left(-\frac{A}{T}t + A\right)e^{-j\omega t}\,dt$$

$$= \frac{A}{T}\left[\int_{-T}^{0} te^{-j\omega t}\,dt - \int_{0}^{T} te^{-j\omega t}\,dt\right] + A\int_{-T}^{T} e^{-j\omega t}\,dt$$

$$= \frac{A}{T}\left[\frac{t}{-j\omega}e^{-j\omega t}\Big|_{-T}^{0} + \frac{1}{j\omega}\int_{-T}^{0} e^{-j\omega t}\,dt - \frac{t}{-j\omega}e^{-j\omega t}\Big|_{0}^{T} - \frac{1}{j\omega}\int_{0}^{T} e^{-j\omega t}\,dt\right] + A\left(\frac{1}{-j\omega}e^{-j\omega t}\Big|_{-T}^{T}\right)$$

$$= \frac{A}{T} \left[\frac{-T}{j\omega} e^{j\omega T} - \frac{1}{(j\omega)^2} e^{-j\omega t} \Big|_{-T}^{0} + \frac{T}{j\omega} e^{-j\omega T} + \frac{1}{(j\omega)^2} e^{-j\omega t} \Big|_{0}^{T} \right] + \frac{A}{j\omega} (e^{j\omega T} - e^{-j\omega T})$$

$$= \frac{A}{T\omega^2} (2 - e^{j\omega T} - e^{-j\omega T})$$

$$= \frac{2A}{T\omega^2} [1 - \tfrac{1}{2} (e^{j\omega T} + e^{-j\omega T})]$$

$$= \frac{2A}{T\omega^2} (1 - \cos \omega T)$$

$$= \frac{4A}{T\omega^2} \sin^2(\omega T/2) = AT \frac{\sin^2(\omega T/2)}{(\omega T/2)^2}$$

which is the same as the result of eq. (b).

PROBLEM 5-14 Show that if $\mathscr{F}[f(t)] = F(\omega)$, then

$$|F(\omega)| \leq \int_{-\infty}^{\infty} |f(t)| \, dt, \qquad |F(\omega)| \leq \frac{1}{|\omega|} \int_{-\infty}^{\infty} \left| \frac{df(t)}{dt} \right| dt, \qquad |F(\omega)| \leq \frac{1}{\omega^2} \int_{-\infty}^{\infty} \left| \frac{d^2 f(t)}{dt^2} \right| dt$$

These inequalities determine the upper bounds of $|F(\omega)|$.

Solution: By definition (5.1),

$$F(\omega) = \int_{-\infty}^{\infty} f(t) e^{-j\omega t} \, dt$$

Hence, by elementary calculus, we have

$$|F(\omega)| = \left| \int_{-\infty}^{\infty} f(t) e^{-j\omega t} \, dt \right| \leq \int_{-\infty}^{\infty} |f(t) e^{-j\omega t}| \, dt = \int_{-\infty}^{\infty} |f(t)| \, dt$$

since

$$|e^{-j\omega t}| = |\cos \omega t - j \sin \omega t| = \sqrt{\cos^2 \omega t + \sin^2 \omega t} = 1$$

$$|f(t) e^{-j\omega t}| = |f(t)|$$

Next, by the differentiation theorem (5.21), we have

$$j\omega F(\omega) = \mathscr{F}[f'(t)] = \int_{-\infty}^{\infty} f'(t) e^{-j\omega t} \, dt$$

Hence, in a similar fashion, we obtain

$$|j\omega F(\omega)| = \left| \int_{-\infty}^{\infty} f'(t) e^{-j\omega t} \, dt \right| \leq \int_{-\infty}^{\infty} |f'(t)| \, dt$$

since

$$|j| = 1, \qquad |j\omega F(\omega)| = |\omega||F(\omega)|$$

Thus,

$$|F(\omega)| \leq \frac{1}{|\omega|} \int_{-\infty}^{\infty} |f'(t)| \, dt$$

Next, by repeated application of (5.21), or by (5.24), we have

$$(j\omega)^2 F(\omega) = \mathscr{F}[f^{(2)}(t)] = \int_{-\infty}^{\infty} \frac{d^2 f(t)}{dt^2} e^{-j\omega t} \, dt$$

Hence,

$$|(j\omega)^2 F(\omega)| = \left| \int_{-\infty}^{\infty} \frac{d^2 f(t)}{dt^2} e^{-j\omega t} \, dt \right| \leq \int_{-\infty}^{\infty} \left| \frac{d^2 f(t)}{dt^2} \right| dt$$

or

$$|F(\omega)| \leq \frac{1}{|\omega^2|} \int_{-\infty}^{\infty} \left| \frac{d^2 f(t)}{dt^2} \right| dt$$

PROBLEM 5-15 If

$$f(t) \leftrightarrow F(\omega)$$

and $F(\omega)$ can be differentiated everywhere n times, show that

$$t^p f(t) \leftrightarrow \frac{1}{(-j)^p} \frac{d^p F(\omega)}{d\omega^p}$$

for every $p \leqslant n$.

Solution: From (5.22), we have

$$(-jt)f(t) \leftrightarrow \frac{dF(\omega)}{d\omega}$$

By repeated application of (5.22), we obtain

$$(-jt)^2 f(t) \leftrightarrow \frac{d^2 F(\omega)}{d\omega^2}$$

$$\cdots$$

$$(-jt)^p f(t) \leftrightarrow \frac{d^p F(\omega)}{d\omega^p}, \qquad p \leqslant n$$

Dividing by $(-j)^p$, we obtain

$$t^p f(t) \leftrightarrow \frac{1}{(-j)^p} \frac{d^p F(\omega)}{d\omega^p}$$

for $p \leqslant n$.

Parseval's Formula

PROBLEM 5-16 Using Parseval's formula (5.27) or (5.28) prove that

$$\int_{-\infty}^{\infty} f(x)g(x)\,dx = \frac{1}{2\pi} \int_{-\infty}^{\infty} F(x)G(-x)\,dx$$

where

$$f(t) \leftrightarrow F(\omega)$$
$$g(t) \leftrightarrow G(\omega)$$

Proof: By (5.28), we have

$$\int_{-\infty}^{\infty} f(\omega)\mathscr{F}[g(t)]\,d\omega = \int_{-\infty}^{\infty} \mathscr{F}[f(t)]g(\omega)\,d\omega$$

Now, by the symmetry property (5.10), we have

$$G(t) \leftrightarrow 2\pi g(-\omega)$$

Using the time-reversal property (5.9), we have

$$G(-t) \leftrightarrow 2\pi g(\omega)$$

Thus, by (5.28), we obtain

$$\int_{-\infty}^{\infty} f(\omega)\mathscr{F}[G(-t)]\,d\omega = \int_{-\infty}^{\infty} \mathscr{F}[f(t)]G(-\omega)\,d\omega$$

or

$$\int_{-\infty}^{\infty} f(\omega)2\pi g(\omega)\,d\omega = \int_{-\infty}^{\infty} F(\omega)G(-\omega)\,d\omega$$

Dividing both sides by 2π and changing the dummy variable, we obtain

$$\int_{-\infty}^{\infty} f(x)g(x)\,dx = \frac{1}{2\pi} \int_{-\infty}^{\infty} F(x)G(-x)\,dx$$

Multidimensional Fourier Transforms

PROBLEM 5-17   If $\mathscr{F}[f(x, y)] = F(u, v)$, show that

$$\mathscr{F}[f(x - a, y - b)] = F(u, v)e^{-j(au + bv)}$$

Solution: By the definition of the two-dimensional Fourier transform (5.30), we have

$$\mathscr{F}[f(x - a, y - b)] = \int_{-\infty}^{\infty}\int_{-\infty}^{\infty} f(x - a, y - b)e^{-j(ux + vy)}\,dx\,dy$$

Letting $x - a = \xi$, $y - b = \eta$, then $x = a + \xi$, $y = b + \eta$ and $dx = d\xi$, $dy = d\eta$; hence,

$$\mathscr{F}[f(x - a, y - b)] = \int_{-\infty}^{\infty}\int_{-\infty}^{\infty} f(\xi, \eta)e^{-j(au + u\xi + bv + v\eta)}\,d\xi\,d\eta$$

$$= e^{-j(au + bv)}\int_{-\infty}^{\infty}\int_{-\infty}^{\infty} f(\xi, \eta)e^{-j(u\xi + v\eta)}\,d\xi\,d\eta$$

$$= e^{-j(au + bv)}F(u, v)$$

Supplementary Exercises

PROBLEM 5-18   If $f(t)$ is purely imaginary, that is, $f(t) = jg(t)$, where $g(t)$ is real, show that the real and imaginary parts of $F(\omega)$ are

$$R(\omega) = \int_{-\infty}^{\infty} g(t)\sin \omega t\,dt, \qquad X(\omega) = \int_{-\infty}^{\infty} g(t)\cos \omega t\,dt$$

Also show that $R(\omega)$ and $X(\omega)$ are odd and even functions of ω; that is,

$$R(-\omega) = -R(\omega), \qquad X(-\omega) = X(\omega), \qquad F(-\omega) = -F^*(\omega)$$

PROBLEM 5-19   If $\mathscr{F}[f(t)] = F(\omega)$, show that $\mathscr{F}[f^*(t)] = F^*(-\omega)$, where $f^*(t)$ is the conjugate of $f(t)$ and $F^*(-\omega)$ is the conjugate of $F(-\omega)$.

PROBLEM 5-20   If $F(\omega) = \mathscr{F}[f(t)]$, show that

$$\mathscr{F}[f(at)e^{j\omega_0 t}] = \frac{1}{|a|} F\left(\frac{\omega - \omega_0}{a}\right)$$

PROBLEM 5-21   The nth moment m_n of a function $f(t)$ is defined by

$$m_n = \int_{-\infty}^{\infty} t^n f(t)\,dt \qquad \text{for} \quad n = 0, 1, 2, \ldots$$

Show that

$$m_n = (j)^n \frac{d^n F(0)}{d\omega^n} \qquad \text{for} \quad n = 0, 1, 2, \ldots$$

where

$$\frac{d^n F(0)}{d\omega^n} = \left.\frac{d^n F(\omega)}{d\omega^n}\right|_{\omega = 0} \qquad \text{and} \quad F(\omega) = \mathscr{F}[f(t)]$$

[*Hint:* Use the differentiation theorem (5.22).]

PROBLEM 5-22   Use the result of Problem 5-21 to show that $F(\omega) = \mathscr{F}[f(t)]$ can be expressed as

$$F(\omega) = \sum_{n=0}^{\infty} (-j)^n m_n \frac{\omega^n}{n!}$$

[*Hint:* Substitute $e^{-j\omega t} = \sum_{n=0}^{\infty} (-j\omega t)^n/n!$ in (5.1) and integrate termwise.]

PROBLEM 5-23   Let $F(\omega)$ be the Fourier transform of $f(t)$ and $f_k(t)$ be defined by

$$f_k(t) = \frac{1}{2\pi}\int_{-k}^{k} F(\omega)e^{j\omega t}\,d\omega$$

Show that

$$f_k(t) = \frac{1}{\pi} \int_{-\infty}^{\infty} f(t - x) \frac{\sin kx}{x} dx$$

PROBLEM 5-24   Let $F(\omega) = \mathscr{F}[f(t)]$ and $G(\omega) = \mathscr{F}[g(t)]$. Prove that

(a) $\int_{-\infty}^{\infty} f(t)g(-t) dt = \frac{1}{2\pi} \int_{-\infty}^{\infty} F(\omega)G(\omega) d\omega$

(b) $\int_{-\infty}^{\infty} f(t)g^*(t) dt = \frac{1}{2\pi} \int_{-\infty}^{\infty} F(\omega)G^*(\omega) d\omega$

where the asterisk denotes complex conjugation. [*Hint:* Use Parseval's formula (5.27).]

PROBLEM 5-25   If $\mathscr{F}[f(x, y)] = F(u, v)$, show that

$$\mathscr{F}[f(ax, by)] = \frac{1}{|ab|} F\left(\frac{u}{a}, \frac{v}{b}\right)$$

PROBLEM 5-26   Prove the Fourier transform theorem

$$\mathscr{F}[\nabla^2 f(x, y)] = -(u^2 + v^2)\mathscr{F}[f(x, y)]$$

where ∇^2 is the Laplacian operator $\nabla^2 = \partial^2/\partial x^2 + \partial^2/\partial y^2$.

6 GENERALIZED FOURIER TRANSFORMS

THIS CHAPTER IS ABOUT

☑ **Functions of Slow Growth**
☑ **Generalized Fourier Transforms**
☑ **The Fourier Transform of a Periodic Function**
☑ **Integral Representation of $\delta(t)$**

6-1. Functions of Slow Growth

The sufficient condition for the existence of a Fourier transform of a function $f(t)$ was given by eq. (5.3):

$$\int_{-\infty}^{\infty} |f(t)| \, dt < \infty$$

In other words, the function $f(t)$ is absolutely integrable.

Functions such as $\sin \omega t$, $\cos \omega t$, the unit step function $u(t)$, etc., do not satisfy this condition. In this chapter we'll find the Fourier transforms of functions that are not absolutely integrable, and we'll define Fourier transforms of generalized functions.

A. Definition of slow growth function

A function $f(t)$ is said to be **of slow growth** if there exist numbers C and R such that

$$|f(t)| \leqslant C|t|^n \qquad \text{whenever} \qquad |t| > R \tag{6.1}$$

EXAMPLE 6-1: Give examples of slow growth function.

Solution: A function of slow growth is one that grows more slowly at infinity than some polynomial does. Functions such as $\sin \omega t$, $\cos \omega t$, the unit step function $u(t)$, and $e^{j\omega t}$ are of slow growth.

EXAMPLE 6-2: Show that $f(t) = e^{-t}$ is not a function of slow growth.

Solution: Since $t^{-n}e^{-t} \to \infty$ as $t \to -\infty$ for any $n \geqslant 0$, e^{-t} is not a function of slow growth.

B. Testing functions of rapid decay

Definition: A continuous function $\phi(t)$ is said to be **a testing function of rapid decay** if

**Testing function
of rapid decay**
$$\lim_{t \to \pm\infty} |t^n \phi^{(r)}(t)| = 0 \qquad \text{for} \quad n, r \geqslant 0 \tag{6.2}$$

where $\phi^{(r)}(t) = d^r\phi(t)/dt^r$.

C. Generalized functions of slow growth

By the definition of generalized functions given in Chapter 4, **a generalized function of slow growth** $g(t)$ is defined as a symbolic function such that to each testing function of rapid decay $\phi(t)$ there is assigned a number

**Generalized function
of slow growth**
$$\langle g, \phi \rangle = \int_{-\infty}^{\infty} g(t)\phi(t) \, dt \tag{6.3}$$

with the property

$$\langle g, a_1\phi_1 + a_2\phi_2 \rangle = a_1\langle g, \phi_1 \rangle + a_2\langle g, \phi_2 \rangle \tag{6.4}$$

Similarly, by the definition of the derivative of a generalized function, the **derivative of a generalized function of slow growth** is defined by

Derivative of a generalized function of slow growth

$$\int_{-\infty}^{\infty} g'(t)\phi(t)\,dt = -\int_{-\infty}^{\infty} g(t)\phi'(t)\,dt \qquad (6.5)$$

EXAMPLE 6-3: As in Chapter 4 (4.1), we define $\delta(t)$ by

δ-Function

$$\int_{-\infty}^{\infty} \delta(t)\phi(t)\,dt = \phi(0) \qquad (6.6)$$

6-2. Generalized Fourier Transforms

If $\phi(t)$ is a testing function of rapid decay, then we can use advanced calculus to show that $\phi(t)$ is absolutely integrable. Hence, by condition (5.3), its Fourier tansform $\Phi(\omega)$ exists. Therefore, all the properties of the ordinary Fourier tansforms that we discussed in Chapter 5 hold for $\Phi(\omega)$.

Since a function of slow growth, $f(t)$ is not absolutely integrable, we cannot define its Fourier transform by definition (5.1). Instead, we shall use Parseval's formula (5.27) to define its Fourier transform.

A. Definition of generalized Fourier transforms of slow growth

The **generalized Fourier transform** $F(\omega)$ of a function of slow growth $f(t)$ is defined by

Generalized Fourier transform

$$\int_{-\infty}^{\infty} F(x)\phi(x)\,dx = \int_{-\infty}^{\infty} f(x)\Phi(x)\,dx \qquad (6.7)$$

EXAMPLE 6-4: Find the Fourier transform of a constant.

Solution: Let $f(t) = 1$. Then $f(t)$ is a function of slow growth, and by definition (6.7),

$$\int_{-\infty}^{\infty} F(\omega)\phi(\omega)\,d\omega = \int_{-\infty}^{\infty} f(t)\Phi(t)\,dt$$

$$= \int_{-\infty}^{\infty} \Phi(t)\,dt = \left[\int_{-\infty}^{\infty} \Phi(t)e^{-j\omega t}\,dt\right]_{\omega=0}$$

By the symmetry property (5.10), we have

$$\left[\int_{-\infty}^{\infty} \Phi(t)e^{-j\omega t}\,dt\right]_{\omega=0} = [\mathscr{F}[\Phi(t)]]_{\omega=0} = [2\pi\phi(-\omega)]_{\omega=0} = 2\pi\phi(0)$$

But

$$\phi(0) = \int_{-\infty}^{\infty} \delta(t)\phi(t)\,dt = \int_{-\infty}^{\infty} \delta(\omega)\phi(\omega)\,d\omega$$

Thus,

$$2\pi\phi(0) = \int_{-\infty}^{\infty} 2\pi\delta(\omega)\phi(\omega)\,d\omega$$

and we have

$$\int_{-\infty}^{\infty} F(\omega)\phi(\omega)\,d\omega = \int_{-\infty}^{\infty} 2\pi\delta(\omega)\phi(\omega)\,d\omega$$

Hence, by the equivalence property (4.2), we conclude that

Fourier transform of a constant

$$\mathscr{F}[1] = F(\omega) = 2\pi\delta(\omega) \qquad (6.8)$$

which indicates that the Fourier transform of a constant is an impulse function.

Remember that $f(t) = A$ means that the function $f(t)$ is constant for all t (see Figure 6-1a). Therefore, we observe that if $f(t) = $ constant, the only frequency we can associate with it is zero frequency (pure dc).

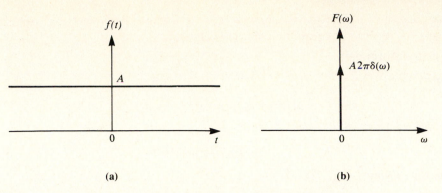

Figure 6-1 (a) Constant function and (b) its Fourier transform.

EXAMPLE 6-5: Find the Fourier transform of $\delta(t)$.

Solution: Let $f(t) = \delta(t)$. Then $f(t)$ is a generalized function of slow growth, and by definition (6.7),

$$\int_{-\infty}^{\infty} F(\omega)\phi(\omega)\,d\omega = \int_{-\infty}^{\infty} f(t)\Phi(t)\,dt$$

$$= \int_{-\infty}^{\infty} \delta(t)\Phi(t)\,dt = \Phi(0)$$

But

$$\Phi(0) = [\mathscr{F}[\phi(t)]]_{\omega=0} = \left[\int_{-\infty}^{\infty} \phi(t)e^{-j\omega t}\,dt\right]_{\omega=0}$$

$$= \int_{-\infty}^{\infty} \phi(t)\,dt = \int_{-\infty}^{\infty} 1\phi(\omega)\,d\omega$$

Thus, we obtain

$$\int_{-\infty}^{\infty} F(\omega)\phi(\omega)\,d\omega = \int_{-\infty}^{\infty} 1\phi(\omega)\,d\omega$$

and—again by the equivalence property (4.2)—we conclude that

Fourier transform of a δ-function

$$\mathscr{F}[\delta(t)] = F(\omega) = 1 \tag{6.9}$$

Hence, the Fourier transform of the unit impulse function is unity. We therefore see that an impulse function has a uniform spectral density over the entire frequency interval (see Figure 6-2b).

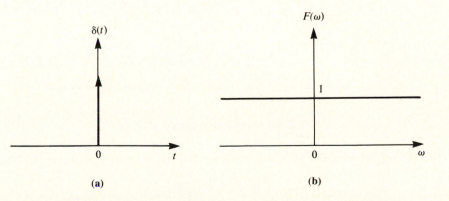

Figure 6-2 (a) Unit impulse function and (b) its Fourier transform.

B. Properties of generalized Fourier transforms

All the properties of the ordinary Fourier transforms (discussed in Chapter 5) also hold for the generalized Fourier transforms of functions of slow growth.

Again, let

$$f(t) \leftrightarrow F(\omega)$$

denote the transform pair.

1. Linearity (see (5.5)):

$$a_1 f_1(t) + a_2 f_2(t) \leftrightarrow a_1 F_1(\omega) + a_2 F_2(\omega) \qquad \textbf{(6.10)}$$

2. Time shifting (see (5.6)):

$$f(t - t_0) \leftrightarrow F(\omega)e^{-j\omega t_0} \qquad \textbf{(6.11)}$$

3. Frequency shifting (see (5.7)):

$$f(t)e^{j\omega_0 t} \leftrightarrow F(\omega - \omega_0) \qquad \textbf{(6.12)}$$

4. Scaling (see (5.8)):

$$f(at) \leftrightarrow \frac{1}{|a|} F\left(\frac{\omega}{a}\right) \qquad \textbf{(6.13)}$$

5. Time-reversal (see (5.9)):

$$f(-t) \leftrightarrow F(-\omega) \qquad \textbf{(6.14)}$$

6. Symmetry (see (5.70)):

$$F(t) \leftrightarrow 2\pi f(-\omega) \qquad \textbf{(6.15)}$$

7. Differentiation (see (5.21) and (5.22)):

$$f'(t) \leftrightarrow j\omega F(\omega) \qquad \textbf{(6.16)}$$

$$(-jt)f(t) \leftrightarrow F'(\omega) \qquad \textbf{(6.17)}$$

EXAMPLE 6–6: Prove the symmetry property (6.15).

Proof: By definition of a generalized Fourier transform (6.7), we know that

$$\int_{-\infty}^{\infty} \mathscr{F}[F(t)]\phi(\omega)\, d\omega = \int_{-\infty}^{\infty} F(\omega)\Phi(\omega)\, d\omega$$

But, again by definition (6.7), we also have

$$\int_{-\infty}^{\infty} F(\omega)\Phi(\omega)\, d\omega = \int_{-\infty}^{\infty} f(\omega)\mathscr{F}[\Phi]\, d\omega$$

Now, by the symmetry property of Fourier transforms (5.10),

$$\mathscr{F}[\Phi] = 2\pi\phi(-\omega)$$

Thus,

$$\int_{-\infty}^{\infty} \mathscr{F}[F(t)]\phi(\omega)\, d\omega = \int_{-\infty}^{\infty} f(\omega)2\pi\phi(-\omega)\, d\omega$$

$$= \int_{-\infty}^{\infty} 2\pi f(-x)\phi(x)\, dx \qquad \text{(by change of variable)}$$

$$= \int_{-\infty}^{\infty} 2\pi f(-\omega)\phi(\omega)\, d\omega$$

Hence, by the equivalence property (4.2), we conclude that the symmetry property (6.15) holds:

$$\mathscr{F}[F(t)] = 2\pi f(-\omega)$$

EXAMPLE 6-7: Prove the differentiation theorem (6.16).

Proof: By definition (6.7),

$$\int_{-\infty}^{\infty} \mathscr{F}[f'(t)]\phi(\omega)\, d\omega = \int_{-\infty}^{\infty} f'(\omega)\Phi(\omega)\, d\omega$$

Now, by the definition of the generalized derivative (6.5), we have

$$\int_{-\infty}^{\infty} f'(\omega)\Phi(\omega)\, d\omega = -\int_{-\infty}^{\infty} f(\omega)\,\Phi'(\omega)\, d\omega$$

and by the differentiation theorem (5.22),

$$\Phi'(\omega) = \mathscr{F}[-jt\phi(t)]$$

Thus,

$$-\int_{-\infty}^{\infty} f(\omega)\,\Phi'(\omega)\, d\omega = -\int_{-\infty}^{\infty} f(\omega)\mathscr{F}[-jt\phi(t)]\, d\omega$$

$$= \int_{-\infty}^{\infty} f(\omega)\mathscr{F}[jt\phi(t)]\, d\omega$$

Again, by definition (6.7), we have

$$\int_{-\infty}^{\infty} f(\omega)\mathscr{F}[jt\phi(t)]\, d\omega = \int_{-\infty}^{\infty} F(\omega)[j\omega\phi(\omega)]\, d\omega$$

$$= \int_{-\infty}^{\infty} [j\omega F(\omega)]\phi(\omega)\, d\omega$$

Hence,

$$\int_{-\infty}^{\infty} \mathscr{F}[f'(t)]\phi(\omega)\, d\omega = \int_{-\infty}^{\infty} [j\omega F(\omega)]\phi(\omega)\, d\omega$$

and, by equivalence property (4.2), we conclude that the differentiation theorem (6.16) holds for generalized Fourier transforms:

$$\mathscr{F}[f'(t)] = j\omega F(\omega)$$

EXAMPLE 6-8: Find the Fourier transform of a complex exponential function $e^{j\omega_0 t}$.

Solution: From the Fourier transform of a constant (6.8), we have

$$1 \leftrightarrow 2\pi\delta(\omega)$$

The function $e^{j\omega_0 t}$ is a function of slow growth; thus, by the frequency-shifting property (6.12), we obtain

Fourier transform of a complex exponential function
$$e^{j\omega_0 t} \leftrightarrow 2\pi\delta(\omega - \omega_0) \tag{6.18}$$

EXAMPLE 6-9: Find the Fourier transform of a cosine function $\cos \omega_0 t$.

Solution: $\cos \omega_0 t$ is a function of slow growth. From (6.18) we have

$$e^{j\omega_0 t} \leftrightarrow 2\pi\delta(\omega - \omega_0)$$

$$e^{-j\omega_0 t} \leftrightarrow 2\pi\delta(\omega + \omega_0)$$

Now, from the identity $\cos \omega_0 t = \frac{1}{2}(e^{j\omega_0 t} + e^{-j\omega_0 t})$, and by the linearity property (6.10), we have

Fourier transform of a cosine function
$$\cos \omega_0 t \leftrightarrow \pi\delta(\omega - \omega_0) + \pi\delta(\omega + \omega_0) \tag{6.19}$$

Note that the function $e^{j\omega_0 t}$ is not a real function of time, and hence it has a spectrum [see eq. (6.18)] which exists only at $\omega = \omega_0$. We have shown that for any real function of time, the amplitude spectrum is an even function of ω (see Problem 5-10). Hence, if there is an impulse at $\omega = \omega_0$, then there must be an impulse at $\omega = -\omega_0$ for any real function of time. This is the case for the function $\cos \omega_0 t$ (see Figure 6-3).

EXAMPLE 6-10: Find the Fourier transform of the unit step function $u(t)$.

Solution: $u(t)$ is a function of slow growth. So, let

$$\mathscr{F}[u(t)] = F(\omega)$$

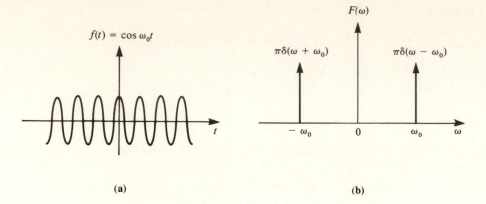

Figure 6-3 **(a) Cosine function and (b) its Fourier transform.**

Then, from the time-reversal property (6.14), we have

$$\mathcal{F}[u(-t)] = F(-\omega)$$

Since

$$u(-t) = \begin{cases} 0 & \text{for } t > 0 \\ 1 & \text{for } t < 0 \end{cases}$$

we have

$$u(t) + u(-t) = 1$$

From the linearity of the Fourier transform (6.10) and from the Fourier transform of a constant (6.8) we have

$$\mathcal{F}[u(t)] + \mathcal{F}[u(-t)] = \mathcal{F}[1]$$

that is

$$F(\omega) + F(-\omega) = 2\pi\delta(\omega)$$

We now assume that

$$F(\omega) = k\delta(\omega) + B(\omega)$$

where $B(\omega)$ is an ordinary function and k is a constant. By $\delta(-t) = \delta(t)$ (4.5) we have $\delta(-\omega) = \delta(\omega)$; thus,

$$\begin{aligned} F(\omega) + F(-\omega) &= k\delta(\omega) + B(\omega) + k\delta(-\omega) + B(-\omega) \\ &= 2k\delta(\omega) + B(\omega) + B(-\omega) \\ &= 2\pi\delta(\omega) \end{aligned}$$

Hence, we conclude that $k = \pi$ and $B(\omega)$ is odd.

To find $B(\omega)$, we proceed as follows: From the derivative of the unit step function (4.18), we know that

$$u'(t) = \frac{du(t)}{dt} = \delta(t)$$

Then, according to the differentiation property (6.16),

$$\begin{aligned} \mathcal{F}[u'(t)] = j\omega F(\omega) &= j\omega[\pi\delta(\omega) + B(\omega)] \\ &= \mathcal{F}[\delta(t)] \\ &= 1 \end{aligned}$$

Now, from the result of Problem 4-1, we have $\omega\delta(\omega) = 0$. Hence, we obtain

$$j\omega B(\omega) = 1$$

and

$$B(\omega) = \frac{1}{j\omega}$$

Finally, we obtain

Fourier transform of a unit step function

$$\mathscr{F}[u(t)] = \pi\delta(\omega) + \frac{1}{j\omega} \qquad (6.20)$$

Equation (6.20) shows that the spectrum of $u(t)$ contains an impulse at $\omega = 0$. Thus, the function $u(t)$ contains a dc component, as expected. Figure 6-4a and b shows the unit step function and its transform, respectively.

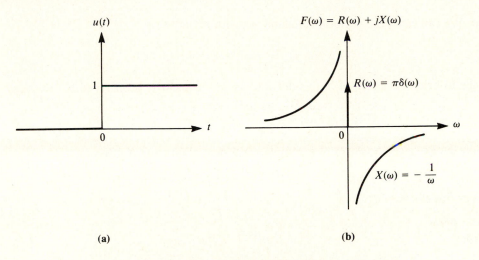

(a)   (b)

Figure 6-4   (a) Unit step function and (b) its Fourier transform.

Note that a superficial application of the differentiation property (6.16) to

$$\delta(t) = \frac{du(t)}{dt}$$

would have resulted in

$$\mathscr{F}[\delta(t)] = j\omega F(\omega)$$

where $F(\omega)$ is the Fourier transform of $u(t)$. Hence, with the Fourier transform of the δ-function (6.9), we have

$$1 = j\omega F(\omega) \qquad (6.21)$$

Therefore,

$$F(\omega) = \frac{1}{j\omega} \qquad (6.22)$$

a result that is not in agreement with (6.20).

In general, if

$$\omega F_1(\omega) = \omega F_2(\omega)$$

it does not follow that

$$F_1(\omega) = F_2(\omega)$$

Instead, the correct conclusion is

$$F_1(\omega) = F_2(\omega) + k\delta(\omega) \qquad (6.23)$$

where k is a constant, because $\omega\delta(\omega) = 0$—as you can see from the property of the δ-function.

Therefore, from (6.21), the correct conclusion is

$$F(\omega) = \frac{1}{j\omega} + k\delta(\omega) \qquad (6.24)$$

6-3. The Fourier Transform of a Periodic Function

Remember that for any periodic function $f(t)$,

$$\int_{-\infty}^{\infty} |f(t)|\,dt = \infty$$

But any periodic function $f(t)$ is a function of slow growth, and its Fourier transform does exist in the sense of a generalized function. This has been demonstrated in finding the generalized Fourier transforms of $\cos \omega_0 t$ and $e^{j\omega_0 t}$.

A. Fourier transform of a periodic function $f(t)$

EXAMPLE 6-11: Find the Fourier transform of a periodic function.

Solution: We can express a periodic function $f(t)$ with period T as

$$f(t) = \sum_{n=-\infty}^{\infty} c_n e^{jn\omega_0 t}, \qquad \omega_0 = 2\pi/T$$

Taking the Fourier transform of both sides, and using the linearity property (6.10), we have

$$\mathcal{F}[f(t)] = F(\omega) = \mathcal{F}\left[\sum_{n=-\infty}^{\infty} c_n e^{jn\omega_0 t}\right] = \sum_{n=-\infty}^{\infty} c_n \mathcal{F}[e^{jn\omega_0 t}]$$

Since from the Fourier transform of a complex exponential function (6.18),

$$\mathcal{F}[e^{jn\omega_0 t}] = 2\pi\delta(\omega - n\omega_0)$$

the Fourier transform of $f(t)$ is

Fourier transform of a periodic function

$$F(\omega) = 2\pi \sum_{n=-\infty}^{\infty} c_n \delta(\omega - n\omega_0) \tag{6.25}$$

Equation (6.25) states that the Fourier transform of a periodic function consists of a sequence of equidistant impulses located at the harmonic frequencies of the function.

B. Fourier transform of the periodic train of unit impulses

EXAMPLE 6-12: Find the Fourier transform of the periodic train of unit impulses.

Solution: The periodic train of unit impulses $\delta_T(t)$ is defined by (4.29); that is,

$$\delta_T(t) = \sum_{n=-\infty}^{\infty} \delta(t - nT)$$

Since $\delta_T(t)$ is a periodic function with period T and from the derivation of the complex Fourier series of $\delta_T(t)$ (4.36) (Example 4-11), the Fourier series for the function $\delta_T(t)$ is given by

$$\delta_T(t) = \frac{1}{T} \sum_{n=-\infty}^{\infty} e^{jn\omega_0 t}$$

where $\omega_0 = 2\pi/T$. Therefore, by the linearity property (6.10), we have

$$\mathcal{F}[\delta_T(t)] = \frac{1}{T} \sum_{n=-\infty}^{\infty} \mathcal{F}[e^{jn\omega_0 t}]$$

Using the Fourier transform of a complex exponential function (6.18), we get

$$\mathcal{F}[\delta_T(t)] = \frac{2\pi}{T} \sum_{n=-\infty}^{\infty} \delta(\omega - n\omega_0)$$

$$= \omega_0 \sum_{n=-\infty}^{\infty} \delta(\omega - n\omega_0)$$

$$= \omega_0 \delta_{\omega_0}(\omega)$$

or

Fourier transform of a unit impulse train

$$\mathcal{F}\left[\sum_{n=-\infty}^{\infty} \delta(t - nT)\right] = \omega_0 \sum_{n=-\infty}^{\infty} \delta(\omega - n\omega_0) \tag{6.26}$$

Equation (6.26) states that the Fourier transform of a unit impulse train is also a similar impulse train (see Figure 6-5).

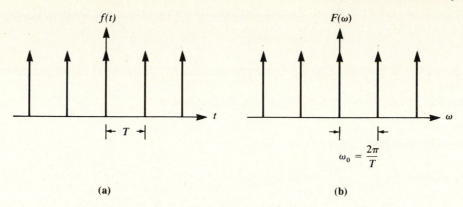

$$\omega_0 = \frac{2\pi}{T}$$

(a)   **(b)**

Figure 6-5 **(a) Unit impulse train and (b) its Fourier transform.**

6-4. Integral Representation of $\delta(t)$

Applying the inverse Fourier transform formula (5.2) to the Fourier transform of $\delta(t)$, we can obtain the following integral representations of $\delta(t)$.

Integral representations of δ-function

$$\delta(t) = \frac{1}{2\pi} \int_{-\infty}^{\infty} e^{j\omega t}\, d\omega \qquad (6.27)$$

$$\delta(t) = \frac{1}{\pi} \int_{0}^{\infty} \cos \omega t\, d\omega \qquad (6.28)$$

Note that the ordinary integration of $\int_{-\infty}^{\infty} e^{j\omega t}\, d\omega$ or $\int_{0}^{\infty} \cos \omega t\, d\omega$ is meaningless here. Instead, we should interpret (6.27) or (6.28) as generalized functions (or symbolic functions); that is, the integration of (6.27) or (6.28) converges to $\delta(t)$ in the sense of a generalized function.

EXAMPLE 6-13: Derive the integral representations of the δ-function (6.27) and (6.28).

Solution: Applying the inverse Fourier transform formula (5.2) to (6.9), we obtain

$$\delta(t) = \mathscr{F}^{-1}[1] = \frac{1}{2\pi} \int_{-\infty}^{\infty} 1 e^{j\omega t}\, d\omega = \frac{1}{2\pi} \int_{-\infty}^{\infty} e^{j\omega t}\, d\omega$$

Next, using the identity $e^{j\omega t} = \cos \omega t + j \sin \omega t$, we have

$$\delta(t) = \frac{1}{2\pi} \int_{-\infty}^{\infty} e^{j\omega t}\, d\omega$$

$$= \frac{1}{2\pi} \int_{-\infty}^{\infty} (\cos \omega t + j \sin \omega t)\, d\omega$$

$$= \frac{1}{2\pi} \int_{-\infty}^{\infty} \cos \omega t\, d\omega + j \frac{1}{2\pi} \int_{-\infty}^{\infty} \sin \omega t\, d\omega$$

$$= \frac{1}{\pi} \int_{0}^{\infty} \cos \omega t\, d\omega$$

with the use of the properties (2.4) and (2.5) of even and odd functions.

SUMMARY

1. A function $f(t)$ of slow growth is defined as

$$|f(t)| \leqslant C|t|^n \quad \text{for} \quad |t| > R$$

2. The generalized Fourier transform $F(\omega)$ of a function of slow growth $f(t)$ is defined by Parseval's formula

$$\int_{-\infty}^{\infty} F(x)\phi(x)\, dx = \int_{-\infty}^{\infty} f(x)\Phi(x)\, dx$$

where $\Phi(\omega) = \mathscr{F}[\phi(t)]$ and $\phi(t)$ is a testing function of rapid decay having the following property:

$$\lim_{t \to \pm\infty} |t^n \phi^{(r)}(t)| = 0 \quad \text{for} \quad n, r \geqslant 0$$

3. All the properties of ordinary Fourier transforms (see Chapter 5) also hold for the generalized Fourier transforms.
4. Generalized Fourier transforms for some special functions:

Constant $\qquad\qquad\qquad\qquad\qquad 1 \leftrightarrow 2\pi\delta(\omega)$

δ-Function (unit impulse function) $\qquad \delta(t) \leftrightarrow 1$

Unit step function $\qquad\qquad\qquad u(t) \leftrightarrow \pi\delta(\omega) + \dfrac{1}{j\omega}$

Complex exponential function $\qquad e^{j\omega_0 t} \leftrightarrow 2\pi\delta(\omega - \omega_0)$

Cosine function $\qquad\qquad\qquad \cos\omega_0 t \leftrightarrow \pi\delta(\omega - \omega_0) + \pi\delta(\omega + \omega_0)$

Unit impulse train $\qquad \displaystyle\sum_{n=-\infty}^{\infty} \delta(t - nT) \leftrightarrow \omega_0 \sum_{n=-\infty}^{\infty} \delta(\omega - n\omega_0), \qquad \omega_0 = 2\pi/T$

RAISE YOUR GRADES

Can you explain . . . ?

☑ what the function of slow growth is
☑ how the generalized Fourier transform of a function of slow growth is defined
☑ how to utilize the properties of the generalized Fourier transforms to derive new transform pairs
☑ how to find the generalized Fourier transform of a periodic function

SOLVED PROBLEMS

Function of Slow Growth

PROBLEM 6-1   Show that e^{-at^2} is a function of rapid decay for any $a > 0$.

Solution: Let

$$\phi(t) = e^{-at^2}$$

Then

$$|t^n \phi^{(r)}(t)| = |(-2a)^r t^{n+r} e^{-at^2}| \to 0 \quad \text{as} \quad t \to \pm\infty$$

for all $n, r \geqslant 0$. Thus, by the definition of the testing function of rapid decay (6.2), $\phi(t)$ is a function of rapid decay.

PROBLEM 6-2   Show that if $|t^n \phi^{(r)}(t)|$ is a bounded function for each $n, r \geqslant 0$, then $\phi(t)$ is a function of rapid decay.

Solution: Since

$$t^n \phi^{(r)}(t) = t^{-1} t^{n+1} \phi^{(r)}(t) = t^{-1} (\text{bounded function}) \to 0$$

as $t \to \pm\infty$ for all n and r, then, by property (6.2), $\phi(t)$ is a function of rapid decay.

Generalized Fourier Transform

PROBLEM 6-3   The Fourier transform of the unit impulse function $\delta(t)$ is often obtained as follows. Applying the ordinary definition (5.1), the Fourier transform of $\delta(t)$ is given by

$$\mathscr{F}[\delta(t)] = \int_{-\infty}^{\infty} \delta(t)e^{-j\omega t}\,dt = e^{-j\omega t}|_{t=0} = 1$$

Is this derivation valid?

Solution: Even though this derivation is very common, strictly speaking, it is not valid. The reason is as follows. By the definition of the unit impulse function (4.1), we have

$$\int_{-\infty}^{\infty} \delta(t)\phi(t)\,dt = \phi(0)$$

where $\phi(t)$ is a testing function. The function $e^{-j\omega t}$, unfortunately, is not a testing function as defined in Chapter 4; neither is it a rapid decay function. Hence, $\int_{-\infty}^{\infty} \delta(t)e^{-j\omega t}\,dt$ is not defined.

PROBLEM 6-4   Find the Fourier transform of e^{-t}.

Solution: e^{-t} does not have a generalized Fourier transform because it is not a function of slow growth (see Example 6-2). Therefore, the definition of the generalized Fourier transform (6.7) is not applicable for e^{-t}.

PROBLEM 6-5   Find the Fourier transform of $\delta(t - t_0)$ by definition (6.7).

Solution: Let $\mathscr{F}[\delta(t - t_0)] = F(\omega)$. Then, by (6.7),

$$\int_{-\infty}^{\infty} F(\omega)\phi(\omega)\,d\omega = \int_{-\infty}^{\infty} \delta(t - t_0)\Phi(t)\,dt$$

$$= \Phi(t_0)$$

$$= [\Phi(\omega)]_{\omega = t_0}$$

$$= \left[\int_{-\infty}^{\infty} \phi(t)e^{-j\omega t}\,dt\right]_{\omega = t_0}$$

$$= \int_{-\infty}^{\infty} e^{-jt_0 x}\phi(x)\,dx$$

$$= \int_{-\infty}^{\infty} e^{-j\omega t_0}\phi(\omega)\,d\omega$$

Hence, by the equivalence property (4.2), we conclude that

$$F(\omega) = \mathscr{F}[\delta(t - t_0)] = e^{-j\omega t_0}$$

PROBLEM 6-6   Prove the time-shifting property $f(t - t_0) \leftrightarrow F(\omega)e^{-j\omega t_0}$ (6.11) by definition (6.7).

Proof: By definition (6.7),

$$\int_{-\infty}^{\infty} \mathscr{F}[f(t - t_0)]\phi(\omega)\,d\omega = \int_{-\infty}^{\infty} f(t - t_0)\Phi(t)\,dt$$

$$= \int_{-\infty}^{\infty} f(t - t_0)\left[\int_{-\infty}^{\infty} \phi(y)e^{-jty}\,dy\right]dt$$

$$= \int_{-\infty}^{\infty} \phi(y)\,dy \int_{-\infty}^{\infty} f(t - t_0)e^{-jty}\,dt$$

$$= \int_{-\infty}^{\infty} \phi(y)\,dy \int_{-\infty}^{\infty} f(x)e^{-j(t_0 + x)y}\,dx$$

$$= \int_{-\infty}^{\infty} \phi(y)e^{-jt_0 y}\,dy \int_{-\infty}^{\infty} f(x)e^{-jxy}\,dx$$

$$= \int_{-\infty}^{\infty} \phi(y)e^{-jyt_0}F(y)\,dy$$

$$= \int_{-\infty}^{\infty} F(\omega)e^{-j\omega t_0}\phi(\omega)\,d\omega$$

Hence, by the equivalence property, we conclude that

$$\mathcal{F}[f(t - t_0)] = F(\omega)e^{-j\omega t_0}$$

PROBLEM 6-7   Rework Problem 6-5 by using the time-shifting property (6.11).

Solution: From the Fourier transform of a δ-function (6.9), $\mathcal{F}[\delta(t)] = 1$. Thus, by (6.11), we have

$$\mathcal{F}[\delta(t - t_0)] = 1 \cdot e^{-j\omega t_0} = e^{-j\omega t_0}$$

PROBLEM 6-8   Find the Fourier transform of $\delta(t)$ by means of the symmetry property (6.15).

Solution: From the Fourier transform of a constant (6.8), we have

$$\mathcal{F}[1] = 2\pi\delta(\omega)$$

Thus, by the symmetry property (6.15),

$$F(t) \leftrightarrow 2\pi f(-\omega)$$

we have

$$\mathcal{F}[2\pi\delta(t)] = 2\pi \cdot 1$$

Hence,

$$\mathcal{F}[\delta(t)] = 1$$

PROBLEM 6-9   Find the Fourier transform of $\sin \omega_0 t$.

Solution: Using the identity

$$\sin \omega_0 t = \frac{1}{2j}(e^{j\omega_0 t} - e^{-j\omega_0 t}),$$

we obtain from linearity (6.10) and the Fourier transform of a complex exponential function (6.18)

$$\mathcal{F}[\sin \omega_0 t] = \mathcal{F}\left[\frac{1}{2j}(e^{j\omega_0 t} - e^{-j\omega_0 t})\right]$$

$$= \frac{1}{2j}[2\pi\delta(\omega - \omega_0) - 2\pi\delta(\omega + \omega_0)]$$

$$= -j\pi\delta(\omega - \omega_0) + j\pi\delta(\omega + \omega_0)$$

PROBLEM 6-10   Find the Fourier transform of $\delta'(t)$.

Solution: From (6.9),

$$\mathcal{F}[\delta(t)] = 1$$

and by the differentiation theorem (6.16), we have

$$\mathcal{F}[\delta'(t)] = j\omega$$

PROBLEM 6-11   Verify the differentiation theorem (6.17); that is,

$$\mathcal{F}[(-jt)f(t)] = F'(\omega) = \frac{dF(\omega)}{d\omega}$$

Solution: By definition (6.7), we have

$$\int_{-\infty}^{\infty} \mathcal{F}[(-jt)f(t)]\phi(\omega)\,d\omega = \int_{-\infty}^{\infty}(-j\omega)f(\omega)\Phi(\omega)\,d\omega$$

$$= -\int_{-\infty}^{\infty} f(\omega)j\omega\Phi(\omega)\,d\omega$$

$$= -\int_{-\infty}^{\infty} f(\omega)\mathcal{F}[\phi'(t)]\,d\omega$$

since $\mathcal{F}[\phi'(t)] = j\omega\Phi(\omega)$ by the differentiation theorem of ordinary Fourier transforms (5.21).

Now, again by definition (6.7), we have

$$\int_{-\infty}^{\infty} f(\omega)\mathscr{F}[\phi'(t)]\,d\omega = \int_{-\infty}^{\infty} F(\omega)\phi'(\omega)\,d\omega$$

$$= -\int_{-\infty}^{\infty} F'(\omega)\phi(\omega)\,d\omega$$

by the definition of generalized derivative (6.5). Thus, we obtain

$$\int_{-\infty}^{\infty} \mathscr{F}[(-jt)f(t)]\phi(\omega)\,d\omega = -\int_{-\infty}^{\infty} f(\omega)\mathscr{F}[\phi'(t)]\,d\omega$$

$$= \int_{-\infty}^{\infty} F'(\omega)\phi(\omega)\,d\omega$$

Hence, by the equivalence property, we conclude that

$$\mathscr{F}[(-jt)f(t)] = F'(\omega)$$

PROBLEM 6-12   Find the Fourier transform of t.

Solution: Function t is a function of slow growth. From the result of Problem 6-11, $\mathscr{F}[(-jt)f(t)] = F'(\omega)$, we have

$$\mathscr{F}[tf(t)] = \frac{1}{-j}F'(\omega) = jF'(\omega)$$

Now, letting $f(t) = 1$ and using the Fourier transform of a constant

$$\mathscr{F}[t] = j2\pi\delta'(\omega)$$

PROBLEM 6-13   Prove that the Fourier transform of the unit step function given by (6.22)—that is, $\mathscr{F}[u(t)] = 1/j\omega$—is incorrect.

Solution: We note that $1/(j\omega) = -j/\omega$ is a purely imaginary function of ω. Then, according to the property of the odd component (5.18), if the Fourier transform of a real function $f(t)$ is purely imaginary, then $f(t)$ is an odd function of t. But $u(t)$ is not an odd function of t, and therefore $1/(j\omega)$ cannot be its Fourier transform.

PROBLEM 6-14   Prove that

$$\mathscr{F}^{-1}\left[\frac{1}{j\omega}\right] = \frac{1}{2}\operatorname{sgn}t$$

where

$$\operatorname{sgn}t = \begin{cases} 1 & \text{for} \quad t > 0 \\ -1 & \text{for} \quad t < 0 \end{cases}$$

Proof: Let $f(t) = \operatorname{sgn}t$ and $\mathscr{F}[\operatorname{sgn}t] = F(\omega)$. Since $\operatorname{sgn}t$ is an odd function of t (Figure 6-6a), according to (5.15) and (5.18), $F(\omega)$ will be purely imaginary and consequently an odd function of ω.

Now, from the result of Problem 4-8,

$$f'(t) = 2\delta(t)$$

Then, from (6.16) and (6.9),

$$\mathscr{F}[f'(t)] = j\omega F(\omega) = \mathscr{F}[2\delta(t)] = 2$$

Therefore, by (6.24), we have

$$F(\omega) = \frac{2}{j\omega} + k\delta(\omega)$$

where k is an arbitrary constant. Since $F(\omega)$ must be purely imaginary and odd, $k = 0$. Hence,

$$F(\omega) = \mathscr{F}[\operatorname{sgn}t] = \frac{2}{j\omega}$$

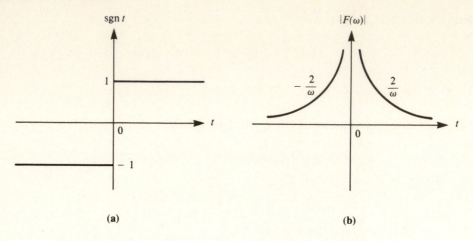

Figure 6-6   (a) The signum function and (b) its magnitude spectrum.

From this, we conclude that

$$\mathscr{F}^{-1}\left[\frac{1}{j\omega}\right] = \frac{1}{2}\,\mathrm{sgn}\,t$$

Alternate Proof: From (6.20), we have

$$\mathscr{F}[u(t)] = \pi\delta(\omega) + \frac{1}{j\omega}$$

Now we note that $u(t)$ may be expressed as (see Figure 6-7)

$$u(t) = f_e(t) + f_o(t)$$

where $f_e(t)$ and $f_o(t)$ are the even and odd components of $u(t)$, respectively. From the result of Problem 2-1, we have

$$f_e(t) = \frac{1}{2}\,[u(t) + u(-t)] = \frac{1}{2}$$

$$f_o(t) = \frac{1}{2}\,[u(t) - u(-t)] = \frac{1}{2}\,\mathrm{sgn}\,t = \begin{cases} \dfrac{1}{2}, & t > 0 \\[2mm] -\dfrac{1}{2}, & t > 0 \end{cases}$$

Hence, from (5.17) and (5.18), we conclude that

$$\mathscr{F}\left[\frac{1}{2}\right] = \pi\delta(\omega)$$

$$\mathscr{F}\left[\frac{1}{2}\,\mathrm{sgn}\,t\right] = \frac{1}{j\omega}$$

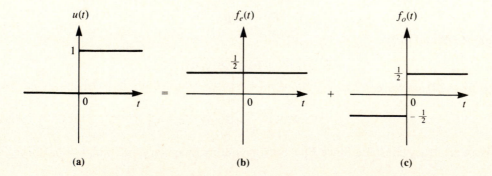

Figure 6-7   (a) Unit step function and its (b) even and (c) odd components.

Therefore,

$$\mathscr{F}^{-1}\left[\frac{1}{j\omega}\right] = \frac{1}{2}\,\text{sgn}\,t$$

The Fourier Transform of a Periodic Function

PROBLEM 6-15   Prove that a sequence of equidistant pulses

$$F(\omega) = \sum_{n=-\infty}^{\infty} A_n \delta(\omega - n\omega_0), \qquad \omega_0 = 2\pi/T$$

is the Fourier transform of a periodic function $f(t)$ with period T.

Proof:

$$f(t) = \mathscr{F}^{-1}[F(\omega)] = \mathscr{F}^{-1}\left[\sum_{n=-\infty}^{\infty} A_n \delta(\omega - n\omega_0)\right]$$

$$= \sum_{n=-\infty}^{\infty} A_n \mathscr{F}^{-1}[\delta(\omega - n\omega_0)]$$

From (6.18), we have

$$\mathscr{F}^{-1}[\delta(\omega - n\omega_0)] = \frac{1}{2\pi}\,e^{jn\omega_0 t}$$

Hence,

$$f(t) = \sum_{n=-\infty}^{\infty} \frac{A_n}{2\pi}\,e^{jn\omega_0 t}$$

Since $e^{jn\omega_0(t + 2\pi/\omega_0)} = e^{jn\omega_0 t}$,

$$f\left[t + \left(\frac{2\pi}{\omega_0}\right)\right] = f(t + T) = f(t)$$

that is, $f(t)$ is a periodic function with period $T = 2\pi/\omega_0$.

PROBLEM 6-16   Show that the complex coefficients c_n of the Fourier series expansion of a periodic function $f(t)$ with period T equal the values of the Fourier transform $F_0(\omega)$ of the function $f_0(t)$ at $\omega = n\omega_0 = n2\pi/T$ multiplied by $1/T$, where $f_0(t)$ is defined by

$$f_0(t) = \begin{cases} f(t), & |t| < \frac{1}{2}T \\[2mm] 0, & |t| > \frac{1}{2}T \end{cases}$$

Solution: The periodic function $f(t)$ with period T can be written as

$$f(t) = \sum_{n=-\infty}^{\infty} c_n e^{jn\omega_0 t}, \qquad \omega_0 = \frac{2\pi}{T}$$

where

$$c_n = \frac{1}{T}\int_{-T/2}^{T/2} f(t)e^{-jn\omega_0 t}\,dt$$

Now,

$$F_0(\omega) = \mathscr{F}[f_0(t)] = \int_{-\infty}^{\infty} f_0(t)e^{-j\omega t}\,dt$$

$$= \int_{-T/2}^{T/2} f(t)e^{-j\omega t}\,dt$$

Since

$$F_0(n\omega_0) = \int_{-T/2}^{T/2} f(t)e^{-jn\omega_0 t}\,dt$$

by comparing to the expression for c_n, we conclude that

$$c_n = \frac{1}{T} F_0(n\omega_0)$$

PROBLEM 6-17   Using the result of Problem 6-16, find the complex Fourier series coefficients of a train of rectangular pulses of width d with period T as shown in Figure 6-8a.

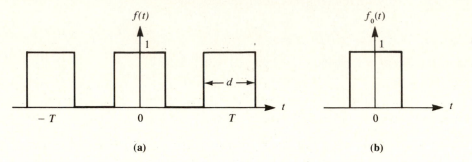

Figure 6-8   (a) A train of rectangular pulses; (b) a single rectangular pulse.

Solution: Let

$$f(t) = \sum_{n=-\infty}^{\infty} c_n e^{jn\omega_0 t}, \qquad \omega_0 = \frac{2\pi}{T}$$

Then from Figure 6-8b,

$$f_0(t) = p_d(t) = \begin{cases} 1, & |t| < d/2 \\ 0, & |t| > d/2 \end{cases}$$

Hence, from the result of Problem 5-2,

$$F_0(\omega) = \mathscr{F}[f_0(t)] = \mathscr{F}[p_d(f)] = \frac{2}{\omega}\sin\left(\frac{\omega d}{2}\right) = d\,\frac{\sin(\omega d/2)}{(\omega d/2)}$$

Therefore, from the result of Problem 6-16, the Fourier series coefficients c_n of $f(t)$ are given by

$$c_n = \frac{1}{T} F_0(n\omega_0) = \frac{d}{T}\,\frac{\sin(n\omega_0 d/2)}{(n\omega_0 d/2)}$$

which is exactly the same as (3.18) (see Example 3-4) except for the factor A of the pulse height.

Integral Representation of $\delta(t)$

PROBLEM 6-18   Using the identity

$$\delta(t) = \frac{1}{2\pi} \int_{-\infty}^{\infty} e^{j\omega t}\, d\omega$$

(6.27) and the relation $\int_{-\infty}^{\infty} \delta(t-t_0)\phi(t)\,dt = \phi(t_0)$ (4.3), prove the inversion formula of the Fourier transform

$$f(t) = \mathscr{F}^{-1}[F(\omega)] = \frac{1}{2\pi} \int_{-\infty}^{\infty} F(\omega)e^{j\omega t}\, d\omega \qquad\qquad \textbf{(a)}$$

where

$$F(\omega) = \int_{-\infty}^{\infty} f(t)e^{-j\omega t}\, dt \qquad\qquad \textbf{(b)}$$

Proof: Substituting eq. (b) into the right-hand side of eq. (a):

$$\frac{1}{2\pi} \int_{-\infty}^{\infty} F(\omega)e^{j\omega t}\, d\omega = \frac{1}{2\pi} \int_{-\infty}^{\infty} \left[\int_{-\infty}^{\infty} f(y)e^{-j\omega y}\, dy\right] e^{j\omega t}\, d\omega$$

Here, in order to avoid confusion, a different dummy variable y is used. Interchanging the order of integration and using (6.27), we obtain

$$\frac{1}{2\pi} \int_{-\infty}^{\infty} F(\omega) e^{j\omega t}\, d\omega = \int_{-\infty}^{\infty} f(y) \left[\frac{1}{2\pi} \int_{-\infty}^{\infty} e^{j\omega(t-y)}\, d\omega\right] dy$$

$$= \int_{-\infty}^{\infty} f(y)\delta(t-y)\, dy = f(t)$$

The last integral was obtained with the use of (4.3).

PROBLEM 6-19 If $f(t)$ is real, then show that

Fourier's
integral
theorem

$$f(t) = \frac{1}{\pi} \int_{0}^{\infty} \int_{-\infty}^{\infty} f(x)\cos \omega(t-x)\, dx\, d\omega$$

which is known as **Fourier's integral theorem**.

Solution: If we substitute the Fourier transform (5.1) into the inverse Fourier transform (5.2), we obtain

$$f(t) = \frac{1}{2\pi} \int_{-\infty}^{\infty} \left[\int_{-\infty}^{\infty} f(x) e^{-j\omega x}\, dx\right] e^{j\omega t}\, d\omega$$

$$= \frac{1}{2\pi} \int_{-\infty}^{\infty} \int_{-\infty}^{\infty} f(x) e^{j\omega(t-x)}\, dx\, d\omega \qquad \textbf{(a)}$$

If $f(t)$ is real, we can equate real parts in eq. (a), which becomes

$$f(t) = \frac{1}{2\pi} \int_{-\infty}^{\infty} \int_{-\infty}^{\infty} f(x)\cos \omega(t-x)\, dx\, d\omega$$

Since $\cos \omega(t-x)$ is even with respect to ω, from (2.4),

$$f(t) = \frac{1}{\pi} \int_{0}^{\infty} \int_{-\infty}^{\infty} f(x)\cos \omega(t-x)\, dx\, d\omega$$

Supplementary Exercises

PROBLEM 6-20 Evaluate the Fourier transforms of the following functions:

(a) $1 - 3\delta(t) + 2\delta'(t-2)$, **(b)** $\sin^3 t$, **(c)** $u(t-1)$

Answer:
(a) $2\pi\delta(\omega) - 3 + 2j\omega\, e^{-j2\omega}$,
(b) $j(\pi/4)[\delta(\omega-3) - 3\delta(\omega-1) + 3\delta(\omega+1) - \delta(\omega+3)]$,
(c) $\pi\delta(\omega) + e^{-j\omega}/j\omega$

PROBLEM 6-21 Show that the unit step function $u(t)$ can be expressed as

$$u(t) = \frac{1}{2} + \frac{1}{\pi} \int_{0}^{\infty} \frac{\sin \omega t}{\omega}\, d\omega$$

[*Hint:* Use (6.20) and (5.19).]

PROBLEM 6-22 Prove that

(a) $\mathscr{F}[\cos \omega_0 t u(t)] = \frac{\pi}{2} [\delta(\omega-\omega_0) + \delta(\omega+\omega_0)] + j\frac{\omega}{\omega_0^2 - \omega^2}$

(b) $\mathscr{F}[\sin \omega_0 t u(t)] = \frac{\omega_0}{\omega_0^2 - \omega^2} - j\frac{\pi}{2} [\delta(\omega-\omega_0) - \delta(\omega+\omega_0)]$

[*Hint:* Use (6.12) and $\cos \omega_0 t u(t) = \frac{1}{2}\{e^{j\omega_0 t} u(t) + e^{-j\omega_0 t} u(t)\}$.]

PROBLEM 6-23 Find the Fourier transform of a finite unit impulse train $f(t) = \sum_{n=0}^{k-1} \delta(t-nT)$

Answer: $e^{-j(k-1)\omega T/2} \dfrac{\sin(k\omega T/2)}{\sin(\omega T/2)}$

PROBLEM 6-24 If $f(t) = e^{-\alpha t}u(t)$, show that $\mathcal{F}[f'(t)] = j\omega\mathcal{F}[f(t)]$. [*Hint:* $f'(t) = \delta(t) - \alpha e^{-\alpha t}u(t)$.]

PROBLEM 6-25 Prove that $\mathcal{F}[1/t] = -\pi j \operatorname{sgn} \omega = \pi j - 2\pi j u(\omega)$. [*Hint:* Apply symmetry property (6.15) to the result of Problem 6-14.]

PROBLEM 6-26 Show that $\mathcal{F}[tu(t)] = j\pi\delta'(\omega) - 1/\omega^2$. [*Hint:* Use (6.17) and (6.20).]

PROBLEM 6-27 Show that $\mathcal{F}[|t|] = -2/\omega^2$. [*Hint:* Use $|t| = 2tu(t) - t$, and the results of Problems 6-12 and 6-26.]

PROBLEM 6-28 Find the particular solution to $x''(t) + 3x'(t) + 2x(t) = 3\delta(t)$ by using the Fourier transform. [*Hint:* Take the Fourier transform of both sides of the equation. Find $X(\omega) = \mathcal{F}[x(t)]$ and take the inverse Fourier transform.]

Answer: $3(e^{-t} - e^{-2t})u(t)$

PROBLEM 6-29 Find the Fourier transform of the shifted unit step function $u(t - t_0)$.

Answer: $\pi\delta(\omega) + \dfrac{e^{-j\omega t_0}}{j\omega}$

PROBLEM 6-30 Using (6.7), show that $\mathcal{F}[e^{j\omega_0 t}] = 2\pi\delta(\omega - \omega_0)$.

PROBLEM 6-31 Find the Fourier integral representation of the function

$$f(t) = \begin{cases} 1 & \text{for} \quad |t| < 1 \\ 0 & \text{for} \quad |t| > 1 \end{cases}$$

Answer: $f(t) = \dfrac{2}{\pi} \displaystyle\int_0^\infty \dfrac{\cos t\omega \sin \omega}{\omega}\, d\omega$

PROBLEM 6-32 Use the result of Problem 6-31 to derive

$$\int_0^\infty \frac{\sin \omega}{\omega}\, d\omega = \frac{\pi}{2}$$

[*Hint:* Set $t = 0$ in the result of Problem 6-31.]

7 CONVOLUTION AND CORRELATION

THIS CHAPTER IS ABOUT

☑ **Convolution**
☑ **Convolution Theorems**
☑ **Parseval's Theorem**
☑ **Correlation Functions**
☑ **The Wiener–Khintchine Theorem**

7-1. Convolution

Let $f_1(t)$ and $f_2(t)$ be two given functions. The **convolution** of $f_1(t)$ and $f_2(t)$ is defined to be the function

Convolution
$$f(t) = \int_{-\infty}^{\infty} f_1(x) f_2(t-x)\, dx \qquad (7.1)$$

which is often expressed symbolically as

$$f(t) = f_1(t) * f_2(t) \qquad (7.2)$$

An important special case is that in which

$$f_1(t) = 0 \quad \text{for} \quad t < 0 \quad \text{and} \quad f_2(t) = 0 \quad \text{for} \quad t < 0$$

In this case, the convolution (7.1) becomes

$$f(t) = f_1(t) * f_2(t) = \int_0^t f_1(x) f_2(t-x)\, dx \qquad (7.3)$$

A. Properties of convolution

Commutative	$f_1(t) * f_2(t) = f_2(t) * f_1(t)$	(7.4)
Associative	$[f_1(t) * f_2(t)] * f_3(t) = f_1(t) * [f_2(t) * f_3(t)]$	(7.5)
Distributive	$f_1(t) * [f_2(t) + f_3(t)] = f_1(t) * f_2(t) + f_1(t) * f_3(t)$	(7.6)

EXAMPLE 7-1: Prove the commutative law (7.4).

Proof: By the definition of convolution (7.1),

$$f_1(t) * f_2(t) = \int_{-\infty}^{\infty} f_1(x) f_2(t-x)\, dx$$

By changing the variable to $t - x = y$, we obtain the commutative law (7.4):

$$f_1(t) * f_2(t) = \int_{-\infty}^{\infty} f_1(t-y) f_2(y)\, dy$$

$$= \int_{-\infty}^{\infty} f_2(y) f_1(t-y)\, dy$$

$$= f_2(t) * f_1(t)$$

EXAMPLE 7-2: Prove the associative law (7.5).

Proof: If we let $f_1(t)*f_2(t) = g(t)$ and $f_2(t)*f_3(t) = h(t)$, then (7.5) can be expressed as

$$g(t)*f_3(t) = f_1(t)*h(t)$$

Since

$$g(t) = \int_{-\infty}^{\infty} f_1(y)f_2(t-y)\,dy$$

we have

$$g(t)*f_3(t) = \int_{-\infty}^{\infty} g(x)f_3(t-x)\,dx$$

$$= \int_{-\infty}^{\infty}\left[\int_{-\infty}^{\infty} f_1(y)f_2(x-y)\,dy\right]f_3(t-x)\,dx$$

Substituting $z = x - y$ and interchanging the order of integration, we have

$$g(t)*f_3(t) = \int_{-\infty}^{\infty} f_1(y)\left[\int_{-\infty}^{\infty} f_2(z)f_3(t-y-z)\,dz\right]dy \tag{7.7}$$

Now, since

$$h(t) = \int_{-\infty}^{\infty} f_2(z)f_3(t-z)\,dz$$

we have

$$h(t-y) = \int_{-\infty}^{\infty} f_2(z)f_3(t-y-z)\,dz$$

Therefore, we identify the integral inside the bracket on the right-hand side of (7.7) as $h(t-y)$. Hence,

$$g(t)*f_3(t) = \int_{-\infty}^{\infty} f_1(y)h(t-y)\,dy = f_1(t)*h(t)$$

that is, the associative law (7.5):

$$[f_1(t)*f_2(t)]*f_3(t) = f_1(t)*[f_2(t)*f_3(t)]$$

B. Convolution with δ-functions

Convolution with δ-functions

$$\phi(t)*\delta(t) = \phi(t) \tag{7.8}$$
$$\phi(t)*\delta(t-T) = \phi(t-T) \tag{7.9}$$

where $\phi(t)$ is a testing function.

EXAMPLE 7-3: Verify (7.8) and (7.9).

Solution: According to the definition of δ-function (4.1),

$$\phi(t)*\delta(t) = \delta(t)*\phi(t) = \int_{-\infty}^{\infty} \delta(x)\phi(t-x)\,dx = \phi(t)$$

And, according to the property (4.3) of δ-functions,

$$\phi(t)*\delta(t-T) = \delta(t-T)*\phi(t) = \int_{-\infty}^{\infty} \delta(x-T)\phi(t-x)\,dx = \phi(t-T)$$

EXAMPLE 7-4: Let $f(t)$ be a periodic function with period T. If a function $f_I(t)$ is defined as

$$f_I(t) = \begin{cases} f(t), & 0 < t < T \\ 0, & \text{otherwise} \end{cases}$$

show that $f(t)$ can be expressed as

$$f(t) = \sum_{n=-\infty}^{\infty} f_I(t-nT) = f_I(t)*\delta_T(t) \tag{7.10}$$

where

$$\delta_T(t) = \sum_{n=-\infty}^{\infty} \delta(t - nT)$$

Solution: From Figure 7-1, it is seen that

$$f(t) = \sum_{n=-\infty}^{\infty} f_I(t - nT)$$

Now, by convolution with a δ-function (7.9),

$$f_I(t) * \delta_T(t) = f_I(t) * \sum_{n=-\infty}^{\infty} \delta(t - nT)$$

$$= \sum_{n=-\infty}^{\infty} f_I(t) * \delta(t - nT)$$

$$= \sum_{n=-\infty}^{\infty} f_I(t - nT)$$

Hence, we conclude that

$$f(t) = f_I(t) * \delta_T(t)$$

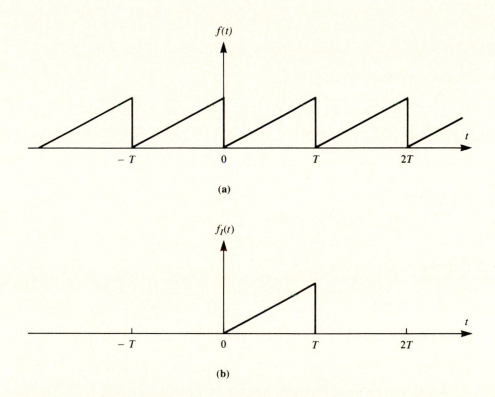

Figure 7-1   (a) A periodic function and (b) the function defined over the first period.

7-2. Convolution Theorems

A. The time convolution theorem

The **time convolution theorem** states that if $\mathscr{F}[f_1(t)] = F_1(\omega)$ and $\mathscr{F}[f_2(t)] = F_2(\omega)$, then

Time convolution theorem $\qquad \mathscr{F}[f_1(t) * f_2(t)] = F_1(\omega)F_2(\omega)$ **(7.11)**

or

$$\mathscr{F}^{-1}[F_1(\omega)F_2(\omega)] = f_1(t) * f_2(t) \qquad\qquad\qquad \textbf{(7.12)}$$

EXAMPLE 7-5: Prove the time convolution theorem (7.11).

Proof: The Fourier transform of $f_1(t) * f_2(t)$ is

$$\mathscr{F}[f_1(t) * f_2(t)] = \int_{-\infty}^{\infty} \left[\int_{-\infty}^{\infty} f_1(x) f_2(t - x) \, dx \right] e^{-j\omega t} \, dt$$

Changing the order of integration,

$$\mathscr{F}[f_1(t) * f_2(t)] = \int_{-\infty}^{\infty} f_1(x) \left[\int_{-\infty}^{\infty} f_2(t - x) e^{-j\omega t} \, dt \right] dx \qquad (7.13)$$

From the time-shifting property of the Fourier transform (5.6), we have

$$\int_{-\infty}^{\infty} f_2(t - x) e^{-j\omega t} \, dt = F_2(\omega) e^{-j\omega x}$$

Substituting this into (7.13), we obtain

$$\mathscr{F}[f_1(t) * f_2(t)] = \int_{-\infty}^{\infty} f_1(x) F_2(\omega) e^{-j\omega x} \, dx$$

$$= \left[\int_{-\infty}^{\infty} f_1(x) e^{-j\omega x} \, dx \right] F_2(\omega) = \left[\int_{-\infty}^{\infty} f_1(t) e^{-j\omega t} \, dt \right] F_2(\omega)$$

$$= F_1(\omega) F_2(\omega)$$

EXAMPLE 7-6: Use time convolution theorem to find $f(t) = \mathscr{F}^{-1}[1/(1 + j\omega)^2]$.

Solution: The Fourier transform of $f(t)$ is

$$F(\omega) = \mathscr{F}[f(t)] = \frac{1}{(1 + j\omega)^2} = \frac{1}{(1 + j\omega)} \times \frac{1}{(1 + j\omega)}$$

From Example 5-1, we recall that

$$\mathscr{F}^{-1}\left[\frac{1}{1 + j\omega} \right] = e^{-t} u(t)$$

Hence, from (7.12),

$$f(t) = \int_{-\infty}^{\infty} e^{-x} u(x) e^{-(t - x)} u(t - x) \, dx$$

In the above integral, the integrand includes the factor $u(x)u(t - x)$. Since $u(x) = 0$ for $x < 0$, and $u(t - x) = 0$ for $x > t$, we have

$$u(x)u(t - x) = \begin{cases} 0 & \text{for } 0 > x \text{ and } x > t \\ 1 & \text{for } 0 < x < t \end{cases}$$

Hence,

$$f(t) = \int_0^t e^{-x} e^{-(t - x)} \, dx = e^{-t} \int_0^t dx = t e^{-t} u(t)$$

B. The frequency convolution theorem

The **frequency convolution theorem** states that if $\mathscr{F}^{-1}[F_1(\omega)] = f_1(t)$ and $\mathscr{F}^{-1}[F_2(\omega)] = f_2(t)$, then

Frequency convolution theorem $\qquad \mathscr{F}^{-1}[F_1(\omega) * F_2(\omega)] = 2\pi f_1(t) f_2(t) \qquad (7.14)$

or

$$\mathscr{F}[f_1(t) f_2(t)] = \frac{1}{2\pi} F_1(\omega) * F_2(\omega) = \frac{1}{2\pi} \int_{-\infty}^{\infty} F_1(y) F_2(\omega - y) \, dy \qquad (7.15)$$

EXAMPLE 7-7: Prove the frequency convolution theorem (7.14).

Proof: From the definition of the inverse Fourier transform (5.2),

$$\mathscr{F}^{-1}[F_1(\omega) * F_2(\omega)] = \mathscr{F}^{-1}\left[\int_{-\infty}^{\infty} F_1(y)F_2(\omega - y)\,dy\right]$$

$$= \frac{1}{2\pi}\int_{-\infty}^{\infty}\left[\int_{-\infty}^{\infty} F_1(y)F_2(\omega - y)\,dy\right]e^{j\omega t}\,d\omega$$

Substituting $\omega - y = x$ and interchanging the order of integration, we get

$$\mathscr{F}^{-1}[F_1(\omega) * F_2(\omega)] = \frac{1}{2\pi}\int_{-\infty}^{\infty} F_1(y)\left[\int_{-\infty}^{\infty} F_2(x)e^{j(x+y)t}\,dx\right]dy$$

$$= \frac{1}{2\pi}\int_{-\infty}^{\infty} F_1(y)e^{jyt}\left[\int_{-\infty}^{\infty} F_2(x)e^{jxt}\,dx\right]dy$$

$$= 2\pi\left[\frac{1}{2\pi}\int_{-\infty}^{\infty} F_1(\omega)e^{j\omega t}\,d\omega\right]\left[\frac{1}{2\pi}\int_{-\infty}^{\infty} F_2(\omega)e^{j\omega t}\,d\omega\right]$$

$$= 2\pi[f_1(t)f_2(t)]$$

by changing the dummy variables of integration.

EXAMPLE 7-8: If $\mathscr{F}[f_1(t)] = F_1(\omega)$ and $\mathscr{F}[f_2(t)] = F_2(\omega)$, show that

$$\int_{-\infty}^{\infty} [f_1(t)f_2(t)]\,dt = \frac{1}{2\pi}\int_{-\infty}^{\infty} F_1(\omega)F_2(-\omega)\,d\omega \qquad (7.16)$$

Solution: From (7.15), we have

$$\mathscr{F}[f_1(t)f_2(t)] = \frac{1}{2\pi}\int_{-\infty}^{\infty} F_1(y)F_2(\omega - y)\,dy$$

that is,

$$\int_{-\infty}^{\infty} [f_1(t)f_2(t)]e^{-j\omega t}\,dt = \frac{1}{2\pi}\int_{-\infty}^{\infty} F_1(y)F_2(\omega - y)\,dy$$

Now, letting $\omega = 0$, we obtain

$$\int_{-\infty}^{\infty} [f_1(t)f_2(t)]\,dt = \frac{1}{2\pi}\int_{-\infty}^{\infty} F_1(y)F_2(-y)\,dy$$

Then, by changing the dummy variable of integration, we get (7.16).

$$\int_{-\infty}^{\infty} [f_1(t)f_2(t)]\,dt = \frac{1}{2\pi}\int_{-\infty}^{\infty} F_1(\omega)F_2(-\omega)\,d\omega$$

EXAMPLE 7-9: If the functions $f_1(t)$ and $f_2(t)$ are real and if $\mathscr{F}[f_1(t)] = F_1(\omega)$ and $\mathscr{F}[f_2(t)] = F_2(\omega)$, show that

$$\int_{-\infty}^{\infty} f_1(t)f_2(t)\,dt = \frac{1}{2\pi}\int_{-\infty}^{\infty} F_1(\omega)F_2^*(\omega)\,d\omega \qquad (7.17)$$

where $F_2^*(\omega)$ denotes the complex conjugate of $F_2(\omega)$.

Solution: If $f(t)$ is real, then property (5.16) applies:

$$F(-\omega) = F^*(\omega)$$

Consequently, from (7.16), we have

$$\int_{-\infty}^{\infty} f_1(t)f_2(t)\,dt = \frac{1}{2\pi}\int_{-\infty}^{\infty} F_1(\omega)F_2(-\omega)\,d\omega$$

$$= \frac{1}{2\pi}\int_{-\infty}^{\infty} F_1(\omega)F_2^*(\omega)\,d\omega$$

7-3. Parseval's Theorem

A. Statement of Parseval's theorem

Parseval's theorem states that if $\mathscr{F}[f(t)] = F(\omega)$, then

Parseval's theorem
$$\int_{-\infty}^{\infty} |f(t)|^2 \, dt = \frac{1}{2\pi} \int_{-\infty}^{\infty} |F(\omega)|^2 \, d\omega \qquad (7.18)$$

B. Energy content of a function

In Section 3-4 we saw that for a periodic function, the power in a signal can be associated with the power contained in each discrete frequency component. The same concept can be extended to nonperiodic functions. A useful concept for a nonperiodic function is its **energy content** E, defined by

Energy content of f(t)
$$E = \int_{-\infty}^{\infty} |f(t)|^2 \, dt \qquad (7.19)$$

Indeed, if we assume $f(t)$ to be the voltage of a source across a 1-Ω resistance, then the quantity $\int_{-\infty}^{\infty} |f(t)|^2 \, dt$ equals the total energy delivered by the source.

Now from Parseval's theorem (7.18),

$$E = \int_{-\infty}^{\infty} |f(t)|^2 \, dt = \frac{1}{2\pi} \int_{-\infty}^{\infty} |F(\omega)|^2 \, d\omega = \int_{-\infty}^{\infty} |F(2\pi v)|^2 \, dv \qquad (7.20)$$

where $\omega = 2\pi v$ and v is expressed in Hertz. Equation (7.20) states that the energy content of $f(t)$ is given by $1/(2\pi)$ times the area under the $|F(\omega)|^2$ curve. For this reason, the quantity $|F(\omega)|^2$ is called the **energy spectrum** or **energy spectral density** of $f(t)$.

EXAMPLE 7-10: Prove Parseval's theorem (7.18).

Proof: If $\mathscr{F}[f(t)] = F(\omega)$, then

$$\mathscr{F}[f^*(t)] = \int_{-\infty}^{\infty} f^*(t) e^{-j\omega t} \, dt = \int_{-\infty}^{\infty} [f(t) e^{j\omega t}]^* \, dt$$

$$= \left[\int_{-\infty}^{\infty} f(t) e^{-j(-\omega)t} \, dt \right]^*$$

$$= F^*(-\omega) \qquad (7.21)$$

Hence, if we let $f_1(t) = f(t)$ and $f_2(t) = f^*(t)$ in (7.16), then

$$\int_{-\infty}^{\infty} f(t) f^*(t) \, dt = \frac{1}{2\pi} \int_{-\infty}^{\infty} F(\omega) F^*[-(-\omega)] \, d\omega$$

$$= \frac{1}{2\pi} \int_{-\infty}^{\infty} F(\omega) F^*(\omega) \, d\omega \qquad (7.22)$$

Since $f(t) f^*(t) = |f(t)|^2$ and $F(\omega) F^*(\omega) = |F(\omega)|^2$, we obtain Parseval's theorem (7.18):

$$\int_{-\infty}^{\infty} |f(t)|^2 \, dt = \frac{1}{2\pi} \int_{-\infty}^{\infty} |F(\omega)|^2 \, d\omega$$

If $f(t)$ is real, then (7.18) can be obtained simply from (7.17).

C. Evaluation of integrals

Some integrals can be evaluated by using Parseval's theorem (7.18) and Fourier transform pairs. This is illustrated in Example 7-11.

EXAMPLE 7-11: Evaluate

$$\int_{-\infty}^{\infty} \frac{dx}{a^2 + x^2} \quad \text{and} \quad \int_{-\infty}^{\infty} \frac{dx}{1 + x^2}$$

Solution: Let

$$f(t) = e^{-at} u(t)$$

Then, finding the Fourier transform from the result of Example 5-1, we have

$$F(\omega) = \mathscr{F}[f(t)] = \frac{1}{a + j\omega}$$

$$|F(\omega)|^2 = \frac{1}{a^2 + \omega^2} \tag{7.23}$$

Now, according to Parseval's theorem (7.18),

$$\int_{-\infty}^{\infty} f^2(t)\,dt = \frac{1}{2\pi} \int_{-\infty}^{\infty} |F(\omega)|^2\,d\omega$$

then multiplying both sides by 2π,

$$\int_{-\infty}^{\infty} |F(\omega)|^2\,d\omega = 2\pi \int_{0}^{\infty} e^{-2at}\,dt$$

Hence, from (7.23),

$$\int_{-\infty}^{\infty} \frac{d\omega}{a^2 + \omega^2} = 2\pi \int_{-\infty}^{\infty} f^2(t)\,dt = 2\pi \int_{0}^{\infty} e^{-2at}\,dt$$

$$= 2\pi \frac{e^{-2at}}{-2a}\bigg|_{0}^{\infty} = \frac{\pi}{a}$$

Thus,

$$\int_{-\infty}^{\infty} \frac{dx}{a^2 + x^2} = \int_{-\infty}^{\infty} \frac{d\omega}{a^2 + \omega^2} = \frac{\pi}{a} \tag{7.24}$$

Setting $a = 1$,

$$\int_{-\infty}^{\infty} \frac{dx}{1 + x^2} = \pi \tag{7.25}$$

7-4. Correlation Functions

A. Definitions of correlation functions

The function defined by

Cross-correlation function $\qquad R_{12}(\tau) = \int_{-\infty}^{\infty} f_1(t)f_2(t - \tau)\,dt \tag{7.26}$

is known as the **cross-correlation function** between functions $f_1(t)$ and $f_2(t)$. In a similar way, we can define

Cross-correlation function $\qquad R_{21}(\tau) = \int_{-\infty}^{\infty} f_2(t)f_1(t - \tau)\,dt \tag{7.27}$

The cross-correlation function $R_{12}(\tau)$ or $R_{21}(\tau)$ provides a measure of the similarity or interdependence between functions $f_1(t)$ and $f_2(t)$ as a function of the parameter τ (the shift of one function with respect to the other). If the cross-correlation function is identically zero for all τ, then the two functions are said to be *uncorrelated*.

If $f_1(t)$ and $f_2(t)$ are identical, then the correlation function

Autocorrelation function $\qquad R_{11}(\tau) = \int_{-\infty}^{\infty} f_1(t)f_1(t - \tau)\,dt \tag{7.28}$

is termed the **autocorrelation function** of $f_1(t)$.

Sometimes, a normalized quantity $\gamma(\tau)$ defined by

$$\gamma(\tau) = \frac{\displaystyle\int_{-\infty}^{\infty} f_1(t)f_1(t - \tau)\,dt}{\displaystyle\int_{-\infty}^{\infty} [f_1(t)]^2\,dt} \tag{7.29}$$

is also called the autocorrelation function of $f(t)$. In this case, it is clear that

$$\gamma(0) = 1 \tag{7.30}$$

EXAMPLE 7-12: By changing the variable t to $(t + \tau)$ in the cross-correlation functions (7.26) and (7.27) and the autocorrelation function (7.28), we obtain

$$R_{12}(\tau) = \int_{-\infty}^{\infty} f_1(t)f_2(t - \tau)\,dt = \int_{-\infty}^{\infty} f_1(t + \tau)f_2(t)\,dt \qquad (7.31)$$

$$R_{21}(\tau) = \int_{-\infty}^{\infty} f_2(t)f_1(t - \tau)\,dt = \int_{-\infty}^{\infty} f_2(t + \tau)f_1(t)\,dt \qquad (7.32)$$

$$R_{11}(\tau) = \int_{-\infty}^{\infty} f_1(t)f_1(t - \tau)\,dt = \int_{-\infty}^{\infty} f_1(t + \tau)f_1(t)\,dt \qquad (7.33)$$

From the results in Example 7-12 we note that it is immaterial whether we shift the function $f_1(t)$ by an amount τ in the negative direction or shift the function $f_2(t)$ by the same amount in the positive direction.

B. Correlation as convolution

The cross-correlation of $f_1(t)$ and $f_2(t)$ is related to the convolution of $f_1(t)$ and $f_2(-t)$.

EXAMPLE 7-13: Let

$$G_{12}(t) = f_1(t) * f_2(-t)$$

Then from the definition (7.1) of convolution, that is,

$$f_1(t) * f_2(t) = \int_{-\infty}^{\infty} f_1(x)f_2(t - x)\,dx$$

we obtain

$$G_{12}(t) = \int_{-\infty}^{\infty} f_1(x)f_2[-(t - x)]\,dx$$

$$= \int_{-\infty}^{\infty} f_1(x)f_2(x - t)\,dx \qquad (7.34)$$

Changing the variable t to τ,

$$G_{12}(\tau) = \int_{-\infty}^{\infty} f_1(x)f_2(x - \tau)\,dx$$

Again, by changing the dummy variable x to t, we have

$$G_{12}(\tau) = \int_{-\infty}^{\infty} f_1(t)f_2(t - \tau)\,dt$$

$$= R_{12}(\tau)$$

Hence,

$$R_{12}(\tau) = G_{12}(\tau) = f_1(t) * f_2(-t)|_{t=\tau} \qquad (7.35)$$

C. Properties of correlation functions

Properties of correlation functions

$$R_{12}(\tau) = R_{21}(-\tau) \qquad (7.36)$$
$$R_{11}(\tau) = R_{11}(-\tau) \qquad (7.37)$$

EXAMPLE 7-14: Verify the properties (7.36) and (7.37) of correlation functions.

Solution: From (7.32),

$$R_{21}(\tau) = \int_{-\infty}^{\infty} f_2(t + \tau)f_1(t)\,dt$$

and hence,

$$R_{21}(-\tau) = \int_{-\infty}^{\infty} f_2(t-\tau)f_1(t)\,dt = \int_{-\infty}^{\infty} f_1(t)f_2(t-\tau)\,dt = R_{12}(\tau)$$

Similarly, from (7.33),

$$R_{11}(\tau) = \int_{-\infty}^{\infty} f_1(t+\tau)f_1(t)\,dt$$

and hence,

$$R_{11}(-\tau) = \int_{-\infty}^{\infty} f_1(t-\tau)f_1(t)\,dt = \int_{-\infty}^{\infty} f_1(t)f_1(t-\tau)\,dt = R_{11}(\tau)$$

7-5. The Wiener–Khintchine Theorem

A. Fourier transforms of correlation functions

If $\mathscr{F}[f_1(t)] = F_1(\omega)$ and $\mathscr{F}[f_2(t)] = F_2(\omega)$, then

Fourier transforms
of correlation
functions

$$\mathscr{F}[R_{12}(\tau)] = F_1(\omega)F_2(-\omega) \tag{7.38}$$

$$\mathscr{F}[R_{21}(\tau)] = F_1(-\omega)F_2(\omega) \tag{7.39}$$

$$\mathscr{F}[R_{11}(\tau)] = F_1(\omega)F_1(-\omega) \tag{7.40}$$

Quantities $S_{12}(\omega) = \mathscr{F}[R_{12}(\omega)]$ and $S_{21}(\omega) = \mathscr{F}[R_{21}(\omega)]$ are referred to as **cross-energy densities**, and $S_{11}(\omega) = \mathscr{F}[R_{11}(\omega)]$ is called the **energy spectrum density** of $f_1(t)$.

EXAMPLE 7-15: Verify the Fourier transforms of the correlation functions (7.38) to (7.40).

Solution: If $\mathscr{F}[f_1(t)] = F_1(\omega)$ and $\mathscr{F}[f_2(t)] = F_2(\omega)$, then by the time-reversal property (5.9) of Fourier transforms, we have

$$\mathscr{F}[f_1(-t)] = F_1(-\omega) \quad \text{and} \quad \mathscr{F}[f_2(-t)] = F_2(-\omega)$$

Now, applying the time convolution theorem (7.11), that is,

$$\mathscr{F}[f_1(t) * f_2(t)] = F_1(\omega)F_2(\omega)$$

to (7.35), we have (7.38):

$$\mathscr{F}[R_{12}(\tau)] = \mathscr{F}[f_1(t) * f_2(-t)] = F_1(\omega)F_2(-\omega)$$

or

$$\int_{-\infty}^{\infty} R_{12}(\tau)e^{-j\omega\tau}\,d\tau = F_1(\omega)F_2(-\omega)$$

In a similar fashion, we get (7.39):

$$\mathscr{F}[R_{21}(\tau)] = \mathscr{F}[f_2(t) * f_1(-t)] = F_2(\omega)F_1(-\omega) = F_1(-\omega)F_2(\omega)$$

or

$$\int_{-\infty}^{\infty} R_{21}(\tau)e^{-j\omega\tau}\,d\tau = F_1(-\omega)F_2(\omega)$$

and (7.40):

$$\mathscr{F}[R_{11}(\tau)] = \mathscr{F}[f_1(t) * f_1(-t)] = F_1(\omega)F_1(-\omega)$$

EXAMPLE 7-16: If $f_1(t)$ is real, show that

$$\mathscr{F}[R_{11}(\tau)] = |F_1(\omega)|^2 \tag{7.41}$$

Solution: Recall property (5.16) of Fourier transforms; that is, if $f_1(t)$ is a real function of t, then $F_1(-\omega) = F_1^*(\omega)$. Hence

$$\mathscr{F}[R_{11}(\tau)] = F_1(\omega)F_1^*(\omega) = |F_1(\omega)|^2$$

or

$$\int_{-\infty}^{\infty} R_{11}(\tau)e^{-j\omega\tau}\,d\tau = |F_1(\omega)|^2 \tag{7.42}$$

B. The Wiener–Khintchine theorem

It follows from (7.41) that the Fourier transform of the autocorrelation function $R_{11}(\tau)$ yields the energy spectrum $|F_1(\omega)|^2$ of $f_1(t)$. In other words, the autocorrelation function $R_{11}(\tau)$ and the energy spectral density $|F_1(\omega)|^2$ constitute a Fourier transform pair; that is,

Wiener–Khintchine theorem

$$|F_1(\omega)|^2 = \mathscr{F}[R_{11}(\tau)] = \int_{-\infty}^{\infty} R_{11}(\tau)e^{-j\omega\tau}\,d\tau \tag{7.43}$$

$$R_{11}(\tau) = \mathscr{F}^{-1}[|\mathscr{F}_1(\omega)|^2] = \frac{1}{2\pi}\int_{-\infty}^{\infty} |F_1(\omega)|^2 e^{j\omega\tau}\,d\omega \tag{7.44}$$

This result is known as the **Wiener–Khintchine theorem**.

SUMMARY

1. The convolution of $f_1(t)$ and $f_2(t)$ is defined by

$$f(t) = f_1(t) * f_2(t) = \int_{-\infty}^{\infty} f_1(x)f_2(t-x)\,dx$$

and it is commutative; that is,

$$f_1(t) * f_2(t) = f_2(t) * f_1(t)$$

2. The time convolution theorem states that

$$\mathscr{F}[f_1(t) * f_2(t)] = F_1(\omega)F_2(\omega)$$

3. The frequency convolution theorem states that

$$\mathscr{F}[f_1(t)f_2(t)] = \frac{1}{2\pi}\int_{-\infty}^{\infty} F_1(y)F_2(\omega - y)\,dy = \frac{1}{2\pi}F_1(\omega) * F_2(\omega)$$

4. Parseval's theorem states that

$$\int_{-\infty}^{\infty} |f(t)|^2\,dt = \frac{1}{2\pi}\int_{-\infty}^{\infty} |F(\omega)|^2\,d\omega$$

5. The cross-correlation function of $f_1(t)$ and $f_2(t)$ is defined by

$$R_{12}(\tau) = \int_{-\infty}^{\infty} f_1(t)f_2(t-\tau)\,dt$$

$$R_{21}(\tau) = \int_{-\infty}^{\infty} f_2(t)f_1(t-\tau)\,dt$$

6. The autocorrelation function of $f_1(t)$ is defined by

$$R_{11}(\tau) = \int_{-\infty}^{\infty} f_1(t)f_1(t-\tau)\,dt$$

7. The Wiener–Khintchine theorem states that the autocorrelation function $R_{11}(\tau)$ and the energy spectral density $|F_1(\omega)|^2$ of $f_1(t)$ constitute a Fourier transform pair; that is,

$$|F_1(\omega)|^2 = \mathscr{F}[R_{11}(\tau)] = \int_{-\infty}^{\infty} R_{11}(\tau)e^{-j\omega\tau}\,d\tau$$

$$R_{11}(\tau) = \mathscr{F}^{-1}[|F_1(\omega)|^2] = \frac{1}{2\pi}\int_{-\infty}^{\infty} |F_1(\omega)|^2 e^{j\omega\tau}\,d\omega$$

RAISE YOUR GRADES

Can you explain ...?

☑ where the differences are between convolution and cross-correlation of two functions
☑ how to use the convolution theorems to obtain new transform pairs
☑ how Parseval's theorem is derived
☑ how to use Parseval's theorem to evaluate some integrals

SOLVED PROBLEMS

Convolution

PROBLEM 7-1 Prove the distributive law of convolution (7.6).

Proof: By the definition of convolution (7.1), we have

$$f_1(t) * [f_2(t) + f_3(t)] = \int_{-\infty}^{\infty} f_1(x)[f_2(t-x) + f_3(t-x)]\, dx$$

$$= \int_{-\infty}^{\infty} f_1(x)f_2(t-x)\, dx + \int_{-\infty}^{\infty} f_1(x)f_3(t-x)\, dx$$

$$= f_1(t) * f_2(t) + f_1(t) * f_3(t)$$

PROBLEM 7-2 Show that

$$f(t) * u(t) = \int_{-\infty}^{t} f(x)\, dx$$

Solution: By convolution (7.1), we have

$$f(t) * u(t) = \int_{-\infty}^{\infty} f(x)u(t-x)\, dx = \int_{-\infty}^{t} f(x)\, dx$$

since

$$u(t-x) = \begin{cases} 1, & x < t \\ 0, & x > t \end{cases}$$

PROBLEM 7-3 Show that

$$f(t - t_1) * \delta(t - t_2) = f(t - t_1 - t_2)$$

Solution:

$$f(t - t_1) * \delta(t - t_2) = \delta(t - t_2) * f(t - t_1) = \int_{-\infty}^{\infty} \delta(x - t_2)f(t - x - t_1)\, dx$$

$$= f(t - t_2 - t_1)$$

$$= f(t - t_1 - t_2)$$

according to property (4.3) of δ-functions.

PROBLEM 7-4 Show that

$$f(t) * \delta'(t) = f'(t)$$
$$u(t) * \delta'(t) = \delta(t)$$

Solution: First, we have to prove that

$$\delta'(-t) = -\delta'(t) \qquad \text{(a)}$$

Now,

$$\int_{-\infty}^{\infty} \delta'(-t)\phi(t)\,dt = \int_{-\infty}^{\infty} \delta'(x)\phi(-x)\,dx \qquad \text{(change of variable)}$$

$$= -\int_{-\infty}^{\infty} \delta(x)\phi'(-x)\,dx \qquad \text{(by (4.15))}$$

$$= \int_{-\infty}^{\infty} \delta(x)\phi'(x)\,dx \qquad \text{(since } \phi'(-x) = -\phi'(x)\text{)}$$

$$= \int_{-\infty}^{\infty} \delta(t)\phi'(t)\,dt \qquad \text{(change of dummy variable)}$$

$$= -\int_{-\infty}^{\infty} \delta'(t)\phi(t)\,dt \qquad \text{(by (4.15))}$$

Hence, by the equivalence property (4.2), we conclude that

$$\delta'(-t) = -\delta'(t)$$

Next, by the definition (7.1) of convolution, we have

$$f(t) * \delta'(t) = \int_{-\infty}^{\infty} f(x)\delta'(t - x)\,dx$$

$$= -\int_{-\infty}^{\infty} f(x)\delta'(x - t)\,dx \qquad \text{(by eq. (a))}$$

$$= \int_{-\infty}^{\infty} f'(x)\delta(x - t)\,dx \qquad \text{(by (4.15))}$$

$$= f'(t) \qquad \text{(by (4.3))}$$

Setting $f(t) = u(t)$, we obtain

$$u(t) * \delta'(t) = u'(t) = \delta(t)$$

since, by relation (4.18), we know that the derivative of the unit step function is $u'(t) = \delta(t)$.

PROBLEM 7-5   Show that if $f(t) * g(t) = h(t)$, then

$$h'(t) = f'(t) * g(t)$$
$$= f(t) * g'(t)$$

Solution: Using the commutative and associative laws of convolution (7.4) and (7.5) and the result of Problem 7-4, we have

$$h(t) * \delta'(t) = f(t) * [g(t) * \delta'(t)] = [f(t) * \delta'(t)] * g(t)$$

that is,

$$h'(t) = f(t) * g'(t) = f'(t) * g(t)$$

Convolution Theorems

PROBLEM 7-6   Using the symmetry property (5.10) of the Fourier transform and the time convolution theorem (7.11), prove the frequency convolution theorem (7.15).

Proof: From the time convolution theorem (7.11), we have

$$\mathscr{F}[f_1(t) * f_2(t)] = F_1(\omega)F_2(\omega)$$

that is,

$$\mathscr{F}\left[\int_{-\infty}^{\infty} f_1(x)f_2(t - x)\,dx \right] = F_1(\omega)F_2(\omega) \qquad \text{(a)}$$

Now from the symmetry property (5.10) of the Fourier transform, we know that if $\mathscr{F}[f(t)] = F(\omega)$, then $\mathscr{F}[F(t)] = 2\pi f(-\omega)$. Applying this result to eq. (a), we obtain

$$\mathscr{F}[F_1(t)F_2(t)] = 2\pi \int_{-\infty}^{\infty} f_1(x)f_2(-\omega - x)\,dx$$

Substituting $x = -y$,

$$\mathscr{F}[F_1(t)F_2(t)] = 2\pi \int_{-\infty}^{\infty} f_1(-y)f_2(-\omega + y)\,dy$$

$$= 2\pi \int_{-\infty}^{\infty} f_1(-y)f_2[-(\omega - y)]\,dy$$

$$= \frac{1}{2\pi} \int_{-\infty}^{\infty} [2\pi f_1(-y)]\{2\pi f_2[-(\omega - y)]\}\,dy \qquad \textbf{(b)}$$

Now, remembering that $2\pi f_1(-\omega) = \mathscr{F}[F_1(t)]$ and $2\pi f_2(-\omega) = \mathscr{F}[F_2(t)]$, we change $F_1(t)$ and $F_2(t)$ to $f_1(t)$ and $f_2(t)$, respectively. Consequently, we can change $2\pi f_1(-\omega)$ and $2\pi f_2(-\omega)$ to $F_1(\omega)$ and $F_2(\omega)$, respectively. Now we can rewrite eq. (b) as

$$\mathscr{F}[f_1(t)f_2(t)] = \frac{1}{2\pi} \int_{-\infty}^{\infty} F_1(y)F_2(\omega - y)\,dy = \frac{1}{2\pi} F_1(\omega) * F_2(\omega)$$

PROBLEM 7-7   In Example 5-14, it was shown that if $\mathscr{F}[f(t)] = F(\omega)$, then

$$\mathscr{F}\left[\int_{-\infty}^{t} f(x)\,dx\right] = \frac{1}{j\omega} F(\omega)$$

provided that

$$\int_{-\infty}^{\infty} f(t)\,dt = F(0) = 0$$

Show that if

$$\int_{-\infty}^{\infty} f(t)\,dt = F(0) \neq 0$$

then

$$\mathscr{F}\left[\int_{-\infty}^{t} f(x)\,dx\right] = \frac{1}{j\omega} F(\omega) + \pi F(0)\delta(\omega)$$

Solution: Let

$$g(t) = \int_{-\infty}^{t} f(x)\,dx$$

From the result of Problem 7-2, we have

$$g(t) = f(t) * u(t) = \int_{-\infty}^{\infty} f(x)u(t - x)\,dx = \int_{-\infty}^{t} f(x)\,dx$$

Therefore, from the time convolution theorem (7.11) and the Fourier transform of a unit step function (6.20), we obtain

$$\mathscr{F}[g(t)] = \mathscr{F}\left[\int_{-\infty}^{t} f(x)\,dx\right] = \mathscr{F}[f(t)]\mathscr{F}[u(t)]$$

$$= F(\omega)\left[\pi\delta(\omega) + \frac{1}{j\omega}\right]$$

$$= \frac{1}{j\omega} F(\omega) + \pi F(\omega)\delta(\omega)$$

From property (4.6) of δ-function, we have

$$F(\omega)\delta(\omega) = F(0)\delta(\omega)$$

Hence,

$$\mathscr{F}\left[\int_{-\infty}^{t} f(x)\,dx\right] = \frac{1}{j\omega} F(\omega) + \pi F(0)\delta(\omega)$$

PROBLEM 7-8   Use convolution to find

$$f(t) = \mathscr{F}^{-1}\left[\frac{1}{(1 + j\omega)(2 + j\omega)}\right]$$

Solution: Let $F_1(\omega) = 1/(1 + j\omega)$ and $F_2(\omega) = 1/(2 + j\omega)$. Then, by the result of Example 5-1, we have

$$f_1(t) = e^{-t}u(t) \quad \text{and} \quad f_2(t) = e^{-2t}u(t)$$

Thus, by the time convolution theorem (7.12), we obtain

$$f(t) = f_1(t) * f_2(t)$$

$$= \int_{-\infty}^{\infty} f_1(x)f_2(t - x)\,dx$$

$$= \int_{-\infty}^{\infty} e^{-x}u(x)e^{-2(t-x)}u(t - x)\,dx$$

$$= e^{-2t}\int_{-\infty}^{\infty} e^{x}u(x)u(t - x)\,dx$$

Since

$$u(x)u(t - x) = \begin{cases} 1, & 0 < x < t \\ 0, & \text{otherwise} \end{cases}$$

we have

$$f(t) = e^{-2t}\left[\int_{0}^{t} e^{x}\,dx\right]u(t)$$

$$= e^{-2t}(e^{t} - 1)u(t) = (e^{-t} - e^{-2t})u(t)$$

PROBLEM 7-9   Let $F(\omega) = \mathscr{F}[f(t)]$ and $G(\omega) = \mathscr{F}[g(t)]$. Prove that

(a) $\displaystyle\int_{-\infty}^{\infty} f(x)g(t - x)\,dx = \frac{1}{2\pi}\int_{-\infty}^{\infty} F(\omega)G(\omega)e^{j\omega t}\,d\omega$

(b) $\displaystyle\int_{-\infty}^{\infty} f(t)g(-t)\,dt = \frac{1}{2\pi}\int_{-\infty}^{\infty} F(\omega)G(\omega)\,d\omega$

(c) $\displaystyle\int_{-\infty}^{\infty} f(t)g^{*}(t)\,dt = \frac{1}{2\pi}\int_{-\infty}^{\infty} F(\omega)G^{*}(\omega)\,d\omega$

where the asterisk (*) denotes complex conjugation. [Note: Equation (c) is sometimes also called Parseval's theorem.]

Proof:
(a) By convolution (7.1),

$$\int_{-\infty}^{\infty} f(x)g(t - x)\,dx = f(t) * g(t)$$

Thus, from the time convolution theorem (7.11), we have

$$\mathscr{F}[f(t) * g(t)] = F(\omega)G(\omega)$$

Taking the inverse Fourier transform (5.2), we obtain

$$\int_{-\infty}^{\infty} f(x)g(t - x)\,dx = \frac{1}{2\pi}\int_{-\infty}^{\infty} F(\omega)G(\omega)e^{j\omega t}\,d\omega$$

(b) Setting $t = 0$ in the result of eq. (a), we have

$$\int_{-\infty}^{\infty} f(x)g(-x)\,dx = \frac{1}{2\pi}\int_{-\infty}^{\infty} F(\omega)G(\omega)\,d\omega$$

or, changing the dummy variable, we have

$$\int_{-\infty}^{\infty} f(t)g(-t)\,dt = \frac{1}{2\pi}\int_{-\infty}^{\infty} F(\omega)G(\omega)\,d\omega$$

(c) Recall (7.21), that is,

$$\mathscr{F}[g^*(t)] = G^*(-\omega)$$

Setting $f_1(t) = f(t)$, and $f_2(t) = g^*(t)$ in (7.16), we obtain

$$\int_{-\infty}^{\infty} f(t)g^*(t)\,dt = \frac{1}{2\pi}\int_{-\infty}^{\infty} F(\omega)G^*\big(-(-\omega)\big)\,d\omega = \frac{1}{2\pi}\int_{-\infty}^{\infty} F(\omega)G^*(\omega)\,d\omega$$

PROBLEM 7-10 If $f(t)$ is an arbitrary function and $F(\omega)$ is its Fourier transform, then prove the following identity:

$$\sum_{n=-\infty}^{\infty} f(t + nT) = \frac{1}{T}\sum_{n=-\infty}^{\infty} e^{jn\omega_0 t} F(n\omega_0)$$

where $\omega_0 = 2\pi/T$.

Proof: From the relation (4.29) for the periodic train of unit impulses, we have

$$\delta_T(t) = \sum_{n=-\infty}^{\infty} \delta(t - nT)$$

By convolution with a δ-function (7.9), we obtain

$$f(t) * \delta_T(t) = f(t) * \sum_{n=-\infty}^{\infty} \delta(t - nT)$$

$$= \sum_{n=-\infty}^{\infty} f(t) * \delta(t - nT)$$

$$= \sum_{n=-\infty}^{\infty} f(t - nT)$$

$$= \sum_{n=-\infty}^{\infty} f(t + nT)$$

since all positive and negative values of n are included in the summation. Hence

$$\sum_{n=-\infty}^{\infty} f(t + nT) = f(t) * \delta_T(t) \qquad\qquad \textbf{(a)}$$

Now, from the Fourier transform of a unit impulse train (6.26), we get

$$\mathscr{F}[\delta_T(t)] = \frac{2\pi}{T}\sum_{n=-\infty}^{\infty} \delta(\omega - n\omega_0), \qquad \omega_0 = \frac{2\pi}{T}$$

Then applying the time convolution theorem (7.11) to eq. (a), we obtain

$$\mathscr{F}\left[\sum_{n=-\infty}^{\infty} f(t + nT)\right] = F(\omega)\mathscr{F}[\delta_T(t)]$$

$$= F(\omega)\frac{2\pi}{T}\sum_{n=-\infty}^{\infty} \delta(\omega - n\omega_0)$$

$$= \frac{2\pi}{T}\sum_{n=-\infty}^{\infty} F(\omega)\delta(\omega - n\omega_0)$$

$$= \frac{2\pi}{T}\sum_{n=-\infty}^{\infty} F(n\omega_0)\delta(\omega - n\omega_0) \qquad\qquad \textbf{(b)}$$

with the use of the δ-function property (4.6).

From the Fourier transform of a complex exponential function (6.18), we have

$$\mathscr{F}^{-1}[\delta(\omega - n\omega_0)] = \frac{1}{2\pi} e^{jn\omega_0 t}$$

Therefore, taking the inverse Fourier transform of eq. (b), we obtain

$$\sum_{n=-\infty}^{\infty} f(t + nT) = \frac{2\pi}{T} \sum_{n=-\infty}^{\infty} F(n\omega_0)\mathscr{F}^{-1}[\delta(\omega - n\omega_0)] = \frac{1}{T} \sum_{n=-\infty}^{\infty} F(n\omega_0)e^{jn\omega_0 t}$$

PROBLEM 7-11   Using the identity of Problem 7-10, prove **Poisson's formula**:

Poisson's formula

$$\sum_{n=-\infty}^{\infty} f(nT) = \frac{1}{T} \sum_{n=-\infty}^{\infty} F(n\omega_0)$$

where $\omega_0 = 2\pi/T$.

Proof: Letting $t = 0$ in the identity of Problem 7-10, we obtain easily the formula

$$\sum_{n=-\infty}^{\infty} f(nT) = \frac{1}{T} \sum_{n=-\infty}^{\infty} F(n\omega_0)$$

PROBLEM 7-12   Prove that

$$\sum_{n=-\infty}^{\infty} e^{-a|n|} = \sum_{n=-\infty}^{\infty} \frac{2a}{a^2 + (2n\pi)^2}, \qquad a > 0$$

Proof: Let

$$f(t) = e^{-a|t|}$$

Then, from the result of Problem 5-3, we have

$$F(\omega) = \mathscr{F}[e^{-a|t|}] = \frac{2a}{a^2 + \omega^2}$$

If we set $T = 1$ (hence, $\omega_0 = 2\pi$) in Poisson's summation formula (Problem 7-11), we obtain

$$\sum_{n=-\infty}^{\infty} f(n) = \sum_{n=-\infty}^{\infty} F(2\pi n)$$

Hence,

$$\sum_{n=-\infty}^{\infty} e^{-a|n|} = \sum_{n=-\infty}^{\infty} \frac{2a}{a^2 + (2\pi n)^2}$$

Parseval's Theorem

PROBLEM 7-13   Evaluate

$$\int_{-\infty}^{\infty} \frac{a^2\, dx}{(a^2 + x^2)^2} \qquad \text{and} \qquad \int_{-\infty}^{\infty} \frac{dx}{(1 + x^2)^2}$$

Solution: Let

$$f(t) = \frac{1}{2} e^{-a|t|}$$

Then from the result of Problem 5-3, we have

$$F(\omega) = \mathscr{F}[f(t)] = \frac{a}{a^2 + \omega^2}$$

Now, using Parseval's theorem (7.18),

$$\int_{-\infty}^{\infty} |F(\omega)|^2\, d\omega = 2\pi \int_{-\infty}^{\infty} f^2(t)\, dt$$

we have

$$\int_{-\infty}^{\infty} \frac{a^2}{(a^2 + \omega^2)^2} \, d\omega = 2\pi \int_{-\infty}^{\infty} \left[\frac{1}{2} e^{-a|t|} \right]^2 dt$$

$$= \frac{\pi}{2} \int_{-\infty}^{\infty} e^{-2a|t|} \, dt$$

$$= \frac{\pi}{2} \left[\int_{-\infty}^{0} e^{2at} \, dt + \int_{0}^{\infty} e^{-2at} \, dt \right]$$

$$= \frac{\pi}{2} \left[\frac{e^{2at}}{2a} \bigg|_{-\infty}^{0} + \frac{e^{-2at}}{-2a} \bigg|_{0}^{\infty} \right] = \frac{\pi}{2a}$$

Thus,

$$\int_{-\infty}^{\infty} \frac{a^2}{(a^2 + x^2)^2} \, dx = \int_{-\infty}^{\infty} \frac{a^2}{(a^2 + \omega^2)^2} \, d\omega = \frac{\pi}{2a}$$

Setting $a = 1$,

$$\int_{-\infty}^{\infty} \frac{dx}{(1 + x^2)^2} = \frac{\pi}{2}$$

Correlation Functions

PROBLEM 7-14 Show that

$$R_{11}(0) = \int_{-\infty}^{\infty} [f_1(t)]^2 \, dt$$

Solution: We use the autocorrelation function (7.28):

$$R_{11}(\tau) = \int_{-\infty}^{\infty} f_1(t) f_1(t - \tau) \, dt$$

Letting $\tau = 0$, we obtain

$$R_{11}(0) = \int_{-\infty}^{\infty} f_1(t) f_1(t) \, dt$$

$$= \int_{-\infty}^{\infty} [f_1(t)]^2 \, dt$$

PROBLEM 7-15 Using the result of Problem 7-14 and $\mathscr{F}[R_{11}(\tau)] = |F_1(\omega)|^2$ (7.41), derive Parseval's theorem, that is,

$$\int_{-\infty}^{\infty} [f_1(t)]^2 \, dt = \frac{1}{2\pi} \int_{-\infty}^{\infty} |F_1(\omega)|^2 \, d\omega$$

Solution: Taking the inverse Fourier transform of (7.41), we have

$$R_{11}(\tau) = \frac{1}{2\pi} \int_{-\infty}^{\infty} |F_1(\omega)|^2 e^{j\omega t} \, d\omega$$

Letting $\tau = 0$, we have

$$R_{11}(0) = \frac{1}{2\pi} \int_{-\infty}^{\infty} |F_1(\omega)|^2 \, d\omega$$

From the result of Problem 7-14, we have

$$R_{11}(0) = \int_{-\infty}^{\infty} [f_1(t)]^2 \, dt$$

Hence,

$$\int_{-\infty}^{\infty} [f_1(t)]^2 \, dt = \frac{1}{2\pi} \int_{-\infty}^{\infty} |F_1(\omega)|^2 \, d\omega$$

PROBLEM 7-16 Using the definition (5.1) of a Fourier transform, derive (7.41), that is,

$$\mathscr{F}[R_{11}(\tau)] = |F_1(\omega)|^2$$

Solution: We use the autocorrelation function (7.28):

$$R_{11}(\tau) = \int_{-\infty}^{\infty} f_1(t) f_1(t - \tau) \, dt$$

Then, by definition (5.1), we have

$$\mathscr{F}[R_{11}(\tau)] = \int_{-\infty}^{\infty} R_{11}(\tau) e^{-j\omega t} \, d\tau$$

$$= \int_{-\infty}^{\infty} \left[\int_{-\infty}^{\infty} f_1(t) f_1(t - \tau) \, dt \right] e^{-j\omega\tau} \, d\tau$$

$$= \int_{-\infty}^{\infty} f_1(t) \left[\int_{-\infty}^{\infty} f_1(t - \tau) e^{-j\omega\tau} \, d\tau \right] dt$$

by interchanging the order of integration.

Changing the variable $(t - \tau)$ to x in the integral within the bracket of the last integral, we obtain

$$\mathscr{F}[R_{11}(\tau)] = \int_{-\infty}^{\infty} f_1(t) \left[\int_{-\infty}^{\infty} f_1(x) e^{-j\omega(t-x)} \, dx \right] dt$$

$$= \int_{-\infty}^{\infty} f_1(t) e^{-j\omega t} \, dt \int_{-\infty}^{\infty} f_1(x) e^{j\omega x} \, dx$$

$$= F_1(\omega) F_1(-\omega) = F_1(\omega) F_1^*(\omega) = |F_1(\omega)|^2$$

PROBLEM 7-17 If $R_{11}(\tau)$ is the autocorrelation function of $f_1(t)$, show that $R_{11}(0) > |R_{11}(\tau)|$ for $\tau \neq 0$.

Solution: Consider the following integral:

$$\int_{-\infty}^{\infty} [f_1(t) \pm f_1(t + \tau)]^2 \, dt \qquad \text{for } \tau \neq 0$$

The integral is always a positive nonzero quantity. It is nonzero because $f_1(t) \neq f_1(t + \tau)$ unless $f_1(t)$ is periodic. Hence,

$$\int_{-\infty}^{\infty} [f_1(t) \pm f_1(t + \tau)]^2 \, dt > 0 \qquad \text{for } \tau \neq 0$$

Expansion of this expression yields

$$\int_{-\infty}^{\infty} f_1^2(t) \, dt + \int_{-\infty}^{\infty} f_1^2(t + \tau) \, dt \pm 2 \int_{-\infty}^{\infty} f_1(t) f_1(t + \tau) \, dt > 0 \qquad \text{for } \tau \neq 0$$

or

$$\int_{-\infty}^{\infty} f_1^2(t) \, dt + \int_{-\infty}^{\infty} f_1^2(t + \tau) \, dt > \pm 2 \int_{-\infty}^{\infty} f_1(t) f_1(t + \tau) \, dt \qquad \text{for } \tau \neq 0$$

By changing the variable in the autocorrelation function, as in (7.33), we have

$$\int_{-\infty}^{\infty} f_1^2(t) \, dt = \int_{-\infty}^{\infty} f_1^2(t + \tau) \, dt = R_{11}(0)$$

$$\int_{-\infty}^{\infty} f_1(t) f_1(t + \tau) \, dt = R_{11}(\tau)$$

Therefore, we obtain

$$R_{11}(0) > \pm R_{11}(\tau) \qquad \text{for } \tau \neq 0$$

Thus,

$$R_{11}(0) > |R_{11}(\tau)| \qquad \text{for } \tau \neq 0$$

PROBLEM 7-18 If $R_{11}(\tau)$ and $R_{22}(\tau)$ are the autocorrelation functions of $f_1(t)$ and $f_2(t)$, and $R_{12}(\tau)$ is the cross-correlation function of $f_1(t)$ and $f_2(t)$, show that

$$R_{11}(0) + R_{22}(0) > 2|R_{12}(\tau)| \qquad \text{for all } \tau.$$

Solution: As shown in Problem 7-17, it is easily seen that

$$\int_{-\infty}^{\infty} [f_1(t) \pm f_2(t + \tau)]^2 \, dt > 0 \qquad \text{for all } \tau$$

Expansion of this expression gives

$$\int_{-\infty}^{\infty} f_1^2(t) \, dt + \int_{-\infty}^{\infty} f_2^2(t + \tau) \, dt > \pm 2 \int_{-\infty}^{\infty} f_1(t) f_2(t + \tau) \, dt \qquad \text{for all } \tau$$

By (7.31) and (7.33), we obtain

$$R_{11}(0) + R_{22}(0) > \pm 2R_{12}(\tau) \qquad \text{for all } \tau$$

which means

$$R_{11}(0) + R_{22}(0) > 2|R_{12}(\tau)| \qquad \text{for all } \tau$$

Wiener–Khintchine Theorem

PROBLEM 7-19 (a) Find the autocorrelation function $R_{11}(\tau)$ of the rectangular pulse $f_1(t)$ defined by

$$f_1(t) = \begin{cases} A & \text{for } |t| < d/2 \\ 0 & \text{for } |t| > d/2 \end{cases}$$

(b) Find the energy spectrum density $S_{11}(\omega)$ of $f_1(t)$ from $R_{11}(\tau)$ obtained in part (a), and also check that $S_{11}(\omega) = |F_1(\omega)|^2$.

Solution:
(a) As shown in Figure 7.2c and by the autocorrelation function (7.28), we have

$$R_{11}(\tau) = \int_{-\infty}^{\infty} f_1(t) f_1(t - \tau) \, dt$$

$$= \int_{-d/2+\tau}^{d/2} A^2 \, dt = A^2(d - \tau) \qquad \text{for } 0 \leqslant \tau \leqslant d$$

Since $R_{11}(\tau)$ is even and vanishes for $\tau > d$, the complete expression of $R_{11}(\tau)$ is given by

$$R_{11}(\tau) = \begin{cases} A^2(d - |\tau|) & \text{for } -d \leqslant \tau \leqslant d \\ 0 & \text{otherwise} \end{cases}$$

which is shown in Figure 7-2d.

(b) Using the result of Problem 5-13 and substituting $A^2 d$ for A and d for T in $F_2(\omega)$, we obtain

$$S_{11}(\omega) = \mathscr{F}[R_{11}(\tau)] = A^2 d^2 \frac{\sin^2(\omega d/2)}{(\omega d/2)^2}$$

Now, from the result of Problem 5-2, we have

$$F_1(\omega) = \mathscr{F}[f_1(t)] = Ad \frac{\sin(\omega d/2)}{(\omega d/2)}$$

Thus,

$$S_{11}(\omega) = |F_1(\omega)|^2$$

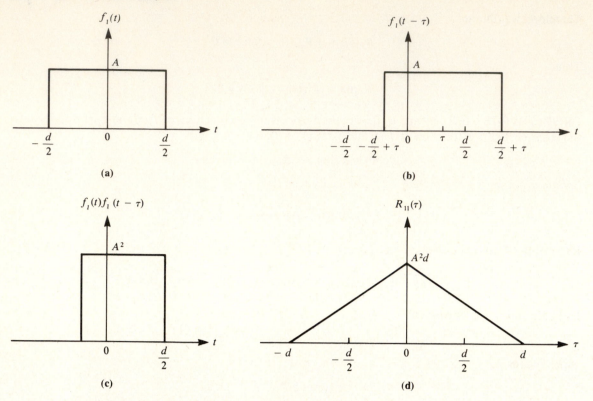

Figure 7-2 **(a) Rectangular pulse; (b) shifted rectangular pulse; (c) product of two rectangular pulses; (d) autocorrelation function of the rectangular pulse.**

Supplementary Exercises:

PROBLEM 7-20   Find $f(t)$ of Problem 7-8 by expanding $F(\omega)$ in partial fractions.

$$\left[Hint: \quad \frac{1}{(j\omega + 1)(j\omega + 2)} = \frac{1}{j\omega + 1} + \frac{-1}{j\omega + 2} \text{ and use the result of Problem 5-1.} \right]$$

PROBLEM 7-21   Using the result of Example 7-4 and the convolution theorem, show that the Fourier transform of a periodic function $f(t)$ with period T and with the complex Fourier coefficients c_n can be expressed as

$$F(\omega) = \frac{2\pi}{T} \sum_{n=-\infty}^{\infty} F_0(n\omega_0)\delta(\omega - n\omega_0) = 2\pi \sum_{n=-\infty}^{\infty} c_n \delta(\omega - n\omega_0)$$

where

$$F_0(\omega) = \mathscr{F}[f_0(t)] \quad \text{and} \quad f_0(t) = \begin{cases} f(t) & \text{for} \quad |t| < t/2 \\ 0 & \text{for} \quad |t| > t/2 \end{cases}$$

[*Hint:* Use the result of Example 6-12.]

PROBLEM 7-22   Use (6.7) to deduce the time convolution theorem

$$\mathscr{F}[f_1(t) * f_2(t)] = F_1(\omega)F_2(\omega)$$

PROBLEM 7-23   Prove the convolution theorem for two dimensions; that is,

$$\mathscr{F}[f(x, y) * g(x, y)] = F(u, v)G(u, v)$$

where

$$f(x, y) * g(x, y) = \int_{-\infty}^{\infty} \int_{-\infty}^{\infty} f(x', y')g(x - x', y - y')\,dx'\,dy'$$

$$F(u, v) = \mathscr{F}[f(x, y)]$$

$$G(u, v) = \mathscr{F}[g(x, y)]$$

PROBLEM 7-24   Prove Parseval's theorem for two dimension; that is,

$$\iint\limits_{-\infty}^{\infty} |f(x, y)|^2 \, dx \, dy = \frac{1}{(2\pi)^2} \iint\limits_{-\infty}^{\infty} |F(u, v)|^2 \, du \, dv$$

PROBLEM 7-25   Let $f_1(t)$ and $f_2(t)$ be two Gaussian functions; that is,

$$f_1(t) = \frac{1}{\sigma_1 \sqrt{2\pi}} e^{-t^2/2\sigma_1^2}, \qquad f_2(t) = \frac{1}{\sigma_2 \sqrt{2\pi}} e^{-t^2/2\sigma_2^2}$$

Show that if $f_3(t) = f_1(t) * f_2(t)$, then $f_3(t)$ is also a Gaussian function and

$$f_3(t) = \frac{1}{\sigma_3 \sqrt{2\pi}} e^{-t^2/2\sigma_3^2}, \qquad \text{where } \sigma_3^2 = \sigma_1^2 + \sigma_2^2$$

PROBLEM 7-26   Show that the correlation function of any two Gaussian functions is itself a Gaussian function.

PROBLEM 7-27   Let $R_{11}(\tau)$ be the autocorrelation function and $S_{11}(\omega) = |F_1(\omega)|^2$ be the energy spectrum density of the function $f_1(t)$. Show that the Wiener-Khintchine theorem (7.43, 7.44) can be rewritten as

$$S_{11}(\tau) = \int_0^\infty R_{11}(\omega) \cos \omega \tau \, d\omega \quad \text{and} \quad R_{11}(\omega) = \frac{1}{\pi} \int_0^\infty S_{11}(\tau) \cos \omega \tau \, d\tau$$

8 APPLICATIONS TO SIGNAL THEORY

THIS CHAPTER IS ABOUT

☑ **Band-Limited and Duration-Limited Signals**
☑ **The Uncertainty Principle**
☑ **Sampling Theorems**
☑ **Modulation**
☑ **Average Correlation Functions**
☑ **Average Power Spectra: Random Signals**
☑ **Analytic Signals and Hilbert Transforms**

8-1. Band-Limited and Duration-Limited Signals

A **band-limited signal** is a time function $f(t)$ for which the Fourier transform of $f(t)$ is identically zero above a certain frequency ω_M:

Band-limited signal
$$F(\omega) = \mathcal{F}[f(t)] = 0 \qquad \text{for } |\omega| > \omega_M = 2\pi f_M \tag{8.1}$$

A **duration-limited signal** is a time function $f(t)$ for which

Duration-limited signal
$$f(t) = 0 \qquad \text{for } |t| > T \tag{8.2}$$

8-2. The Uncertainty Principle

A. Definitions used in expressing the uncertainty principle

Consider a pulse signal $f(t)$ with a single maximum, taken at $t = 0$ for convenience, as shown in Figure 8-1a. We define the **equivalent pulse duration** T_D by

Equivalent pulse duration
$$T_D = \frac{1}{f(0)} \int_{-\infty}^{\infty} |f(t)| \, dt \tag{8.3}$$

Similarly, we define the **equivalent spectral bandwidth** W_B of $f(t)$ by

Equivalent spectral bandwidth
$$W_B = \frac{1}{F(0)} \int_{-\infty}^{\infty} |F(\omega)| \, d\omega \tag{8.4}$$

B. Statement of the uncertainty principle

The **uncertainty principle** in spectral analysis can be stated as follows: *The product of the spectral bandwidth and the time duration of a signal cannot be less than a certain minimum value.*

Using the definitions of the equivalent pulse duration (8.3) and the equivalent spectral bandwidth (8.4), the uncertainty principle can be expressed as

Uncertainty principle
$$W_B T_D \geqslant 2\pi \tag{8.5}$$

EXAMPLE 8-1: Verify the uncertainty principle (8.5).

Solution: By the definition (8.3) of T_D, we have

$$f(0) T_D = \int_{-\infty}^{\infty} |f(t)| \, dt \geqslant \int_{-\infty}^{\infty} f(t) \, dt = F(0) \tag{8.6}$$

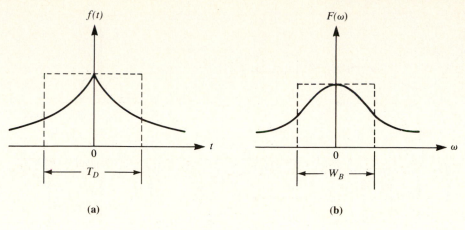

Figure 8-1

since

$$F(0) = F(\omega)\big|_{\omega=0} = \left[\int_{-\infty}^{\infty} f(t)e^{-j\omega t}\,dt\right]_{\omega=0} = \int_{-\infty}^{\infty} f(t)\,dt$$

Similarly, by the definition (8.4) of W_B, we have

$$F(0)W_B = \int_{-\infty}^{\infty} |F(\omega)|\,d\omega \geqslant \int_{-\infty}^{\infty} F(\omega)\,d\omega = 2\pi f(0) \tag{8.7}$$

since

$$f(0) = f(t)\big|_{t=0} = \left[\frac{1}{2\pi}\int_{-\infty}^{\infty} F(\omega)e^{j\omega t}\,d\omega\right]_{t=0} = \frac{1}{2\pi}\int_{-\infty}^{\infty} F(\omega)\,d\omega$$

Thus, from (8.7), we have

$$F(0) \geqslant 2\pi f(0)/W_B \tag{8.8}$$

Substituting (8.8) into (8.6), we obtain

$$f(0)T_D \geqslant F(0) \geqslant 2\pi f(0)/W_B \tag{8.9}$$

From (8.9), we conclude that

$$W_B T_D \geqslant 2\pi$$

8-3. Sampling Theorems

A. The sampling theorem in the time domain

The **uniform sampling theorem in the time domain** states that if a time function $f(t)$ is a band-limited signal satisfying definition (8.1), then $f(t)$ can be uniquely determined by its sampled values $f_n = f(n\pi/\omega_M)$ at uniform intervals π/ω_M or less apart. In fact, $f(t)$ is given by

Sampling theorem in the time domain
$$f(t) = \sum_{n=-\infty}^{\infty} f\left(\frac{n\pi}{\omega_M}\right)\frac{\sin(\omega_M t - n\pi)}{(\omega_M t - n\pi)} \tag{8.10}$$

EXAMPLE 8-2: Prove the sampling theorem and its expression (8.10).

Proof: Consider a sampled function $f_s(t)$ defined by the product of the function $f(t)$ and a periodic unit impulse train function $\delta_T(t)$ (see Figure 8-2c):

$$f_s(t) = f(t)\delta_T(t) = f(t)\sum_{n=-\infty}^{\infty} \delta(t - nT) \tag{8.11}$$

Using the result of Problem 4-2, we have

$$f_s(t) = f(t)\sum_{n=-\infty}^{\infty} \delta(t - nT)$$

$$= \sum_{n=-\infty}^{\infty} f(t)\delta(t - nT)$$

$$= \sum_{n=-\infty}^{\infty} f(nT)\delta(t - nT) \tag{8.12}$$

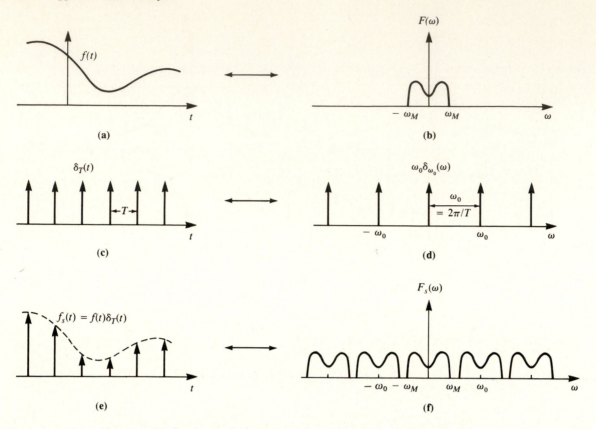

Figure 8-2 **(a) A band-limited signal $f(t)$; (b) the spectrum of $f(t)$; (c) the unit impulse train; (d) the spectrum of the unit impulse train; (e) the sampled function $f_s(t)$; (f) the spectrum of $f_s(t)$.**

Equation (8.12) shows that the function $f_s(t)$ is a sequence of impulses located at regular intervals of T seconds and having strength equal to the values of $f(t)$ at the sampling instants (see Figure 8-2e).

From the expression (6.26) of a Fourier transform of a unit impulse train, we have

$$\mathscr{F}[\delta_T(t)] = \omega_0 \delta_{\omega_0}(\omega) = \omega_0 \sum_{n=-\infty}^{\infty} \delta(\omega - n\omega_0)$$

Now according to the frequency convolution theorem (7.15), we have

$$\mathscr{F}[f_s(t)] = F_s(\omega) = \frac{1}{2\pi}[F(\omega) * \omega_0 \delta_{\omega_0}(\omega)]$$

Substituting $\omega_0 = 2\pi/T$,

$$F_s(\omega) = \frac{1}{T}[F(\omega) * \delta_{\omega_0}(\omega)]$$

$$= \frac{1}{T}\left[F(\omega) * \sum_{n=-\infty}^{\infty} \delta(\omega - n\omega_0)\right]$$

$$= \frac{1}{T}\sum_{n=-\infty}^{\infty} F(\omega) * \delta(\omega - n\omega_0) \tag{8.13}$$

From convolution with δ-function (7.9), we have

$$F(\omega) * \delta(\omega - n\omega_0) = F(\omega - n\omega_0)$$

Hence, we can rewrite (8.13) as

$$F_s(\omega) = \frac{1}{T}\sum_{n=-\infty}^{\infty} F(\omega - n\omega_0) \tag{8.14}$$

Equation (8.14) shows that the Fourier transform of $f_s(t)$ repeats itself every ω_0 rad/sec, as shown in Figure 8-2f. Note that $F(\omega)$ will repeat periodically without overlapping as long as $\omega_0 \geq 2\omega_M$, or

$2\pi/T \geqslant 2\omega_M$; that is,

$$T \leqslant \pi/\omega_M \quad \text{or} \quad T \leqslant 1/(2f_M) \tag{8.15}$$

where $\omega_M = 2\pi f_M$. Therefore, as long as we sample $f(t)$ at uniform intervals less than π/ω_M or $1/(2f_M)$ sec apart, the Fourier spectrum of $f_s(t)$ will be a periodic replica of $F(\omega)$, and will contain all the information about $f(t)$.

Since $F_s(\omega)$ is a periodic function of ω with period ω_0, it can be expanded into the Fourier series

$$F_s(\omega) = \sum_{n=-\infty}^{\infty} c_n e^{jn2\pi\omega/\omega_0} \tag{8.16}$$

where, by the formula (3.10) for complex Fourier coefficients,

$$c_n = \frac{1}{\omega_0} \int_{-\omega_0/2}^{\omega_0/2} F_s(\omega) e^{-jn2\pi\omega/\omega_0} \, d\omega \tag{8.17}$$

Since $F_s(\omega) = F(\omega)$ for $-\omega_M < \omega < \omega_M$ and $\frac{1}{2}\omega_0 > \omega_M$, we can rewrite (8.17) as

$$c_n = \frac{1}{\omega_0} \int_{-\omega_M}^{\omega_M} F(\omega) e^{-jn2\pi\omega/\omega_0} \, d\omega \tag{8.18}$$

Now by the inverse Fourier transform (5.2),

$$f(t) = \mathscr{F}^{-1}[F(\omega)] = \frac{1}{2\pi} \int_{-\infty}^{\infty} F(\omega) e^{j\omega t} \, d\omega \tag{8.19}$$

Since $f(t)$ is band-limited, i.e., $F(\omega) = 0$ for $|\omega| > \omega_M$, eq. (8.19) becomes

$$f(t) = \frac{1}{2\pi} \int_{-\omega_M}^{\omega_M} F(\omega) e^{j\omega t} \, d\omega \tag{8.20}$$

Designating the sampling points to be at $t = -nT = -n2\pi/\omega_0$, from (8.20), we have

$$f(-nT) = f\left(-\frac{n2\pi}{\omega_0}\right) = \frac{1}{2\pi} \int_{-\omega_M}^{\omega_M} F(\omega) e^{-jn2\pi\omega/\omega_0} \, d\omega \tag{8.21}$$

Comparing (8.21) with (8.18),

$$c_n = \frac{2\pi}{\omega_0} f\left(-\frac{n2\pi}{\omega_0}\right) = Tf(-nT) \tag{8.22}$$

Now if we select $T = \pi/\omega_M$, then $\omega_0 = 2\pi/T = 2\omega_M$. Hence, (8.16) becomes

$$F_s(\omega) = \sum_{n=-\infty}^{\infty} c_n e^{jn2\pi\omega/2\omega_M} = \sum_{n=-\infty}^{\infty} c_n e^{jnT\omega} \tag{8.23}$$

From (8.22),

$$c_n = Tf(-nT) = \frac{\pi}{\omega_M} f(-nT) \tag{8.24}$$

Substituting (8.24) into (8.23), we have

$$F_s(\omega) = \sum_{n=-\infty}^{\infty} \frac{\pi}{\omega_M} f(-nT) e^{jnT\omega} \tag{8.25}$$

Since $F_s(\omega) = F(\omega)$ for $-\omega_M < \omega < \omega_M$, substituting (8.25) into (8.20), we obtain

$$f(t) = \frac{1}{2\pi} \int_{-\omega_M}^{\omega_M} \left[\sum_{n=-\infty}^{\infty} \frac{\pi}{\omega_M} f(-nT) e^{jnT\omega} \right] e^{j\omega t} \, d\omega \tag{8.26}$$

Interchanging the order of integration and summation,

$$f(t) = \sum_{n=-\infty}^{\infty} \left[f(-nT) \int_{-\omega_M}^{\omega_M} \frac{1}{2\omega_M} e^{j\omega(t+nT)} \, d\omega \right] = \sum_{n=-\infty}^{\infty} f(-nT) \frac{\sin \omega_M(t+nT)}{\omega_M(t+nT)}$$

$$= \sum_{n=-\infty}^{\infty} f(nT) \frac{\sin \omega_M(t-nT)}{\omega_M(t-nT)} \tag{8.27}$$

In the last equation, $(-n)$ was replaced by n because all positive and negative values of n are included in the summation. Since $T = \pi/\omega_M$, eq. (8.27) also can be written as (8.10):

$$f(t) = \sum_{n=-\infty}^{\infty} f\left(\frac{n\pi}{\omega_M}\right) \frac{\sin(\omega_M t - n\pi)}{(\omega_M t - n\pi)}$$

Note that the maximum sampling interval $T = 1/(2f_M)$ is sometimes called the **Nyquist interval** and the minimum sampling frequency $2f_M$ is called the **Nyquist rate**.

B. The sampling theorem in the frequency domain

The **sampling theorem in the frequency domain** states that if a time function $f(t)$ is a duration-limited signal satisfying definition (8.2), then its Fourier transform $F(\omega)$ can be uniquely determined from its values $F(n\pi/T)$ at a series of equidistant points spaced π/T apart. In fact, $F(\omega)$ is given by

Sampling theorem in the frequency domain

$$F(\omega) = \sum_{n=-\infty}^{\infty} F\left(\frac{n\pi}{T}\right) \frac{\sin(\omega T - n\pi)}{(\omega T - n\pi)} \tag{8.28}$$

EXAMPLE 8-3: Verify the sampling theorem in the frequency domain and its expression (8.28).

Solution: Since $f(t) = 0$ for $|t| > T$, then, in the interval $-T < t < T$, the function $f(t)$ can be expanded in a Fourier series

$$f(t) = \sum_{n=-\infty}^{\infty} c_n e^{j2\pi n t/2T} = \sum_{n=-\infty}^{\infty} c_n e^{jn\pi t/T} \tag{8.29}$$

where

$$c_n = \frac{1}{2T} \int_{-T}^{T} f(t) e^{-j2\pi n t/2T} dt = \frac{1}{2T} \int_{-T}^{T} f(t) e^{-jn\pi t/T} dt \tag{8.30}$$

Since $f(t) = 0$ for $t > T$ and $t < -T$, eq. (8.30) can be written as

$$c_n = \frac{1}{2T} \int_{-\infty}^{\infty} f(t) e^{-jn\pi t/T} dt = \frac{1}{2T} F\left(\frac{n\pi}{T}\right) \tag{8.31}$$

where

$$F(\omega) = \mathscr{F}[f(t)] = \int_{-\infty}^{\infty} f(t) e^{-j\omega t} dt$$

and $\omega = n\pi/T$.

Substituting (8.31) into (8.29), we have

$$f(t) = \sum_{n=-\infty}^{\infty} \frac{1}{2T} F\left(\frac{n\pi}{T}\right) e^{jn\pi t/T} \tag{8.32}$$

Now

$$F(\omega) = \int_{-\infty}^{\infty} f(t) e^{-j\omega t} dt = \int_{-T}^{T} f(t) e^{-j\omega t} dt \tag{8.33}$$

in view of the definition (8.2) of a duration-limited signal.

Substituting (8.32) into (8.33) and interchanging the order of summation and integration, we get

$$F(\omega) = \int_{-T}^{T} \left[\sum_{n=-\infty}^{\infty} \frac{1}{2T} F\left(\frac{n\pi}{T}\right) e^{jn\pi t/T} \right] e^{-j\omega t} dt$$

$$= \sum_{n=-\infty}^{\infty} \left[F\left(\frac{n\pi}{T}\right) \frac{1}{2T} \int_{-T}^{T} e^{-j(\omega - n\pi/T)t} dt \right]$$

$$= \sum_{n=-\infty}^{\infty} F\left(\frac{n\pi}{T}\right) \frac{\sin(\omega T - n\pi)}{(\omega T - n\pi)}$$

Hence, we complete the proof of the frequency sampling theorem.

8-4. Modulation

The method of processing a signal for more efficient transmission is called **modulation**. One commonly used type of modulation is based on the **frequency translation theorem** (sometimes called the **modulation theorem**) of the Fourier transform discussed in Chapter 5 (see (5.11)).

A. Frequency translation theorem

The **frequency translation** (or **modulation**) **theorem** states that the multiplication of a signal $f(t)$ by a sinusoidal signal of the frequency ω_c translates its spectrum by $\pm\omega_c$.

EXAMPLE 8-4: Verify the frequency translation theorem.

Solution: Let $\mathscr{F}[f(t)] = F(\omega)$. From the Fourier transform of a cosine function (6.19) and the result of Problem 6-9, we have

$$\mathscr{F}[\cos \omega_c t] = \pi\delta(\omega - \omega_c) + \pi\delta(\omega + \omega_c)$$
$$\mathscr{F}[\sin \omega_c t] = -j\pi\delta(\omega - \omega_c) + j\pi\delta(\omega + \omega_c)$$

Therefore, according to the frequency convolution theorem (7.15),

$$\mathscr{F}[f(t)\cos \omega_c t] = \frac{1}{2\pi} F(\omega) * [\pi\delta(\omega - \omega_c) + \pi\delta(\omega + \omega_c)]$$

Frequency translation theorem (cosine)

$$= \frac{1}{2} F(\omega) * \delta(\omega - \omega_c) + \frac{1}{2} F(\omega) * \delta(\omega + \omega_c)$$

$$= \frac{1}{2} F(\omega - \omega_c) + \frac{1}{2} F(\omega + \omega_c) \tag{8.34}$$

with the use of convolution with δ-function (7.9). Similarly, we have

$$\mathscr{F}[f(t)\sin \omega_c t] = \frac{1}{2\pi} F(\omega) * [-j\pi\delta(\omega - \omega_c) + j\pi\delta(\omega + \omega_c)]$$

Frequency translation theorem (sine)

$$= -\frac{1}{2} j F(\omega) * \delta(\omega - \omega_c) + \frac{1}{2} j F(\omega) * \delta(\omega + \omega_c)$$

$$= -\frac{1}{2} j F(\omega - \omega_c) + \frac{1}{2} j F(\omega + \omega_c) \tag{8.35}$$

B. Modulated signal

Let $m(t)$ be a band-limited signal. Then the signal $f(t)$ given by

$$f(t) = m(t)\cos \omega_c t \tag{8.36}$$

is called the **modulated signal**, where the sinusoid $\cos \omega_c t$ is the **carrier** and $m(t)$ is the **modulating signal**.

EXAMPLE 8-5: Find the spectrum of the modulated signal of (8.36).

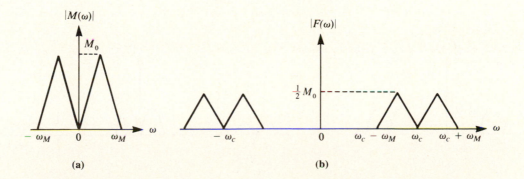

Figure 8-3 (a) The spectrum of $m(t)$; (b) the spectrum of the modulated signal.

Solution: If $\mathscr{F}[m(t)] = M(\omega)$, then

$$F(\omega) = \mathscr{F}[f(t)] = \mathscr{F}[m(t)\cos \omega_c t] = \tfrac{1}{2}[M(\omega - \omega_c) + M(\omega + \omega_c)] \qquad (8.37)$$

in view of the frequency translation theorem (8.34). The spectrum of the modulated signal is shown in Figure 8-3b.

C. Demodulation

The process of separating a modulating signal from a modulated signal is called **demodulation** or **detection**.

EXAMPLE 8-6: Show that the spectrum of the modulated signal can be conveniently retranslated to the original position by multiplying the modulated signal by $\cos \omega_c t$ at the receiving end.

Solution: Let the modulated signal be expressed as

$$f(t) = m(t)\cos \omega_c t$$

Then, as shown in Figure 8-4a, at the receiver we multiply the received signal $f(t)$ by $\cos \omega_c t$ to obtain, by the use of a trigonometric identity,

$$f(t)\cos \omega_c t = m(t)\cos^2 \omega_c t$$

$$= m(t)\frac{1}{2}(1 + \cos 2\omega_c t)$$

$$= \frac{1}{2} m(t) + \frac{1}{2} m(t)\cos 2\omega_c t \qquad (8.38)$$

Now if $\mathscr{F}[m(t)] = M(\omega)$ and $M(\omega) = 0$ for $|\omega| > \omega_M$, then

$$\mathscr{F}[f(t)\cos \omega_c t] = \mathscr{F}[m(t)\cos^2 \omega_c t]$$

$$= \mathscr{F}\left[\frac{1}{2} m(t)\right] + \mathscr{F}\left[\frac{1}{2} m(t)\cos 2\omega_c t\right]$$

$$= \frac{1}{2} M(\omega) + \frac{1}{4} M(\omega - 2\omega_c) + \frac{1}{4} M(\omega - 2\omega_c) \qquad (8.39)$$

The spectrum of $f(t)\cos \omega_c t = m(t)\cos^2 \omega_c t$ is shown in Figure 8-4c. From this you can see that the original signal $m(t)$ can be recovered by using a low-pass filter that passes the spectrum up to ω_M. The demodulation process is diagrammed in Figure 8-4a.

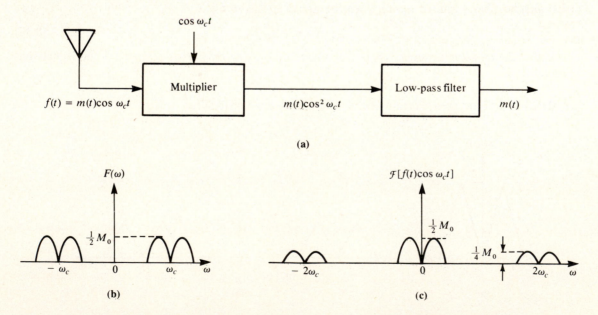

(a)

(b)

(c)

Figure 8-4   (a) Demodulation system; (b) the spectrum of the modulated signal *f*(*t*); (c) the spectrum of *f*(*t*)cos *ω_c t*.

8-5. Average Correlation Functions

A. Definitions used in average correlation functions

The concept of correlation functions was introduced in Section 7-4. For periodic or random noise signals that exist over the entire time interval $(-\infty, \infty)$, the energy content will be infinite; that is,

$$\int_{-\infty}^{\infty} [f(t)]^2 \, dt = \infty$$

Thus, it is obvious that the correlation functions as defined in Section 7-4 do not exist. In such cases, we consider the following average correlation functions.

The **average autocorrelation function** of $f_1(t)$, denoted by $\bar{R}_{11}(\tau)$, is defined as the limit

Average autocorrelation functions

$$\bar{R}_{11}(\tau) = \lim_{T \to \infty} \frac{1}{T} \int_{-T/2}^{T/2} f_1(t) f_1(t - \tau) \, dt \tag{8.40}$$

Similarly, the **average cross-correlation function** of $f_1(t)$ and $f_2(t)$, denoted by $\bar{R}_{12}(\tau)$, is defined as the limit

Average cross-correlation function

$$\bar{R}_{12}(\tau) = \lim_{T \to \infty} \frac{1}{T} \int_{-T/2}^{T/2} f_1(t) f_2(t - \tau) \, dt \tag{8.41}$$

B. Periodic signals

For periodic signals (with period T_1), eqs. (8.40) and (8.41) reduce to

$$\bar{R}_{11}(\tau) = \frac{1}{T_1} \int_{-T_1/2}^{T_1/2} f_1(t) f_1(t - \tau) \, dt \tag{8.42}$$

$$\bar{R}_{12}(\tau) = \frac{1}{T_1} \int_{-T_1/2}^{T_1/2} f_1(t) f_2(t - \tau) \, dt \tag{8.43}$$

EXAMPLE 8-7: Derive (8.42) and (8.43).

Solution: Let $f_1(t)$ and $f_2(t)$ both be periodic functions of period T_1. Then

$$f_1(t) = f_1(t + T_1) \tag{8.44}$$

$$f_1(t - \tau) = f_1(t - \tau + T_1) \tag{8.45}$$

$$f_2(t - \tau) = f_2(t - \tau + T_1) \tag{8.46}$$

Hence, the integrands in (8.40) and (8.41) are periodic functions in variable t with period T_1. The integral of such a function over each period is the same; therefore, it is immaterial whether the correlation functions are averaged over a very large interval $T \to \infty$, or over one period T_1. Thus, for periodic signals,

$$\bar{R}_{11}(\tau) = \lim_{T \to \infty} \frac{1}{T} \int_{-T/2}^{T/2} f_1(t) f_1(t - \tau) \, dt = \frac{1}{T_1} \int_{-T_1/2}^{T_1/2} f_1(t) f_1(t - \tau) \, dt$$

$$\bar{R}_{12}(\tau) = \lim_{T \to \infty} \frac{1}{T} \int_{-T/2}^{T/2} f_1(t) f_2(t - \tau) \, dt = \frac{1}{T_1} \int_{-T_1/2}^{T_1/2} f_1(t) f_2(t - \tau) \, dt$$

EXAMPLE 8-8: Show that the average autocorrelation and cross-correlation functions of periodic signals of period T_1 are also periodic with the same period.

Solution: From (8.42),

$$\bar{R}_{11}(\tau - T_1) = \frac{1}{T_1} \int_{-T_1/2}^{T_1/2} f_1(t) f_1[t - (\tau - T_1)] \, dt$$

$$= \frac{1}{T_1} \int_{-T_1/2}^{T_1/2} f_1(t) f_1(t - \tau + T_1) \, dt$$

But, from (8.45), we obtain

$$\bar{R}_{11}(\tau - T_1) = \frac{1}{T_1} \int_{-T_1/2}^{T_1/2} f_1(t) f_1(t - \tau) \, dt = \bar{R}_{11}(\tau) \tag{8.47}$$

Similarly, from (8.43) and (8.46), we obtain

$$\bar{R}_{12}(\tau - T_1) = \frac{1}{T_1} \int_{-T_1/2}^{T_1/2} f_1(t) f_2[t - (\tau - T_1)] \, dt$$

$$= \frac{1}{T_1} \int_{-T_1/2}^{T_1/2} f_1(t) f_2(t - \tau + T_1) \, dt$$

$$= \frac{1}{T_1} \int_{-T_1/2}^{T_1/2} f_1(t) f_2(t - \tau) \, dt$$

$$= \bar{R}_{12}(\tau) \tag{8.48}$$

Equations (8.47) and (8.48) show that $\bar{R}_{11}(\tau)$ and $\bar{R}_{12}(\tau)$ are periodic with the period T_1.

EXAMPLE 8-9: Find the average autocorrelation function of the sine wave given by

$$f(t) = A \sin(\omega_1 t + \phi), \qquad \omega_1 = \frac{2\pi}{T_1}$$

Solution: Since $f(t)$ is periodic, from (8.42),

$$\bar{R}_{ff}(\tau) = \lim_{T \to \infty} \frac{1}{T} \int_{-T/2}^{T/2} f(t) f(t - \tau) \, dt$$

$$= \frac{1}{T_1} \int_{-T_1/2}^{T_1/2} f(t) f(t - \tau) \, dt$$

$$= \frac{A^2}{T_1} \int_{-T_1/2}^{T_1/2} \sin(\omega_1 t + \phi) \sin[\omega_1(t - \tau) + \phi] \, dt$$

$$= \frac{A^2}{T_1} \int_{-T_1/2}^{T_1/2} \sin(\omega_1 t + \phi) \sin(\omega_1 t + \phi - \omega_1 \tau) \, dt$$

Using the trigonometric identity $\sin A \sin B = \frac{1}{2}[\cos(A - B) - \cos(A + B)]$, we obtain

$$\bar{R}_{ff}(\tau) = \frac{A^2}{2T_1} \int_{-T_1/2}^{T_1/2} [\cos \omega_1 \tau - \cos(2\omega_1 t + 2\phi - \omega_1 \tau)] \, dt$$

$$= \frac{A^2}{2T_1} \cos \omega_1 \tau \int_{-T_1/2}^{T_1/2} dt$$

$$= \frac{A^2}{2} \cos(\omega_1 \tau) \tag{8.49}$$

Equation (8.49) shows that $\bar{R}_{ff}(\tau)$ is independent of the phase ϕ of $f(t)$ and periodic with period T_1.

C. Uncorrelated signals

In general, two signals $f_1(t)$ and $f_2(t)$ are said to be **uncorrelated** if

Uncorrelated signals

$$\bar{R}_{12}(\tau) = \lim_{T \to \infty} \frac{1}{T} \int_{-T/2}^{T/2} f_1(t) f_2(t - \tau) \, dt$$

$$= \left[\lim_{T \to \infty} \frac{1}{T} \int_{-T/2}^{T/2} f_1(t) \, dt \right] \left[\lim_{T \to \infty} \frac{1}{T} \int_{-T/2}^{T/2} f_2(t) \, dt \right] \tag{8.50}$$

Thus, if one signal, say $f_2(t)$, is assumed to have a zero average value, that is

$$\lim_{T \to \infty} \frac{1}{T} \int_{-T/2}^{T/2} f_2(t) \, dt = 0 \tag{8.51}$$

then

$$\bar{R}_{12}(\tau) = 0 \qquad \text{for all } \tau \tag{8.52}$$

8-6. Average Power Spectra: Random Signals

A. Random signals

In Section 7-3 we introduced the idea of the energy spectrum, or energy-density function of $f(t)$. There we assumed that the energy content of $f(t)$ is finite, that is,

$$\int_{-\infty}^{\infty} [f(t)]^2 \, dt = \text{finite} \tag{8.53}$$

For such functions, the average power over the interval T approaches zero as T approaches infinity, thus,

$$\lim_{T \to \infty} \frac{1}{T} \int_{-T/2}^{T/2} [f(t)]^2 \, dt = 0 \tag{8.54}$$

For signals that do not have finite energy content, we define the average power of $f(t)$ by

$$\lim_{T \to \infty} \frac{1}{T} \int_{-T/2}^{T/2} [f(t)]^2 \, dt \tag{8.55}$$

When this limit exists, the quantity

Power spectral density
$$P(\omega) = \lim_{T \to \infty} \frac{1}{T} \left| \int_{-T/2}^{T/2} f(t) e^{-j\omega t} \, dt \right|^2 \tag{8.56}$$

is called the **power spectrum**, or the **power spectral density**, of the function $f(t)$.

If the power spectral density of a function $f(t)$ only is specified, we do not know the waveform because only a time-average spectrum is known. Signals specified in this fashion may be called **random signals**. Random signals are usually described in terms of their statistical properties, but we won't discuss these properties here.

B. Definition of the power spectral density

Although the quantity (8.56) is referred to as the power spectral density of the function $f(t)$, the **power spectral density** (or simply spectral density) of the function $f(t)$ is usually defined as the Fourier transform of the average autocorrelation function of $f(t)$. Thus, we define

Power spectral density
$$P(\omega) = \mathscr{F}[\bar{R}_{ff}(\tau)] = \int_{-\infty}^{\infty} \bar{R}_{ff}(\tau) e^{-j\omega t} \, d\tau \tag{8.57}$$

Then,

$$\bar{R}_{ff}(\tau) = \mathscr{F}^{-1}[P(\omega)] = \frac{1}{2\pi} \int_{-\infty}^{\infty} P(\omega) e^{j\omega \tau} \, d\omega \tag{8.58}$$

With these definitions, we can show that

$$\lim_{T \to \infty} \frac{1}{T} \int_{-T/2}^{T/2} [f(t)]^2 \, dt = \frac{1}{2\pi} \int_{-\infty}^{\infty} P(\omega) \, d\omega = \int_{-\infty}^{\infty} P(2\pi v) \, dv \tag{8.59}$$

where $\omega = 2\pi v$.

Equation (8.59) states that the total average power (or the mean-square value) of a function $f(t)$ is given by the integration of $P(\omega)$ over the entire frequency range. For this reason the quantity $P(\omega)$ is called the power spectrum or power spectral density of $f(t)$.

EXAMPLE 8-10: Prove (8.59)

Proof: It follows from (8.58) that

$$\bar{R}_{ff}(0) = \frac{1}{2\pi} \int_{-\infty}^{\infty} P(\omega) d\omega = \int_{-\infty}^{\infty} P(2\pi v) dv \tag{8.60}$$

Now from the average autocorrelation function (8.40), we have

$$\bar{R}_{ff}(\tau) = \lim_{T \to \infty} \frac{1}{T} \int_{-T/2}^{T/2} f(t) f(t - \tau) \, dt$$

Hence,

$$\bar{R}_{ff}(0) = \lim_{T \to \infty} \frac{1}{T} \int_{-T/2}^{T/2} [f(t)]^2 \, dt \tag{8.61}$$

Comparing (8.61) and (8.60), we prove that

$$\lim_{T \to \infty} \frac{1}{T} \int_{-T/2}^{T/2} [f(t)]^2 \, dt = \frac{1}{2\pi} \int_{-\infty}^{\infty} P(\omega) \, d\omega = \int_{-\infty}^{\infty} P(2\pi v) \, dv$$

C. White noise

White noise is defined as any random signal whose power spectral density is a constant (independent of frequency).

EXAMPLE 8-11: Find the average autocorrelation function of white noise.

Solution: From the definition of white noise

$$P(\omega) = K \tag{8.62}$$

It follows from (8.58) that

$$\bar{R}(\tau) = \mathcal{F}^{-1}[P(\omega)] = \frac{1}{2\pi} \int_{-\infty}^{\infty} P(\omega) e^{j\omega\tau} \, d\omega = K \frac{1}{2\pi} \int_{-\infty}^{\infty} e^{j\omega\tau} \, d\omega$$

From the integral representation of the δ-function (6.27),

$$\frac{1}{2\pi} \int_{-\infty}^{\infty} e^{j\omega\tau} \, d\omega = \delta(\tau)$$

we have

$$\bar{R}(\tau) = K\delta(\tau) \tag{8.63}$$

Hence, the average autocorrelation function for white noise is found to be an impulse.

8-7. Analytic Signals and Hilbert Transforms

A. Analytic signals

An **analytic signal** is a complex-valued signal whose Fourier spectrum is one-sided. Consider a given real signal $f(t)$. Then the analytic signal $f_+(t)$ associated with $f(t)$ is defined by

Analytic signal
$$f_+(t) = f(t) + j\hat{f}(t) \tag{8.64}$$

(where $\hat{f}(t)$ is yet to be determined), such that

$$\mathcal{F}[f_+(t)] = F_+(\omega) = 0 \quad \text{for} \quad \omega < 0 \tag{8.65}$$

EXAMPLE 8-12: Find the Fourier transform of $\hat{f}(t)$ of (8.64) such that (8.65) is satisfied.

Solution: Let $\mathcal{F}[f(t)] = F(\omega)$ and $\mathcal{F}[\hat{f}(t)] = \hat{F}(\omega)$. Then from definition (8.64) and the linearity property of the Fourier transform (5.5), we have

$$F_+(\omega) = F(\omega) + j\hat{F}(\omega) \tag{8.66}$$

Since, by definition (8.65),

$$F_+(\omega) = 0 \quad \text{for} \quad \omega < 0$$

we obtain

$$j\hat{F}(\omega) = -F(\omega) \quad \text{for} \quad \omega < 0$$

or

$$\hat{F}(\omega) = jF(\omega) \quad \text{for} \quad \omega < 0 \tag{8.67}$$

Since $j = e^{j\pi/2}$ and $e^{-j\pi/2} = -j$, to maintain a phase characteristic that is an odd function of ω (see

Problem 5-10) then requires that

$$\hat{F}(\omega) = -jF(\omega) \qquad \text{for} \quad \omega > 0 \tag{8.68}$$

Combining (8.67) and (8.68), we obtain

$$\hat{F}(\omega) = \begin{cases} -jF(\omega), & \omega > 0 \\ jF(\omega), & \omega < 0 \end{cases}$$

$$= -j\,\text{sgn}\,\omega\,F(\omega) \tag{8.69}$$

where

$$\text{sgn}\,\omega = \begin{cases} 1, & \omega > 0 \\ -1, & \omega < 0 \end{cases}$$

EXAMPLE 8-13: Show that if $\mathcal{F}[f(t)] = F(\omega)$, then

$$\mathcal{F}[f_+(t)] = F_+(\omega) = 2F(\omega)u(\omega) = \begin{cases} 2F(\omega), & \omega > 0 \\ 0, & \omega < 0 \end{cases} \tag{8.70}$$

where $u(\omega)$ is the unit step function.

Solution: From (8.66) and (8.69), we have

$$F_+(\omega) = F(\omega) + j\hat{F}(\omega)$$

$$= F(\omega) - j^2\,\text{sgn}\,\omega\,F(\omega)$$

$$= F(\omega)(1 + \text{sgn}\,\omega)$$

Since

$$1 + \text{sgn}\,\omega = \begin{cases} 2, & \omega > 0 \\ 0, & \omega < 0 \end{cases}$$

we obtain

$$F_+(\omega) = \begin{cases} 2F(\omega), & \omega > 0 \\ 0, & \omega < 0 \end{cases}$$

$$= 2F(\omega)u(\omega)$$

B. Hilbert transforms

The function $\hat{f}(t)$ of the analytic signal (8.64) can be expressed as

Hilbert transform
$$\hat{f}(t) = \frac{1}{\pi}\int_{-\infty}^{\infty}\frac{f(\tau)}{t-\tau}\,d\tau \tag{8.71}$$

The function $\hat{f}(t)$ defined by (8.71) is called the **Hilbert transform** of $f(t)$.

EXAMPLE 8-14: Verify (8.71).

Solution: Let $\mathcal{F}[\hat{f}(t)] = \hat{F}(\omega)$. Then, from (8.69), we have

$$\hat{F}(\omega) = -j\,\text{sgn}\,\omega\,F(\omega)$$

where $F(\omega) = \mathcal{F}[f(t)]$.

Now, from the result of Problem 6-14, we have

$$\text{sgn}\,t \leftrightarrow \frac{2}{j\omega}$$

Applying the symmetry property of the Fourier transform (5.10), we have

$$\frac{2}{jt} \leftrightarrow 2\pi\,\text{sgn}(-\omega) = -2\pi\,\text{sgn}\,\omega$$

Thus,

$$\frac{1}{\pi t} \leftrightarrow -j\,\text{sgn}\,\omega$$

Application of the time convolution theorem (7.12) yields (8.71):

$$\hat{f}(t) = \mathscr{F}^{-1}[\hat{F}(\omega)] = \mathscr{F}^{-1}[-j \operatorname{sgn} \omega F(\omega)]$$

$$= \frac{1}{\pi t} * f(t) = \frac{1}{\pi} \int_{-\infty}^{\infty} \frac{f(\tau)}{t - \tau} d\tau$$

EXAMPLE 8-15: Show that $f(t)$ and $\hat{f}(t)$ are orthogonal; that is,

$$\int_{-\infty}^{\infty} f(t)\hat{f}(t) \, dt = 0 \qquad (8.72)$$

Solution: Using (7.17), we have

$$\int_{-\infty}^{\infty} f(t)\hat{f}(t) \, dt = \frac{1}{2\pi} \int_{-\infty}^{\infty} F(\omega)\hat{F}^*(\omega) \, d\omega \qquad (8.73)$$

where $\hat{F}^*(\omega)$ is the complex conjugate of $\hat{F}(\omega)$.

From (8.69), we have

$$\hat{F}^*(\omega) = +j \operatorname{sgn} \omega F^*(\omega)$$

and (8.73) reduces to

$$\int_{-\infty}^{\infty} f(t)\hat{f}(t) \, dt = \frac{1}{2\pi} \int_{-\infty}^{\infty} (+j \operatorname{sgn} \omega)|F(\omega)|^2 \, d\omega \qquad (8.74)$$

But the integrand of the right side of (8.74) is odd, being the product of the even function $|F(\omega)|^2$ and the odd function of $+j \operatorname{sgn} \omega$. Therefore, the integral is zero; that is,

$$\int_{-\infty}^{\infty} f(t)\hat{f}(t) \, dt = 0$$

SUMMARY

1. The uncertainty principle in spectral analysis is stated by

$$W_B T_D \geqslant K$$

where W_B is the spectral bandwidth and T_D is the time duration of a signal, and K is a constant whose value depends on the selection of W_B and T_D.

2. The sampling theorem in the time domain states that if $F(\omega) = \mathscr{F}[f(t)] = 0$ for $|\omega| > \omega_M$, then it is uniquely determined by

$$f(t) = \sum_{n=-\infty}^{\infty} f\left(\frac{n\pi}{\omega_M}\right) \frac{\sin(\omega_M t - n\pi)}{(\omega_M t - n\pi)}$$

3. The sampling theorem in the frequency domain states that if $f(t) = 0$ for $|t| > T$, then $F(\omega)$ can be uniquely determined by

$$F(\omega) = \sum_{n=-\infty}^{\infty} F\left(\frac{n\pi}{T}\right) \frac{\sin(\omega T - n\pi)}{(\omega T - n\pi)}$$

4. The average autocorrelation function of $f_1(t)$ is defined by

$$\bar{R}_{11}(\tau) = \lim_{T \to \infty} \frac{1}{T} \int_{-T/2}^{T/2} f_1(t)f_1(t - \tau) \, dt$$

and the average cross-correlation function of $f_1(t)$ and $f_2(t)$ is defined by

$$\bar{R}_{12}(\tau) = \lim_{T \to \infty} \frac{1}{T} \int_{-T/2}^{T/2} f_1(t)f_2(t - \tau) \, dt$$

5. The power spectral density of a random signal $f(t)$ is defined by

$$P(\omega) = \mathscr{F}[\bar{R}_{ff}(\tau)] = \int_{-\infty}^{\infty} \bar{R}_{ff}(\tau)e^{-j\omega\tau} \, d\tau$$

6. The analytic signal $f_+(t)$ associated with a real signal $f(t)$ is defined by

$$f_+(t) = f(t) + j\hat{f}(t)$$

where $\hat{f}(t)$ is the Hilbert transform of $f(t)$ defined by

$$\hat{f}(t) = \frac{1}{\pi} \int_{-\infty}^{\infty} \frac{f(\tau)}{t - \tau} \, d\tau$$

and

$$F_+(\omega) = \mathscr{F}[f_+(t)] = 2F(\omega)u(\omega)$$

RAISE YOUR GRADES

Can you explain...?

☑ how to select the sampling interval
☑ how the modulation and demodulation of a signal are accomplished
☑ how to find the power spectral density of a random signal
☑ how to find an analytic signal associated with a given real signal

SOLVED PROBLEMS

Band-Limited and Duration-Limited Signals

PROBLEM 8-1   Show that if $f(t)$ is a band-limited signal with no spectral components above the frequency ω_M, then the signal $f(t)\cos\omega_c t$ is also band-limited.

Solution: Since the signal $f(t)$ is a band-limited signal, we know, by definition (8.1), that

$$\mathscr{F}[f(t)] = F(\omega) = 0 \qquad \text{for} \quad |\omega| > \omega_M$$

From the results (8.34) of Example 8-4 and Figure 8-3, it follows, therefore, that the signal $f(t)\cos\omega_c t$ is also band-limited and that its spectrum is zero for $|\omega| > \omega_c + \omega_M$.

PROBLEM 8-2   Show that if $f(t)$ is band-limited and

$$F(\omega) = 0 \qquad \text{for} \quad |\omega| > \omega_M$$

then

$$f(t) * \frac{\sin at}{\pi t} = f(t)$$

for every $a > \omega_M$.

Solution: From the result of Problem 5-6, we have

$$\mathscr{F}\left(\frac{\sin at}{\pi t}\right) = p_{2a}(\omega)$$

where

$$p_{2a}(\omega) = \begin{cases} 1 & \text{for} \quad |\omega| < a \\ 0 & \text{for} \quad |\omega| > a \end{cases}$$

Thus, applying the time convolution theorem (7.11), we have

$$\mathscr{F}\left[f(t) * \frac{\sin at}{\pi t}\right] = F(\omega)p_{2a}(\omega)$$

Now, if $a > \omega_M$, then

$$F(\omega)p_{2a}(\omega) = F(\omega)$$

Hence, by taking the inverse Fourier transform, we obtain

$$f(t) * \frac{\sin at}{\pi t} = f(t)$$

for every $a > \omega_M$.

Uncertainty Principle

PROBLEM 8-3 Consider the rectangular pulse $p_d(t)$ of Problem 5-2. Show that the product of spectral bandwidth and pulse duration is a constant with "appropriate" selection of some measure of the bandwidth.

Solution: Referring to Figure 5-2, it is common to select the spectral bandwidth W_B of $p_d(t)$ as the frequency range to the first zero of $F(\omega)$, since most of the energy of the pulse is included within this range. Thus,

$$W_B = \frac{4\pi}{d}$$

Since the pulse duration of $p_d(t)$ is d, we obtain

$$W_B d = 4\pi$$

that is, the product of the bandwidth and the pulse duration is a constant.

PROBLEM 8-4 Using $f(t) = e^{-a|t|}$, $a > 0$, verify the relation (8.5) of the uncertainty principle.

Solution: From the result of Problem 5-3, we have

$$F(\omega) = \mathscr{F}[e^{-a|t|}] = \frac{2a}{a^2 + \omega^2}$$

Thus,

$$\int_{-\infty}^{\infty} |f(t)|\, dt = \int_{-\infty}^{\infty} e^{-a|t|}\, dt = \int_{-\infty}^{0} e^{at}\, dt + \int_{0}^{\infty} e^{-at}\, dt = \frac{2}{a}$$

From (7.24) of Example 7-11, we have

$$\int_{-\infty}^{\infty} |F(\omega)|\, d\omega = \int_{-\infty}^{\infty} \frac{2a}{a^2 + \omega^2}\, d\omega = 2\pi$$

Since $f(0) = 1$ and $F(0) = 2/a$, by definitions (8.3) and (8.4), we have

$$T_D = \frac{1}{f(0)} \int_{-\infty}^{\infty} |f(t)|\, dt = \frac{2}{a}$$

and

$$W_B = \frac{1}{F(0)} \int_{-\infty}^{\infty} |F(\omega)|\, d\omega = a\pi$$

Hence,

$$W_B T_D = 2\pi$$

Sampling Theorems

PROBLEM 8-5 Consider the sampling functions

$$\phi_n(t) = \frac{\sin \omega_M(t - nT)}{\omega_M(t - nT)}, \qquad n = 0, \pm 1, \pm 2, \ldots$$

where $T = \pi/\omega_M$. Show that $\phi_n(t)$ are orthogonal over the interval $-\infty < t < \infty$, and

$$\int_{-\infty}^{\infty} \phi_n(t)\phi_m(t)\,dt = T\delta_{nm}$$

where δ_{nm} is Kronecker's delta.

Solution: From the result of Problem 5-6, we have

$$\frac{\sin at}{\pi t} \leftrightarrow p_{2a}(\omega) = \begin{cases} 1 & \text{for } |\omega| < a \\ 0 & \text{for } |\omega| > a \end{cases}$$

Thus,

$$\frac{\sin \omega_M t}{\omega_M t} \leftrightarrow \frac{\pi}{\omega_M} p_{2\omega_M}(\omega) = \begin{cases} \dfrac{\pi}{\omega_M}, & |\omega| < \omega_M \\ 0, & |\omega| > \omega_M \end{cases}$$

Then by the time-shifting property (5.6), we have

$$\phi_n(t) = \frac{\sin \omega_M(t - nT)}{\omega_M(t - nT)} \leftrightarrow \begin{cases} \dfrac{\pi}{\omega_M} e^{-j\omega nT}, & |\omega| < \omega_M \\ 0, & |\omega| > \omega_M \end{cases}$$

$$\phi_m(t) = \frac{\sin \omega_M(t - mT)}{\omega_M(t - mT)} \leftrightarrow \begin{cases} \dfrac{\pi}{\omega_M} e^{-j\omega mT}, & |\omega| < \omega_M \\ 0, & |\omega| > \omega_M \end{cases}$$

Thus, by (7.17) of Example 7-9, that is

$$\int_{-\infty}^{\infty} f_1(t)f_2(t)\,dt = \frac{1}{2\pi} \int_{-\infty}^{\infty} F_1(\omega)F_2^*(\omega)\,d\omega$$

we obtain

$$\int_{-\infty}^{\infty} \phi_n(t)\phi_m(t)\,dt = \frac{1}{2\pi} \int_{-\omega_M}^{\omega_M} \left(\frac{\pi}{\omega_M}\right)^2 e^{-j\omega nT} e^{j\omega mT}\,d\omega$$

$$= \frac{\pi}{2\omega_M^2} \int_{-\omega_M}^{\omega_M} e^{-j\omega(n-m)T}\,d\omega$$

$$= \begin{cases} 0, & n \neq m \\ \dfrac{\pi}{\omega_M} = T, & n = m \end{cases}$$

Hence, we conclude that

$$\int_{-\infty}^{\infty} \phi_n(t)\phi_m(t)\,dt = T\delta_{nm}$$

where

$$\delta_{nm} = \begin{cases} 1, & n = m \\ 0, & n \neq m \end{cases}$$

PROBLEM 8-6 If $f(t)$ is band-limited, that is, $F(\omega) = \mathscr{F}[f(t)] = 0$ for $|\omega| > \omega_M$, show that

$$\int_{-\infty}^{\infty} f(t)\phi_n(t)\,dt = Tf(nT)$$

where $\phi_n(t)$ is the sampling function of Problem 8-5 and $T = \pi/\omega_M$.

Solution: From (8.27), we can express $f(t)$ as

$$f(t) = \sum_{n=-\infty}^{\infty} f(nT)\phi_n(t) = \sum_{m=-\infty}^{\infty} f(mT)\phi_m(t)$$

Thus, with the result of Problem 8-5, we obtain

$$\int_{-\infty}^{\infty} f(t)\phi_n(t)\,dt = \int_{-\infty}^{\infty}\left[\sum_{m=-\infty}^{\infty} f(mT)\phi_m(t)\right]\phi_n(t)\,dt$$

$$= \sum_{m=-\infty}^{\infty} f(mT)\int_{-\infty}^{\infty}\phi_m(t)\phi_n(t)\,dt$$

$$= \sum_{m=-\infty}^{\infty} f(mT)T\delta_{nm} = Tf(nT)$$

PROBLEM 8-7   If $f(t)$ is band-limited, that is, $F(\omega) = \mathscr{F}[f(t)] = 0$ for $|\omega| > \omega_M$, prove that

$$T\sum_{n=-\infty}^{\infty}[f(nT)]^2 = \int_{-\infty}^{\infty}[f(t)]^2\,dt$$

where $T = \pi/\omega_M$.

Solution: Using the result of Problem 8-6, we have

$$\int_{-\infty}^{\infty}[f(t)]^2\,dt = \int_{-\infty}^{\infty}\left[\sum_{n=-\infty}^{\infty} f(nT)\phi_n(t)\right]\left[\sum_{m=-\infty}^{\infty} f(mT)\phi_m(t)\right]dt$$

$$= \sum_{n=-\infty}^{\infty} f(nT)\left[\sum_{m=-\infty}^{\infty} f(mT)\int_{-\infty}^{\infty}\phi_n(t)\phi_m(t)\,dt\right]$$

$$= \sum_{n=-\infty}^{\infty} f(nT)\left[\sum_{m=-\infty}^{\infty} f(mT)T\delta_{nm}\right]$$

$$= T\sum_{n=-\infty}^{\infty}[f(nT)]^2$$

PROBLEM 8-8   Show that a band-limited periodic function, with no harmonics of order higher than N, is uniquely specified by its values at $2N + 1$ instants in one period.

Solution: A band-limited periodic function $f(t)$ with no harmonics of order higher than N can be expressed as a trigonometric Fourier series (see Section 1-2):

$$f(t) = C_0 + \sum_{n=1}^{N} C_n\cos(n\omega_0 t - \theta_n)$$

Now there are $2N + 1$ unknowns; that is,

$$C_0, C_1, \ldots, C_N \quad \text{and} \quad \theta_1, \theta_2, \ldots, \theta_N$$

Thus, if we substitute the values of $f(t)$ at $2N + 1$ instants in one period into the trigonometric Fourier series, we obtain $2N + 1$ simultaneous algebraic equations from which we can find the $2N + 1$ unknowns and $f(t)$ is uniquely determined.

Modulation

PROBLEM 8-9   An ordinary amplitude-modulated (AM) signal is usually written in the form

$$f(t) = K[1 + m(t)]\cos\omega_c t$$

where $m(t)$ is a band-limited signal, such that

$$\mathscr{F}[m(t)] = M(\omega) = 0 \quad \text{for} \quad |\omega| > \omega_M, \omega_c > \omega_M, \quad \text{and} \quad |m(t)| < 1$$

Find the frequency spectrum of an ordinary AM signal.

Solution: Using the Fourier transform of the cosine function (6.19) and the frequency translation theorem (8.34), the Fourier transform of $f(t)$ is given by

$$F(\omega) = \mathscr{F}[f(t)]$$

$$= \mathscr{F}\{K[1 + m(t)]\cos\omega_c t\}$$

$$= \mathscr{F}[K\cos\omega_c t] + \mathscr{F}[Km(t)\cos\omega_c t]$$

$$= K\pi\delta(\omega - \omega_c) + K\pi\delta(\omega + \omega_c) + \tfrac{1}{2}KM(\omega - \omega_c) + \tfrac{1}{2}KM(\omega + \omega_c)$$

where $\mathscr{F}[m(t)] = M(\omega)$.

Thus, the spectrum of $f(t)$ consists of impulses at the carrier frequency ω_c and the frequency-translated spectrum of $m(t)$ (see Figure 8-5).

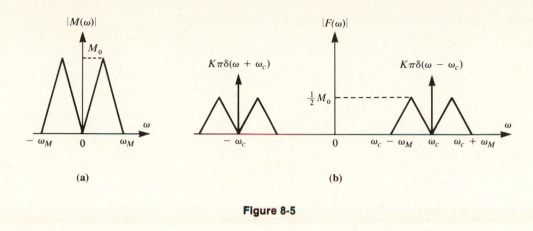

(a) (b)

Figure 8-5

PROBLEM 8-10 Let $m(t)$ be a band-limited signal with $M(\omega) = \mathscr{F}[m(t)] = 0$ for $|\omega| > \omega_M$ and let $g(t)$ be a periodic pulse train with period T. Then the product $f(t) = m(t)g(t)$ is called a **pulse amplitude modulated (PAM)** signal, provided that $T \leqslant \pi/\omega_M$. Find the spectrum of $f(t)$.

Solution: Since $g(t)$ is a periodic function, it can be expanded into a Fourier series; thus

$$g(t) = \sum_{n=-\infty}^{\infty} c_n e^{jn\omega_0 t}, \qquad \omega_0 = \frac{2\pi}{T}$$

Then the PAM signal $f(t)$ can be written as

$$f(t) = m(t)\left(\sum_{n=-\infty}^{\infty} c_n e^{jn\omega_0 t} \right) = \sum_{n=-\infty}^{\infty} c_n m(t) e^{jn\omega_0 t}$$

Thus,

$$F(\omega) = \mathscr{F}[f(t)] = \mathscr{F}\left[\sum_{n=-\infty}^{\infty} c_n m(t) e^{jn\omega_0 t} \right] = \sum_{n=-\infty}^{\infty} c_n \mathscr{F}[m(t) e^{jn\omega_0 t}]$$

Now according to the frequency-shifting property of the Fourier transform (5.7), if $\mathscr{F}[m(t)] = M(\omega)$, then

$$\mathscr{F}[m(t) e^{jn\omega_0 t}] = M(\omega - n\omega_0)$$

Hence,

$$F(\omega) = \sum_{n=-\infty}^{\infty} c_n M(\omega - n\omega_0)$$

Figure 8-6b illustrates the amplitude spectrum of the PAM signal, which consists of periodically spaced pulses with amplitude modified by the Fourier coefficients of $g(t)$. In the figure, ω_0 is selected such that $T < \pi/\omega_M$.

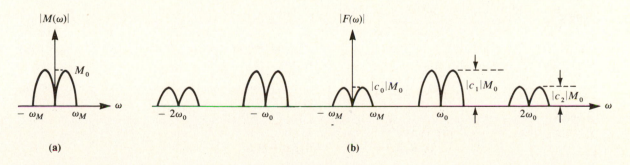

(a) (b)

Figure 8-6

PROBLEM 8-11 Show that demodulation may also be accomplished by multiplying the modulated signal $f(t) = m(t)\cos\omega_c t$ by any periodic signal of frequency ω_c.

Solution: If $p(t)$ is a periodic signal of frequency ω_c and of the form

$$p(t) = \sum_{n=-\infty}^{\infty} c_n e^{jn\omega_c t}$$

then from the Fourier transform of a complex exponential function (6.18), its Fourier transform can be written as

Now from (8.37),

$$\mathscr{F}[f(t)] = \frac{1}{2}M(\omega - \omega_c) + \frac{1}{2}M(\omega + \omega_c)$$

Hence, according to the frequency convolution theorem (7.15), the Fourier transform of $f(t)p(t)$ is given by

$$\mathscr{F}[f(t)p(t)] = \frac{1}{2}[M(\omega - \omega_c) + M(\omega + \omega_c)] * \sum_{n=-\infty}^{\infty} c_n \delta(\omega - n\omega_c)$$

$$= \frac{1}{2} \sum_{n=-\infty}^{\infty} c_n [M(\omega - \omega_c) + M(\omega + \omega_c)] * \delta(\omega - n\omega_c)$$

$$= \frac{1}{2} \sum_{n=-\infty}^{\infty} c_n \{M[\omega - (n+1)\omega_c] + M[\omega - (n-1)\omega_c]\}$$

with the use of (7.9). It is obvious that this spectrum contains a term $M(\omega)$, the spectrum of $m(t)$. This can be recovered by using a low-pass filter that passes only up to $\omega = \omega_M$.

Average Correlation Functions

PROBLEM 8-12 Show that if $f_1(t)$ and $f_2(t)$ are real periodic functions having the same period T_1, then

$$\bar{R}_{12}(\tau) = \sum_{n=-\infty}^{\infty} [c_{1n}^* c_{2n}] e^{-jn\omega_1 \tau}$$

where $\omega_1 = 2\pi/T_1$ and c_{1n}, c_{2n} are the complex Fourier coefficients of $f_1(t)$ and $f_2(t)$, respectively, and c_{1n}^* denotes complex conjugate of c_{1n}.

Solution: In the case of periodic functions, from (8.43), we have

$$\bar{R}_{12}(\tau) = \frac{1}{T_1} \int_{-T_1/2}^{T_1/2} f_1(t) f_2(t - \tau) \, dt$$

Let the complex Fourier series expansions for $f_1(t)$ and $f_2(t)$ be

$$f_1(t) = \sum_{n=-\infty}^{\infty} c_{1n} e^{jn\omega_1 t}$$

$$f_2(t) = \sum_{n=-\infty}^{\infty} c_{2n} e^{jn\omega_1 t}$$

where

$$c_{1n} = \frac{1}{T_1} \int_{-T_1/2}^{T_1/2} f_1(t) e^{-jn\omega_1 t} \, dt \qquad \textbf{(a)}$$

$$c_{2n} = \frac{1}{T_1} \int_{-T_1/2}^{T_1/2} f_2(t) e^{-jn\omega_1 t} \, dt \qquad \textbf{(b)}$$

Then

$$\bar{R}_{12}(\tau) = \frac{1}{T_1} \int_{-T_1/2}^{T_1/2} f_1(t) f_2(t - \tau) \, dt$$

$$= \frac{1}{T_1} \int_{-T_1/2}^{T_1/2} f_1(t) \left[\sum_{n=-\infty}^{\infty} c_{2n} e^{jn\omega_1(t-\tau)} \right] dt$$

Interchanging the order of summation and integration,

$$\bar{R}_{12}(\tau) = \sum_{n=-\infty}^{\infty} c_{2n} e^{-jn\omega_1\tau} \left[\frac{1}{T_1} \int_{-T_1/2}^{T_1/2} f_1(t) e^{jn\omega_1 t}\, dt \right]$$

The integral in the bracket is recognized, by comparison with eq. (a), as the complex conjugate of c_{1n}. Hence,

$$\bar{R}_{12}(\tau) = \sum_{n=-\infty}^{\infty} [c_{1n}^* c_{2n}] e^{-jn\omega_1\tau}$$

Note that $\bar{R}_{12}(\tau)$ is also a periodic function of τ with period T_1.

PROBLEM 8-13   Show that if $f(t)$ is a real periodic function having the period T, then

$$\bar{R}_{ff}(\tau) = \sum_{n=-\infty}^{\infty} |c_n|^2 e^{jn\omega_0\tau}$$

where $\omega_0 = 2\pi/T$ and c_n are the complex Fourier coefficients of $f(t)$.

Solution: In Problem 8-12, if we let $f_1(t) = f_2(t) = f(t)$ and $T_1 = T$, we obtain

$$\bar{R}_{ff}(\tau) = \sum_{n=-\infty}^{\infty} c_n^* c_n e^{-jn\omega_0\tau} = \sum_{n=-\infty}^{\infty} |c_n|^2 e^{-jn\omega_0\tau} = \sum_{n=-\infty}^{\infty} |c_n|^2 e^{jn\omega_0\tau}$$

since $|c_{-n}|^2 = |c_n|^2$.

PROBLEM 8-14   Show that the average autocorrelation function of the sum of a message signal $s(t)$ and a noise signal $n(t)$ is the sum of the individual autocorrelation functions of signal and noise, respectively, if the noise signal $n(t)$ is assumed to have a zero average value and $s(t)$ and $n(t)$ are uncorrelated.

Solution: Let $f(t) = s(t) + n(t)$. Then

$$\bar{R}_{ff}(\tau) = \lim_{T\to\infty} \frac{1}{T} \int_{-T/2}^{T/2} f(t) f(t-\tau)\, dt$$

$$= \lim_{T\to\infty} \frac{1}{T} \int_{-T/2}^{T/2} [s(t) + n(t)][s(t-\tau) + n(t-\tau)]\, dt$$

$$= \bar{R}_{ss}(\tau) + \bar{R}_{nn}(\tau) + \bar{R}_{sn}(\tau) + \bar{R}_{ns}(\tau)$$

Since the message signal $s(t)$ and the noise signal $n(t)$ are uncorrelated and $n(t)$ is assumed to have a zero average value, by (8.52) we have

$$\bar{R}_{sn}(\tau) = \bar{R}_{ns}(\tau) = 0$$

Thus,

$$\bar{R}_{ff}(\tau) = \bar{R}_{ss}(\tau) + \bar{R}_{nn}(\tau)$$

Average Power Spectra: Random Signals

PROBLEM 8-15   Find the power spectral density of a periodic function $f(t)$ with period T.

Solution: Let the Fourier series of a periodic function $f(t)$ be given by

$$f(t) = \sum_{n=-\infty}^{\infty} c_n e^{jn\omega_0 t}, \qquad \omega_0 = \frac{2\pi}{T}$$

Then from the result of Problem 8-13, we have

$$\bar{R}_{ff}(\tau) = \sum_{n=-\infty}^{\infty} |c_n|^2 e^{jn\omega_0\tau}$$

Taking the Fourier transform of $\bar{R}_{ff}(\tau)$, we obtain

$$P(\omega) = \mathscr{F}[\bar{R}_{ff}(\tau)] = \sum_{n=-\infty}^{\infty} |c_n|^2 \mathscr{F}[e^{jn\omega_0\tau}]$$

$$= \sum_{n=-\infty}^{\infty} 2\pi |c_n|^2 \delta(\omega - n\omega_0)$$

with the use of the Fourier transform of a complex exponential function (6.18).

Hence, $P(\omega)$ consists of a series of impulses at the harmonic frequencies of $f(t)$. Each impulse has a strength equal to the power contained in that component frequency, and is clearly a measure of the distribution of the power in $f(t)$.

PROBLEM 8-16   The average autocorrelation function of the thermal-noise current $i_n(t)$ is given by

$$\bar{R}_{ii}(\tau) = kTG\alpha e^{-\alpha|\tau|}$$

where

      k = Boltzmann's constant = $1.38 \times 10^{-23}\,J/°K$

      T = ambient temperature (°K)

      G = conductance of the resistor (℧)

      α = average number of collisions per second of an electron

Find the average power spectral density of the thermal-noise current.

Solution: Taking the Fourier transform of $\bar{R}_{ii}(\tau)$, we have

$$P(\omega) = \mathscr{F}[\bar{R}_{ii}(\tau)]$$

$$= kTG\alpha \int_{-\infty}^{\infty} e^{-\alpha|\tau|} e^{-j\omega\tau}\, d\tau$$

$$= kTG\alpha \int_{-\infty}^{0} e^{\alpha\tau} e^{-j\omega\tau}\, d\tau + \int_{0}^{\infty} e^{-\alpha\tau} e^{-j\omega\tau}\, d\tau$$

$$= \frac{2kTG\alpha^2}{\alpha^2 + \omega^2} = \frac{2kTG}{1 + (\omega^2/\alpha^2)}$$

Since α, the number of collisions per second, is of the order of 10^{12}, the factor $1 + \omega^2/\alpha^2$ is close to unity for frequencies below $\sim 10^{10}$ Hz. Therefore, for frequencies below $\sim 10^{10}$ Hz, the average power spectral density of the thermal-noise current may be approximated by

$$P(\omega) = 2kTG$$

Analytic Signals and Hilbert Transforms

PROBLEM 8-17   Find the analytic signal associated with the signal $f(t) = \cos \omega_m t$.

Solution: From the Fourier transform of the cosine function (6.19),

$$\mathscr{F}[f(t)] = F(\omega) = \pi\delta(\omega - \omega_m) + \pi\delta(\omega + \omega_m)$$

Now, by (8.69), we have

$$\mathscr{F}[\hat{f}(t)] = \hat{F}(\omega) = -j\,\mathrm{sgn}\,\omega F(\omega)$$

Since

$$\mathrm{sgn}\,\omega = \begin{cases} 1, & \omega > 0 \\ -1, & \omega < 0 \end{cases}$$

$$\hat{F}(\omega) = -j\pi\delta(\omega - \omega_m) + j\pi\delta(\omega + \omega_m)$$

Thus, from the result of Problem 6-9, we conclude that

$$\hat{f}(t) = \sin \omega_m t$$

Hence, from the definition of the analytic signal (8.64),

$$f_+(t) = f(t) + j\hat{f}(t)$$

$$= \cos \omega_m t + j\sin \omega_m t = e^{j\omega_m t}$$

which is the required analytic signal associated with $\cos \omega_m t$. Note that by the Fourier transform of a complex exponential function (6.18),

$$F_+(\omega) = 2\pi\delta(\omega - \omega_m)$$

PROBLEM 8-18   Let $f(t)$ be a given band-limited real signal whose Fourier spectrum is shown in Figure 8-7a. Find the Fourier spectrum of the signal $g(t)$ given by

$$g(t) = f(t)\cos \omega_c t - \hat{f}(t)\sin \omega_c t$$

in terms of $F(\omega) = \mathscr{F}[f(t)]$, where $\hat{f}(t)$ is the Hilbert transform of $f(t)$. Signal $g(t)$ is often referred as the **single sideband (SSB)** signal.

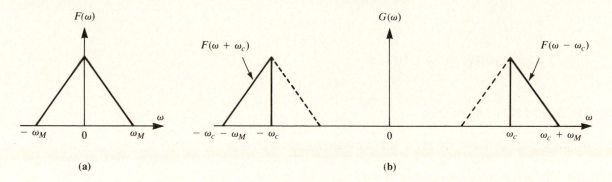

(a)   (b)

Figure 8-7

Solution: From frequency translation theorem (8.34), we have

$$\mathscr{F}[f(t)\cos \omega_c t] = \frac{1}{2}F(\omega - \omega_c) + \frac{1}{2}F(\omega + \omega_c)$$

Now, from (8.69),

$$\mathscr{F}[\hat{f}(t)] = -j\,\mathrm{sgn}\,\omega F(\omega)$$

Thus, from frequency translation theorem (8.35), we have

$$\mathscr{F}[\hat{f}(t)\sin \omega_c t] = -\frac{1}{2}j[-j\mathrm{sgn}(\omega - \omega_c)F(\omega - \omega_c)]$$

$$+ \frac{1}{2}j[-j\,\mathrm{sgn}(\omega + \omega_c)F(\omega + \omega_c)]$$

$$= -\frac{1}{2}\mathrm{sgn}(\omega - \omega_c)F(\omega - \omega_c) + \frac{1}{2}\mathrm{sgn}(\omega + \omega_c)F(\omega + \omega_c)$$

Hence, we obtain

$$G(\omega) = \frac{1}{2}F(\omega - \omega_c)[1 + \mathrm{sgn}(\omega - \omega_c)] + \frac{1}{2}F(\omega + \omega_c)[1 - \mathrm{sgn}(\omega + \omega_c)]$$

Since

$$1 + \mathrm{sgn}(\omega - \omega_c) = \begin{cases} 2, & \omega > \omega_c \\ 0, & \omega < \omega_c \end{cases}$$

and

$$1 - \mathrm{sgn}(\omega + \omega_c) = \begin{cases} 2, & \omega < -\omega_c \\ 0, & \omega > -\omega_c \end{cases}$$

we have

$$G(\omega) = 0 \qquad \text{for} \quad |\omega| < \omega_c$$
$$G(\omega) = F(\omega + \omega_c) \qquad \text{for} \quad \omega < -\omega_c$$
$$G(\omega) = F(\omega - \omega_c) \qquad \text{for} \quad \omega > \omega_c$$

which is plotted in Figure 8-7b.

PROBLEM 8-19   Show that the Hilbert transform of $\hat{f}(t)$ is $-f(t)$; that is,

$$\hat{\hat{f}}(t) = -f(t)$$

Solution: Let

$$f(t) \leftrightarrow F(\omega)$$

Then by (8.69),

$$\hat{f}(t) \leftrightarrow -j \operatorname{sgn} \omega F(\omega)$$

and

$$\hat{\hat{f}}(t) \leftrightarrow -j \operatorname{sgn} \omega [-j \operatorname{sgn} \omega F(\omega)]$$
$$= j^2 (\operatorname{sgn} \omega)^2 F(\omega)$$
$$= -F(\omega)$$

since $j^2 = -1$ and $(\operatorname{sgn} \omega)^2 = 1$. Hence,

$$\hat{\hat{f}}(t) = \mathscr{F}^{-1}[-F(\omega)] = -\mathscr{F}^{-1}[F(\omega)] = -f(t)$$

PROBLEM 8-20 Show that

$$\int_{-\infty}^{\infty} [f(t)]^2 \, dt = \int_{-\infty}^{\infty} [\hat{f}(t)]^2 \, dt$$

where $\hat{f}(t)$ is the Hilbert transform of $f(t)$.

Solution: By (8.69), we have

$$\hat{F}(\omega) = -j \operatorname{sgn} \omega F(\omega)$$

Thus,

$$|\hat{F}(\omega)|^2 = |-j \operatorname{sgn} \omega|^2 |F(\omega)|^2 = |F(\omega)|^2$$

since $|-j \operatorname{sgn} \omega|^2 = 1$. Hence, by Parseval's theorem (7.18), we have

$$\int_{-\infty}^{\infty} [f(t)]^2 \, dt = \frac{1}{2\pi} \int_{-\infty}^{\infty} |F(\omega)|^2 \, d\omega = \frac{1}{2\pi} \int_{-\infty}^{\infty} |\hat{F}(\omega)|^2 \, d\omega = \int_{-\infty}^{\infty} [\hat{f}(t)]^2 \, dt$$

Supplementary Exercises

PROBLEM 8-21 Let $f(t)$ be a real signal and define

$$(\Delta t)^2 = \frac{1}{\|f\|^2} \int_{-\infty}^{\infty} (t - \bar{t})^2 f^2(t) \, dt < \infty$$

where

$$\|f\|^2 = \int_{-\infty}^{\infty} f^2(t) \, dt < \infty$$

$$\bar{t} = \frac{1}{\|f\|^2} \int_{-\infty}^{\infty} t f^2(t) \, dt$$

and (Δt) is a measure of how much the signal is spread about $\bar{t}$, and is called the **dispersion** in time of the signal. Find the time dispersion Δt of the exponentially decaying pulse signal shown in Figure 8-8.

Answer: $\Delta t = \frac{1}{2} T$.

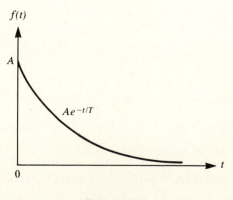

Figure 8-8

PROBLEM 8-22   Let $F(\omega) = \mathscr{F}[f(t)]$ where $f(t)$ is real, and define

$$(\Delta\omega)^2 = \frac{1}{\|F\|^2} \int_{-\infty}^{\infty} (\omega - \bar{\omega})^2 |F(\omega)|^2 \, d\omega$$

where

$$\|F\|^2 = \int_{-\infty}^{\infty} |F(\omega)|^2 \, d\omega$$

$$\bar{\omega} = \frac{1}{\|F\|^2} \int_{-\infty}^{\infty} \omega |F(\omega)|^2 \, d\omega$$

Show that ω is equal to zero and $(\Delta\omega)^2$ reduces to

$$(\Delta\omega)^2 = \frac{1}{\|F\|^2} \int_{-\infty}^{\infty} \omega^2 |F(\omega)|^2 \, d\omega$$

Here $\Delta\omega$ can also be defined as the **spectral bandwidth** of the signal. [*Hint:* Note that $\omega|F(\omega)|^2$ is an odd function of ω and use (2.5).]

PROBLEM 8-23   Show that the spectral bandwidth $\Delta\omega$ of a signal $f(t)$ defined in Problem 8-22 will be finite only if the following integral is finite; that is,

$$\int_{-\infty}^{\infty} [f'(t)]^2 \, dt = \text{finite}$$

where $f'(t) = df(t)/dt$. [*Hint:* Use Parseval's theorem (7.18).]

PROBLEM 8-24   Show that if the signal $f(t)$ is such that $(\Delta t)^2$ and $(\Delta\omega)^2$ defined in Problems 8-21 and 8-22 are finite and $\lim_{t\to\infty} \sqrt{t} f(t) = 0$, then $\Delta t \Delta\omega \geq \frac{1}{2}$, which is an another form of the uncertainty principle in spectral analysis.

PROBLEM 8-25   Determine the minimum sampling rate for the following signal $\cos\omega_0 t + \cos 8\omega_0 t$.

Answer: $16 f_0/\text{sec}$, $\omega_0 = 2\pi f_0$

PROBLEM 8-26   Show that the product of an ordinary AM signal, with a periodic waveform whose fundamental frequency is the carrier frequency of the ordinary AM signal, includes a term proportional to the signal $m(t)$.

PROBLEM 8-27   Find the spectrum of the PAM signal (see Problem 8-10) if $g(t)$ is the periodic symmetric rectangular pulse shown in Figure 8-9. This special PAM signal is called a **chopped** signal.

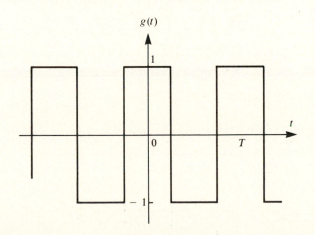

Figure 8-9

Answer: $\displaystyle\sum_{n=1}^{\infty} \frac{1}{2} a_{2n-1} [M\{\omega - (2n-1)\omega_0\} + M\{\omega + (2n-1)\omega_0\}]$

where

$$a_{2n-1} = \begin{cases} \dfrac{4}{(2n-1)\pi} & \text{for} \quad (2n-1) = 1, 5, \dots \\[2ex] \dfrac{-4}{(2n-1)\pi} & \text{for} \quad (2n-1) = 3, 7, \dots \end{cases}$$

PROBLEM 8-28   Show that the average autocorrelation function $\bar{R}_{11}(\tau)$ is an even function of τ.

PROBLEM 8-29   Show that the derivative of the average autocorrelation function of $f(t)$ is the negative of the average cross-correlation of $f(t)$ and df/dt; that is,

$$d\bar{R}_{ff}/d\tau = -\bar{R}_{f\,df/dt}$$

PROBLEM 8-30   Two periodic signals $f_1(t)$ and $f_2(t)$ with period T are said to be **uncorrelated** or **noncoherent** if for all τ,

$$\bar{R}_{12}(\tau) = \frac{1}{T}\int_{-T/2}^{T/2} f_1(t) f_2(t-\tau)\,dt = \frac{1}{T}\int_{-T/2}^{T/2} f_1(t)\,dt \times \frac{1}{T}\int_{-T/2}^{T/2} f_2(t)\,dt$$

that is, the average cross-correlation function of $f_1(t)$ and $f_2(t)$ is equal to the product of the average of $f_1(t)$ and $f_2(t)$ over one period. Show that the mean-squared value of the sum of two periodic non-coherent signals is the sum of the mean-squared values of the two signals when the average value of each signal is zero.

PROBLEM 8-31   It is often convenient to represent an arbitrary real signal $f(t)$ as an amplitude- and angle-modulated sinusoid of the form $f(t) = A(t)\cos\theta(t)$, where $A(t)$ is called the **envelope function**, $\theta(t)$ the **phase function**, and $\omega_i = d\theta(t)/dt$ the **instantaneous frequency** of the signal $f(t)$. Let $\hat{f}(t)$ be the signal defined in (8.71). Then the envelope function $A(t)$ can be defined by

$$A(t) = \frac{f(t)}{\cos\{\tan^{-1}[\hat{f}(t)/f(t)]\}}$$

and the phase function $\theta(t)$ can be defined by

$$\theta(t) = \tan^{-1}[\hat{f}(t)/f(t)]$$

Using the above definitions, express $f(t) = A\sin\omega t$, where A and ω are constant, in the form of an amplitude- and angle-modulated sinusoid.

Answer: $f(t) = A\cos\left(\omega t - \dfrac{\pi}{2}\right)$

PROBLEM 8-32   Find the instantaneous frequency of the signal $f(t) = 1/(1+t^2)$.

Answer: $\omega_i = 1/(1+t^2)$

9 APPLICATIONS TO LINEAR SYSTEMS

THIS CHAPTER IS ABOUT

☑ **Linear Systems**
☑ **Unit Impulse Response and System Function**
☑ **Operational System Function**
☑ **Causal Systems**
☑ **Response to Random Signals**

9-1. Linear Systems

A. System representation

A system is a mathematical model of a physical process that relates the **input function** (or *source* function) to the **output function** (or *response* function).

Now suppose that $f_i(t)$ and $f_o(t)$, as shown in Figure 9-1, are the input and output, respectively, of a system. Then the system can be considered as a mapping of the input $f_i(t)$ into the output $f_o(t)$. Using a symbol L to indicate this mapping, we express the relationship as

$$L\{f_i(t)\} = f_o(t) \tag{9.1}$$

The symbol or operator L in (9.1) designates the law for determining the response function $f_o(t)$ from a given source function $f_i(t)$.

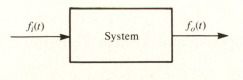

Figure 9-1   Input and output of a system.

B. Linear systems

Definition: If the operator L in (9.1) satisfies the following two conditions, then L is called the **linear operator** and the system represented by L is called the **linear system**:

Linear system

$$L\{f_{i1}(t) + f_{i2}(t)\} = L\{f_{i1}(t)\} + L\{f_{i2}(t)\} = f_{o1}(t) + f_{o2}(t) \quad \text{(additivity)} \tag{9.2}$$

$$L\{af_i(t)\} = aL\{f_i(t)\} = af_o(t) \quad \text{(homogeneity)} \tag{9.3}$$

where a is any constant.

C. Time-invariant systems

Definition: If the system satisfies the following condition, then the system is called the **time-invariant system**:

Time-invariant system

$$L\{f_i(t + t_0)\} = f_o(t + t_0) \tag{9.4}$$

where t_0 is an arbitrary constant.

D. Response to exponential source functions

The responses of linear systems to source functions that are exponential functions of time are particularly important in the analysis of linear systems.

The responses of a linear time-invariant system to an exponential function $e^{j\omega t}$ is also an exponential function and is proportional to the input; that is,

$$L\{e^{j\omega t}\} = ke^{j\omega t} \tag{9.5}$$

In mathematical language, a function $f(t)$ satisfying the equation

Eigenfunction
$$L\{f(t)\} = kf(t) \tag{9.6}$$

is called an **eigenfunction** (or **characteristic function**) of the operator L, and the corresponding value of k is called an **eigenvalue** (or **characteristic value**) of L. From (9.5), we can say that the eigenfunction of a linear time invariant system is an exponential function.

EXAMPLE 9-1: Verify (9.5).

Solution: Let $f_o(t)$ be the response to $e^{j\omega t}$. Then

$$L\{e^{j\omega t}\} = f_o(t) \tag{9.7}$$

Since the system is time-invariant, from definition (9.4) we have

$$L\{e^{j\omega(t + t_0)}\} = f_o(t + t_0) \tag{9.8}$$

But from the homogeneity condition (9.3), we have

$$L\{e^{j\omega(t + t_0)}\} = L\{e^{j\omega t_0}e^{j\omega t}\} = e^{j\omega t_0} \cdot L\{e^{j\omega t}\} \tag{9.9}$$

Hence,

$$f_o(t + t_0) = e^{j\omega t_0}f_o(t) \tag{9.10}$$

Setting $t = 0$, we have

$$f_o(t_0) = f_o(0)e^{j\omega t_0} \tag{9.11}$$

Since t_0 is arbitrary, by changing t_0 to t, we can rewrite (9.11) as

$$f_o(t) = f_o(0)e^{j\omega t} = ke^{j\omega t}$$

Thus, the output is proportional to the input, with $k = f_o(0)$ as the proportionality constant. In general, k is complex and depends on ω.

9-2. Unit Impulse Response and System Function

A. Unit impulse response

We denote the response of a linear system to a unit impulse $\delta(t)$ by $h(t)$ and call the $h(t)$ the **unit impulse response** of the system. Symbolically, this is expressed as

Unit impulse response
$$L\{\delta(t)\} = h(t) \tag{9.12}$$

If the system is time-invariant, then we see from definition (9.4) that its response to $\delta(t - \tau)$ is given by $h(t - \tau)$; that is,

$$L\{\delta(t - \tau)\} = h(t - \tau) \tag{9.13}$$

EXAMPLE 9-2: Show that the response $f_o(t)$ of a time-invariant linear system to an arbitrary input $f_i(t)$ can be expressed as the convolution of the input $f_i(t)$ and the unit impulse response $h(t)$ of the system; that is,

Response to an arbitrary input
$$f_o(t) = \int_{-\infty}^{\infty} f_i(\tau)h(t - \tau)\,d\tau = f_i(t) * h(t) \tag{9.14}$$

$$= \int_{-\infty}^{\infty} f_i(t - \tau)h(\tau)\,d\tau = h(t) * f_i(t) \tag{9.15}$$

Solution: From the property (4.3) of the δ-function, we can express $f_i(t)$ as

$$f_i(t) = \int_{-\infty}^{\infty} f_i(\tau)\delta(t - \tau)\,d\tau \tag{9.16}$$

Then from the linearity of the *L* operator, in view of (9.13), we obtain

$$f_o(t) = L\{f_i(t)\} = \int_{-\infty}^{\infty} f_i(\tau)L\{\delta(t-\tau)\}\,d\tau = \int_{-\infty}^{\infty} f_i(\tau)h(t-\tau)\,d\tau \tag{9.17}$$

From the definition (7.1) and commutative property (7.4) of convolution, (9.17) can be expressed as

$$f_o(t) = f_i(t) * h(t) = h(t) * f_i(t) = \int_{-\infty}^{\infty} f_i(t-\tau)h(\tau)\,d\tau$$

Equations (9.14) and (9.15) indicate a very interesting result: the response of a linear system is uniquely determined from the knowledge of the unit impulse response $h(t)$ of the system.

B. System function

Definition: The Fourier transform of the unit impulse response of a linear system is called the **system function**:

System function $$H(\omega) = \mathscr{F}[h(t)] = \int_{-\infty}^{\infty} h(t)e^{-j\omega t}\,dt \tag{9.18}$$

Unit impulse response $$h(t) = \mathscr{F}^{-1}[H(\omega)] = \frac{1}{2\pi}\int_{-\infty}^{\infty} H(\omega)e^{j\omega t}\,d\omega \tag{9.19}$$

Equations (9.18) and (9.19) indicate that the system function and the unit impulse response together constitute the Fourier transform pair.

Now, if $F_i(\omega) = \mathscr{F}[f_i(t)]$ and $F_o(\omega) = \mathscr{F}[f_o(t)]$, where $f_i(t)$ and $f_o(t)$ are the input and output of a linear time-invariant system, respectively, then

$$F_o(\omega) = F_i(\omega)H(\omega) \tag{9.20}$$

$$f_o(t) = \frac{1}{2\pi}\int_{-\infty}^{\infty} F_i(\omega)H(\omega)e^{j\omega t}\,d\omega \tag{9.21}$$

where $H(\omega)$ is the system function defined by (9.18). Then, from (9.20), we have

System function $$H(\omega) = \frac{F_o(\omega)}{F_i(\omega)} = \frac{\mathscr{F}[f_o(t)]}{\mathscr{F}[f_i(t)]} \tag{9.22}$$

Equation (9.22) indicates that the system function $H(\omega)$ is also the ratio of the response transform to the source transform.

EXAMPLE 9-3: Prove (9.20) and (9.21).

Proof: From (9.14), we have

$$f_o(t) = f_i(t) * h(t)$$

Hence, applying the time convolution theorem (7.11), we obtain

$$F_o(\omega) = F_i(\omega)H(\omega)$$

Applying the Fourier inverse transform formula (5.2), we have

$$f_o(t) = \mathscr{F}^{-1}[F_o(\omega)] = \frac{1}{2\pi}\int_{-\infty}^{\infty} F_i(\omega)H(\omega)e^{j\omega t}\,d\omega$$

Relations (9.14) and (9.20) are illustrated in Figure 9-2.

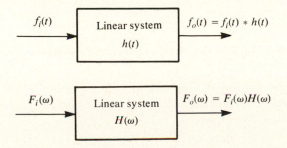

Figure 9-2 Unit impulse response and system function.

9-3. Operational System Function

A. Definition of the operational system function

Another definition of a linear system is that the input function and the output function of the system are related by a linear differential equation; that is,

$$a_n \frac{d^n f_o(t)}{dt^n} + a_{n-1} \frac{d^{n-1} f_o(t)}{dt^{n-1}} + \cdots + a_1 \frac{df_o(t)}{dt} + a_0 f_o(t)$$

$$= b_m \frac{d^m f_i(t)}{dt^m} + b_{m-1} \frac{d^{m-1} f_i(t)}{dt^{m-1}} + \cdots + b_1 \frac{df_i(t)}{dt} + b_0 f_i(t) \qquad (9.23)$$

If we denote d/dt by the operator p, such that

$$pf(t) = \frac{df(t)}{dt}, \qquad p^n f(t) = \frac{d^n f(t)}{dt^n}$$

then (9.23) can be rewritten as

$$\sum_{n=0}^{\infty} a_n p^n f_o(t) = \sum_{m=0}^{\infty} b_m p^m f_i(t) \qquad (9.24)$$

or

$$A(p) f_o(t) = B(p) f_i(t) \qquad (9.25)$$

where

$$A(p) = a_n p^n + a_{n-1} p^{n-1} + \cdots + a_1 p + a_0$$
$$B(p) = b_m p^m + b_{m-1} p^{m-1} + \cdots + b_1 p + b_0$$

In a linear system the coefficients a_n and b_m are *independent* of the output response. In the time-invariant (or **constant-parameter**) system, the coefficients a_n and b_m are *constants*.

Equation (9.25) can be written symbolically in the form

Operational system function

$$f_0(t) = \frac{B(p)}{A(p)} f_i(t) = H(p) f_i(t) \qquad (9.26)$$

where $H(p) = B(p)/A(p)$. It is understood that (9.26) is an operational expression of the differential equation (9.23). The operator $H(p)$, which operates on the source function to produce a response, is termed the **operational system function**. Comparing (9.26) and (9.1), you can see that $L = H(p)$.

EXAMPLE 9-4: Obtain the operational expression for the current response $i(t)$ to the voltage source $v(t)$ in the *RLC* circuit of Figure 9-3a.

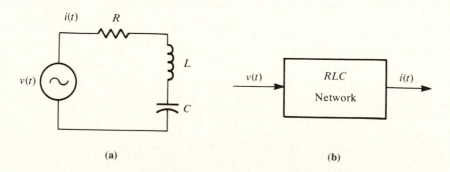

Figure 9-3 (a) *RLC* circuit; (b) system representation of the circuit.

Solution: The source is the applied voltage $v(t)$, and the response is the current $i(t)$, as shown in Figure 9-3b. Applying Kirchhoff's law, the differential equation relating $i(t)$ and $v(t)$ can be obtained as

$$Ri(t) + L \frac{di(t)}{dt} + \frac{1}{C} \int_{-\infty}^{t} i(t)\, dt = v(t)$$

Differentiating both sides,

$$L\frac{d^2i(t)}{dt^2} + R\frac{di(t)}{dt} + \frac{1}{C}i(t) = \frac{dv(t)}{dt} \tag{9.27}$$

where the symbol L stands for inductance and is *not* an operator.

Using the operator $p = d/dt$, eq. (9.27) can be written as

$$\left(Lp^2 + Rp + \frac{1}{C}\right)i(t) = pv(t)$$

Hence,

$$i(t) = \frac{p}{Lp^2 + Rp + 1/C}v(t) = H(p)v(t) \tag{9.28}$$

where

$$H(p) = \frac{p}{(Lp^2 + Rp + 1/C)} = \frac{1}{\left(R + Lp + \dfrac{1}{Cp}\right)}$$

B. Response to exponential source functions

EXAMPLE 9-5: Using the symbolic expression (9.26), show that the response of a linear time-invariant system to an exponential function $e^{j\omega t}$ is also an exponential function and proportional to the input.

Solution: Let the input source function in (9.25) be $f_i(t) = e^{j\omega t}$. Then

$$A(p)f_o(t) = B(p)e^{j\omega t} \tag{9.29}$$

where $f_o(t)$ is the output response function. Now

$$B(p) = b_m p^m + b_{m-1}p^{m-1} + \cdots + b_1 p + b_0$$
$$B(p)e^{j\omega t} = B(j\omega)e^{j\omega t}$$

since

$$p^m e^{j\omega t} = \frac{d^m}{dt^m}(e^{j\omega t}) = (j\omega)^m e^{j\omega t}$$

Hence, the response $f_o(t)$ is defined by the ordinary linear differential equation

$$A(p)f_o(t) = B(j\omega)e^{j\omega t} \tag{9.30}$$

Now the forcing function of (9.30) is $B(j\omega)e^{j\omega t}$, an exponential function, and from the theory of differential equations we can assume that the response $f_o(t)$ is also exponential. Hence, if $f_o(t) = k_1 e^{j\omega t}$, then

$$A(p)f_o(t) = A(p)[k_1 e^{j\omega t}] = k_1 A(p)[e^{j\omega t}] = k_1 A(j\omega)e^{j\omega t} = A(j\omega)f_o(t) \tag{9.31}$$

Substituting (9.31) in (9.30), we obtain

$$A(j\omega)f_o(t) = B(j\omega)e^{j\omega t} \tag{9.32}$$

Hence, if $A(j\omega) \neq 0$,

$$f_o(t) = \frac{B(j\omega)}{A(j\omega)}e^{j\omega t} = H(j\omega)e^{j\omega t} \tag{9.33}$$

In view of (9.5), we can rewrite (9.33) symbolically as

$$L\{e^{j\omega t}\} = H(j\omega)e^{j\omega t} \tag{9.34}$$

and we recognize that the eigenvalue of the operator L is equal to $H(j\omega)$. The block diagram in Figure 9-4 illustrates the relationship between input and output given by (9.33).

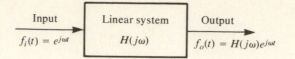

Figure 9-4   System function.

C. The system function

The eigenvalue $H(j\omega)$ of the operator L is defined as the system function represented by L.

EXAMPLE 9-6: Verify that the system function $H(\omega)$ defined by (9.18) is exactly the same system function $H(j\omega)$ defined by (9.34)

Solution: If $f_i(t) = e^{j\omega_0 t}$, then from the Fourier transform of a complex exponential function (6.18),

$$F_i(\omega) = \mathscr{F}[f_i(t)] = \mathscr{F}[e^{j\omega_0 t}] = 2\pi\delta(\omega - \omega_0)$$

Hence,

$$F_i(\omega)H(\omega) = 2\pi\delta(\omega - \omega_0)H(\omega) = 2\pi H(\omega_0)\delta(\omega - \omega_0) \tag{9.35}$$

in view of the result of Problem 4-2 (a property of the δ-function). Then from (9.21), we obtain

$$f_o(t) = L\{e^{j\omega_0 t}\} = \frac{1}{2\pi}\int_{-\infty}^{\infty} 2\pi H(\omega_0)\delta(\omega - \omega_0)e^{j\omega t}d\omega$$

$$= H(\omega_0)\int_{-\infty}^{\infty} \delta(\omega - \omega_0)e^{j\omega t}\,d\omega$$

$$= H(\omega_0)e^{j\omega_0 t} \tag{9.36}$$

Since (9.36) holds for any value of ω_0, we can change ω_0 to ω and obtain

$$f_o(t) = L\{e^{j\omega t}\} = H(\omega)e^{j\omega t} \tag{9.37}$$

Recalling (9.34), we have

$$f_o(t) = L\{e^{j\omega t}\} = H(j\omega)e^{j\omega t}$$

Comparing (9.34) with (9.37), we conclude that

$$H(\omega) = H(j\omega)$$

9-4. Causal Systems

A. Definition of causal systems and causal functions

A physical passive system has the property that if the input source is zero for $t < t_0$, then the output response is also zero for $t < t_0$, that is, if

$$f_i(t) = 0 \qquad \text{for } t < t_0 \tag{9.38}$$

then

Causal system $$f_o(t) = L\{f_i(t)\} = 0 \qquad \text{for } t < t_0 \tag{9.39}$$

A system that satisfies (9.38) and (9.39) is called a **causal system**.

A function $f(t)$ will be called **causal** if it has zero values for $t < 0$, that is,

Causal function $$f(t) = 0 \qquad \text{for } t < 0 \tag{9.40}$$

B. Response of causal systems

The response $f_o(t)$ of a linear causal system to any source $f_i(t)$ is given by

$$f_o(t) = \int_{-\infty}^{t} f_i(\tau)h(t - \tau)\,d\tau \tag{9.41}$$

or

$$f_o(t) = \int_{0}^{\infty} f_i(t - \tau)h(\tau)\,d\tau \tag{9.42}$$

If the source function $f_i(t)$ is causal, that is, if the source $f_i(t)$ is impressed at $t = 0$, then the response $f_o(t)$ of a linear causal system is given by

$$f_o(t) = \int_0^t f_i(\tau) h(t - \tau) \, d\tau \qquad (9.43)$$

EXAMPLE 9-7: Verify (9.41) to (9.43).

Solution: From (9.38) and (9.39), it follows that the unit impulse response $h(t)$ is causal; that is,

$$h(t) = 0 \qquad \text{for } t < 0 \qquad (9.44)$$

This means that

$$h(\tau) = 0 \qquad \text{for } \tau < 0 \qquad (9.45)$$

and

$$h(t - \tau) = 0 \qquad \text{for} \quad t - \tau < 0 \qquad \text{or} \qquad \tau > t \qquad (9.46)$$

Therefore, in view of (9.46), the integrand in (9.14) is zero in the interval $\tau = t$ to $\tau = \infty$. Thus, from (9.14), we have

$$f_o(t) = \int_{-\infty}^{\infty} f_i(\tau) h(t - \tau) \, d\tau = \int_{-\infty}^{t} f_i(\tau) h(t - \tau) \, d\tau$$

Similarly, from (9.45), the integrand in (9.15) is zero in the interval $\tau = -\infty$ to $\tau = 0$. Thus, from (9.15), we obtain

$$f_o(t) = \int_{-\infty}^{\infty} f_i(t - \tau) h(\tau) \, d\tau = \int_0^{\infty} f_i(t - \tau) h(\tau) \, d\tau$$

In (9.41), if $f_i(\tau) = 0$ for $\tau < 0$, then the lower limit of the integral may be changed to 0, since in the interval $\tau = -\infty$ to $\tau = 0$, the integrand is zero. Thus,

$$f_o(t) = \int_{-\infty}^{t} f_i(\tau) h(t - \tau) \, d\tau = \int_0^t f_i(\tau) h(t - \tau) \, d\tau$$

C. Unit step response

The response of a system to a unit step function $u(t)$ is called the **unit step response** of the system and is denoted by $a(t)$; that is,

Unit step response $\qquad\qquad L\{u(t)\} = a(t) \qquad (9.47)$

The unit step response $a(t)$ of a linear system can be expressed as

$$a(t) = \int_{-\infty}^{t} h(\tau) \, d\tau \qquad (9.48)$$

and

$$a(\infty) = a(t)\big|_{t=\infty} = H(0) \qquad (9.49)$$

where $H(\omega)$ is the system function of the system and $h(t)$ is its unit impulse response. If the system is causal, then

$$a(t) = \int_0^t h(\tau) \, d\tau \qquad (9.50)$$

EXAMPLE 9-8: Verify (9.48) to (9.50).

Solution: Since $f_i(t) = u(t)$ and $f_o(t) = a(t)$, it follows from (9.15) that

$$a(t) = \int_{-\infty}^{\infty} u(t - \tau) h(\tau) \, d\tau \qquad (9.51)$$

Since

$$u(t - \tau) = \begin{cases} 0 & \text{for } \tau > t \\ 1 & \text{for } \tau < t \end{cases}$$

we have

$$a(t) = \int_{-\infty}^{t} h(\tau) \, d\tau$$

With $t = -\infty$,

$$a(\infty) = a(t)|_{t=\infty} = \int_{-\infty}^{\infty} h(\tau) \, d\tau$$

Certainly, this integral can be written as

$$a(\infty) = \int_{-\infty}^{\infty} h(\tau) e^{-j\omega t} \, d\tau|_{\omega=0} = H(\omega)|_{\omega=0} = H(0)$$

For a causal system, since $h(\tau) = 0$ for $\tau < 0$, eq. (9.48) becomes

$$a(t) = \int_{0}^{t} h(\tau) \, d\tau$$

9-5. Response to Random Signals

For random signals, we do not have—nor can we get—an explicit expression for an input noise source or for the response of a system to such a source. Consequently, a relationship such as (9.20) is not available for random signals. In this section, we'll study the application of correlation functions and power spectral densities to system analysis problems involving random signals.

Let $x(t)$ and $y(t)$ be the random input and output signals, respectively, for a linear time-invariant system characterized by the system function $H(\omega)$. Then the average autocorrelation of the input and the output is related by

$$\bar{R}_{yy}(\tau) = \int_{-\infty}^{\infty} h(\lambda) \int_{-\infty}^{\infty} h(\sigma) \bar{R}_{xx}(\tau + \sigma - \lambda) \, d\sigma \, d\lambda \qquad (9.52)$$

where $h(t) = \mathcal{F}^{-1}[H(\omega)]$ is the unit impulse response of the system. The output power spectral density $P_o(\omega)$ and the input power spectral density $P_i(\omega)$ of a linear system are related by

**Power spectral
densities relationship**
$$P_o(\omega) = |H(\omega)|^2 P_i(\omega) \qquad (9.53)$$

EXAMPLE 9-9: Prove (9.52).

Proof: It was shown in (9.15) that the output $y(t)$ is related to the input $x(t)$ by the convolution integral, that is,

$$y(t) = \int_{-\infty}^{\infty} h(\tau) x(t - \tau) \, d\tau \qquad (9.54)$$

Now from the average autocorrelation function (8.40), we have

$$\bar{R}_{yy}(\tau) = \lim_{T \to \infty} \frac{1}{T} \int_{-T/2}^{T/2} y(t) y(t - \tau) \, dt \qquad (9.55)$$

From (9.54), we may write $y(t)$ and $y(t - \tau)$ as

$$y(t) = \int_{-\infty}^{\infty} h(\lambda) x(t - \lambda) \, d\lambda \qquad (9.56)$$

$$y(t - \tau) = \int_{-\infty}^{\infty} h(\sigma) x(t - \tau - \sigma) \, d\sigma \qquad (9.57)$$

Substituting (9.56) and (9.57) into (9.55), we have

$$\bar{R}_{yy}(\tau) = \lim_{T \to \infty} \frac{1}{T} \int_{-T/2}^{T/2} \left[\int_{-\infty}^{\infty} h(\lambda) x(t - \lambda) \, d\lambda \int_{-\infty}^{\infty} h(\sigma) x(t - \tau - \sigma) \, d\sigma \right] dt \qquad (9.58)$$

By interchanging the order of integration, we can write (9.58) as

$$\bar{R}_{yy}(\tau) = \int_{-\infty}^{\infty} h(\lambda) \int_{-\infty}^{\infty} h(\sigma) \left[\lim_{T \to \infty} \frac{1}{T} \int_{-T/2}^{T/2} x(t - \lambda) x(t - \tau - \sigma) \, dt \right] d\sigma \, d\lambda \qquad (9.59)$$

Since, by (8.40),

$$\bar{R}_{xx}(\tau + \sigma - \lambda) = \lim_{T \to \infty} \frac{1}{T} \int_{-T/2}^{T/2} x(t - \lambda) x(t - \tau - \sigma) \, dt \qquad \textbf{(9.60)}$$

(9.59) becomes

$$\bar{R}_{yy}(\tau) = \int_{-\infty}^{\infty} h(\lambda) \int_{-\infty}^{\infty} h(\sigma) \bar{R}_{xx}(\tau + \sigma - \lambda) \, d\sigma \, d\lambda$$

EXAMPLE 9-10: Prove (9.53).

Proof: From the definition (8.57) of the power spectral density, $P_o(\omega)$ is given by

$$P_o(\omega) = \mathscr{F}[\bar{R}_{yy}(\tau)] = \int_{-\infty}^{\infty} \bar{R}_{yy}(\tau) e^{-j\omega\tau} \, d\tau \qquad \textbf{(9.61)}$$

Substituting (9.52) into (9.61), we have

$$P_o(\omega) = \int_{-\infty}^{\infty} \left[\int_{-\infty}^{\infty} h(\lambda) \int_{-\infty}^{\infty} h(\sigma) \bar{R}_{xx}(\tau + \sigma - \lambda) \, d\sigma \, d\lambda \right] e^{-j\omega\tau} \, d\tau \qquad \textbf{(9.62)}$$

With the change of variable $\mu = \tau + \sigma - \lambda$, followed by a separation of variables,

$$P_o(\omega) = \int_{-\infty}^{\infty} h(\lambda) \, d\lambda \int_{-\infty}^{\infty} h(\sigma) \, d\sigma \int_{-\infty}^{\infty} \bar{R}_{xx}(\mu) e^{-j\omega(\mu - \sigma + \lambda)} \, d\mu$$

$$= \int_{-\infty}^{\infty} h(\lambda) e^{-j\omega\lambda} \, d\lambda \int_{-\infty}^{\infty} h(\sigma) e^{j\omega\sigma} \, d\sigma \int_{-\infty}^{\infty} \bar{R}_{xx}(\mu) e^{-j\omega\mu} \, d\mu \qquad \textbf{(9.63)}$$

Since by (8.57) and (9.18),

$$P_i(\omega) = \int_{-\infty}^{\infty} R_{xx}(\tau) e^{-j\omega\tau} \, d\tau$$

$$H(\omega) = \int_{-\infty}^{\infty} h(\tau) e^{-j\omega\tau} \, d\tau$$

and $h(t)$ is always real,

$$H^*(\omega) = \int_{-\infty}^{\infty} h(\tau) e^{j\omega\tau} \, d\tau$$

Then (9.63) can be written as

$$P_o(\omega) = H(\omega) H^*(\omega) P_i(\omega) \qquad \textbf{(9.64)}$$

Since $H(\omega)H^*(\omega) = |H(\omega)|^2$, we obtain

$$P_o(\omega) = |H(\omega)|^2 P_i(\omega)$$

which is shown in Figure 9-5.

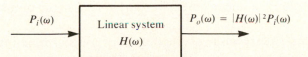

Figure 9-5   Input and output power spectral density.

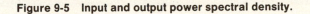

SUMMARY

1. If a system is represented by the operator L such that

$$L\{f_i(t)\} = f_o(t)$$

where $f_i(t)$ is the input and $f_o(t)$ is the output of the system, then the system is linear if

$$L\{f_{i1}(t) + f_{i2}(t)\} = f_{o1}(t) + f_{o2}(t)$$

and

$$L\{af_i(t)\} = af_o(t)$$

2. The system is time-invariant if

$$L\{f_i(t + t_0)\} = f_o(t + t_0) \qquad \text{for any} \quad t_0$$

3. The output $f_o(t)$ of a linear time-invariant system is given by the convolution of the input $f_i(t)$ and the unit impulses response $h(t)$; that is,

$$f_o(t) = f_i(t) * h(t)$$

where

$$h(t) = L\{\delta(t)\}$$

4. The system function $H(\omega)$ of a linear time-invariant system is defined as

$$H(\omega) = \mathscr{F}[h(t)]$$

or

$$F_o(\omega) = F_i(\omega)H(\omega)$$

where $F_i(\omega) = \mathscr{F}[f_i(t)]$ and $F_o(\omega) = \mathscr{F}[f_o(t)]$.

5. A causal system is defined by

$$L\{f_i(t)\} = f_o(t) = 0 \qquad \text{for} \quad t < t_0$$

if $f_i(t) = 0$ for $t < t_0$.

6. The unit step response $a(t)$ of a linear time-invariant system is given by

$$a(t) = \int_{-\infty}^{t} h(\tau)\, dt$$

7. The average autocorrelation of the input $x(t)$ and the output $y(t)$ of a linear time-invariant system is related by

$$\bar{R}_{yy}(\tau) = \int_{-\infty}^{\infty} h(\lambda) \int_{-\infty}^{\infty} h(\sigma) \bar{R}_{xx}(\tau + \sigma - \lambda)\, d\sigma\, d\lambda$$

8. The output power spectral density $P_o(\omega)$ and the input power spectral density $P_i(\omega)$ of a linear time-invariant system are related by

$$P_o(\omega) = |H(\omega)|^2 P_i(\omega)$$

RAISE YOUR GRADES

Can you explain ···?

☑ under what conditions a system is linear and time-invariant
☑ how to find the unit impulse response of a given linear system
☑ how to find the output of a linear time-invariant system if the input of the system is given
☑ how to find the mean-square value of the output of a linear time-invariant system if the input is a random signal

SOLVED PROBLEMS

Linear Systems

PROBLEM 9-1 Suppose a system has an input–output relation given by the linear equation

$$y(t) = ax(t) + b$$

where $x(t)$ and $y(t)$ are the input and output of the system, respectively, and a and b are constants. Is this system linear?

Solution: We can represent the input–output relation of the system by an operator T such that

$$y(t) = T\{x(t)\} = ax(t) + b$$

Consider two inputs $x_1(t)$ and $x_2(t)$. The corresponding outputs are

$$T\{x_1(t)\} = ax_1(t) + b$$
$$T\{x_2(t)\} = ax_2(t) + b$$

Now apply an input $x_1(t)$ and $x_2(t)$. Then the output is given by

$$T\{x_1(t) + x_2(t)\} = a[x_1(t) + x_2(t)] + b$$

But

$$T\{x_1(t)\} + T\{x_2(t)\} = a[x_1(t) + x_2(t)] + 2b \neq T\{x_1(t) + x_2(t)\}$$

which indicates that the additivity condition (9.2) is not satisfied. Thus, the system is not linear.

Note that

$$T\{2x(t)\} = 2ax(t) + b \neq 2T\{x(t)\}$$

Hence, the system also does not satisfy the homogeneity condition (9.3).

PROBLEM 9-2   Give an example of a system that satisfies the condition of additivity (9.2) but not the condition of homogeneity (9.3).

Solution: Let the system be represented by an operator T such that

$$T\{f(t)\} = f^*(t)$$

where $f^*(t)$ is the complex conjugate of $f(t)$. Then, clearly,

$$T\{f_1(t) + f_2(t)\} = [f_1(t) + f_2(t)]^* = f_1^*(t) + f_2^*(t)$$
$$= T\{f_1(t)\} + T\{f_2(t)\}$$

Now, if α is an arbitrary complex-valued constant, then

$$T\{\alpha f(t)\} = [\alpha f(t)]^* = \alpha^* f^*(t) = \alpha^* T\{f(t)\} \neq \alpha T\{f(t)\}$$

Thus, the system represented by T is additive but not homogeneous.

PROBLEM 9-3   Given an example of a system that satisfies the condition of homogeneity (9.3) but not the condition of additivity (9.2).

Solution: Let the system be represented by an operator T such that

$$T\{f(t)\} = \left[\int_a^b [f(\tau)]^2 \, d\tau \right]^{1/2}$$

Then,

$$T\{f_1(t) + f_2(t)\} = \left[\int_a^b [f_1(\tau) + f_2(\tau)]^2 \, d\tau \right]^{1/2}$$
$$= \left[\int_a^b \{[f_1(\tau)]^2 + [f_2(\tau)]^2 + 2f_1(\tau)f_2(\tau)f_2(\tau)\} \, d\tau \right]^{1/2}$$
$$\neq \left[\int_a^b [f_1(\tau)]^2 \, d\tau \right]^{1/2} + \left[\int_a^b [f_2(\tau)]^2 \, d\tau \right]^{1/2} = T\{f_1(t)\} + T\{f_2(t)\}$$

and

$$T\{\alpha f(t)\} = \left[\int_a^b [\alpha f(\tau)]^2 \, d\tau \right]^{1/2}$$
$$= \left[\alpha^2 \int_a^b [f(\tau)]^2 \, d\tau \right]^{1/2} = \alpha \left[\int_a^b [f(\tau)]^2 \, d\tau \right]^{1/2} = \alpha T\{f(t)\}$$

Thus, the system represented by T is homogeneous but not additive.

Unit Impulse Response and System Function

PROBLEM 9-4   If the input function of the linear system specified by $H(j\omega)$ is a periodic time function with period T, find the output response of the system.

Solution: Since the input function $f_i(t)$ is periodic, then

$$f_i(t) = \sum_{n=-\infty}^{\infty} c_n e^{jn\omega_0 t}, \qquad \omega_0 = \frac{2\pi}{T}$$

where

$$c_n = \frac{1}{T} \int_{-T/2}^{T/2} f_i(t) e^{-jn\omega_0 t} \, dt$$

It follows from (9.34) that

$$f_{on}(t) = H(jn\omega_0) c_n e^{jn\omega_0 t}$$

is the output in response to the input component

$$f_{in}(t) = c_n e^{jn\omega_0 t}$$

Since the system is linear, its total response to $f_i(t)$ is the sum of the component output $f_{on}(t)$. Thus,

$$f_o(t) = \sum_{n=-\infty}^{\infty} c_n H(jn\omega_0) e^{jn\omega_0 t}$$

which indicates that if the input to a linear system is periodic, then the output is also periodic.

PROBLEM 9-5   Consider the simple spring and mass mechanical system shown in Figure 9-6a. Obtain the operational system function for the displacement $x(t)$ of a mass m from the equilibrium position.

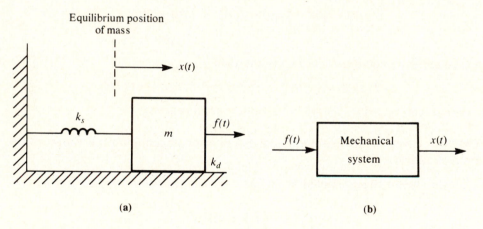

Figure 9-6

Solution: The source is the impressed force $f(t)$, and the response is the displacement $x(t)$ of a mass m from its equilibrium position (Figure 9-6b).

The forces acting on the mass are as follows:
(1) The impressed force $f(t)$
(2) The inertial reaction $(-m d^2x/dt^2)$
(3) The damping (frictional resistance) force $(-k_d dx/dt)$
(4) The elastic restoring force $(-k_s x)$

In items (3) and (4), k_d and k_s are the frictional coefficient and the spring constant, respectively.

By applying d'Alembert's principle, we have

$$m \frac{d^2x(t)}{dt^2} + k_d \frac{dx(t)}{dt} + k_s x(t) = f(t)$$

In operator form, this becomes

$$(mp^2 + k_d p + k_s)x(t) = f(t)$$

where $p = d/dt$. Hence,

$$x(t) = \frac{1}{mp^2 + k_d p + k_s} f(x) = H(p)f(t)$$

and we obtain the operational system function $H(p)$, which is given by

$$H(p) = \frac{1}{mp^2 + k_d p + k_s}$$

PROBLEM 9-6   Find the unit impulse response of the RC network shown in Figure 9-7a.

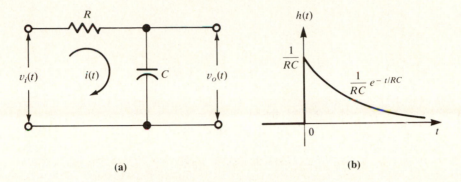

(a)   (b)

Figure 9-7   (a) RC network; (b) unit impulse response.

Solution: Applying Kirchhoff's law, the input source is given by

$$v_i(t) = Ri(t) + \frac{1}{C}\int_{-\infty}^{t} i(t)\,dt = \left(R + \frac{1}{pC}\right)i(t) \tag{a}$$

and the output response is given by

$$v_o(t) = \frac{1}{C}\int_{-\infty}^{t} i(t)\,dt = \frac{1}{pC}i(t) \tag{b}$$

Dividing eq. (b) by eq. (a), we have

$$\frac{v_o(t)}{v_i(t)} = \frac{1/(pC)}{R + 1/(pC)} = \frac{1}{1 + pRC}$$

Hence, the output response $v_o(t)$ and the input source $v_i(t)$ are related by

$$v_o(t) = \frac{1/(pC)}{R + 1/(pC)}v_i(t) = H(p)v_i(t)$$

where

$$H(p) = \frac{1/(pC)}{R + 1/(pC)} = \frac{1}{1 + pRC}$$

Thus, the system function $H(j\omega)$ is given by

$$H(j\omega) = \frac{1/(j\omega C)}{R + 1/(j\omega C)} = \frac{1}{1 + j\omega RC} = \frac{1}{RC(j\omega + 1/(RC))}$$

Therefore, from the result of Example 5-1, we obtain

$$h(t) = \mathscr{F}^{-1}[H(j\omega)] = \frac{1}{RC}\mathscr{F}^{-1}\left[\frac{1}{j\omega + 1/(RC)}\right] = \frac{1}{RC}e^{-t/RC}u(t)$$

The unit impulse response $h(t)$ is plotted in Figure 9-7b.

PROBLEM 9-7   A voltage source $v_i(t) = e^{-t}u(t)$ is applied to the RC network of Figure 9-7a of Problem 9-6. Find the response output voltage $v_0(t)$ when $R = \frac{1}{2}\,\Omega$ and $C = 1$ F.

Solution: Substituting $R = \frac{1}{2}\,\Omega$ and $C = 1$ F in the result of Problem 9-6, we have

$$h(t) = 2e^{-2t}u(t)$$

Hence, from the relation (9.14) defining the response to an arbitrary input, we obtain

$$v_o(t) = v_i(t) * h(t)$$

$$= \int_{-\infty}^{\infty} v_i(\tau)h(t - \tau)\,d\tau$$

$$= \int_{-\infty}^{\infty} e^{-\tau}u(\tau)2e^{-2(t-\tau)}u(t - \tau)\,d\tau$$

$$= 2e^{-2t}\int_{-\infty}^{\infty} e^{\tau}u(\tau)u(t - \tau)\,d\tau$$

Since

$$u(\tau)u(t - \tau) = \begin{cases} 0 & \text{for} \quad \tau < 0, \tau > t \\ 1 & \text{for} \quad 0 < \tau < t \end{cases}$$

we have

$$v_o(t) = \left(2e^{-2t}\int_0^t e^{\tau}\,d\tau\right)u(t)$$

$$= 2e^{-2t}(e^t - 1)u(t)$$

$$= 2(e^{-t} - e^{-2t})u(t)$$

PROBLEM 9-8   Suppose the system function $H(j\omega)$ of a linear time-invariant system is given by

$$H(j\omega) = Ke^{-j\omega t_0}$$

where K and t_0 are positive constants. Find the response $f_o(t)$ of the system to an input source $f_i(t)$.

Solution: Let

$$\mathscr{F}[f_i(t)] = F_i(j\omega), \qquad \mathscr{F}[f_o(t)] = F_o(j\omega)$$

From (9.20), $F_i(j\omega)$ and $F_o(j\omega)$ are related by

$$F_o(j\omega) = F_i(j\omega)H(j\omega)$$
$$= KF_i(j\omega)e^{-j\omega t_0}$$

Hence,

$$f_o(t) = \mathscr{F}^{-1}[F_o(j\omega)]$$

$$= \frac{K}{2\pi}\int_{-\infty}^{\infty} [F_i(j\omega)e^{-j\omega t_0}]e^{j\omega t}\,d\omega$$

$$= \frac{K}{2\pi}\int_{-\infty}^{\infty} F_i(j\omega)e^{j\omega(t-t_0)}\,d\omega$$

In view of the fact that

$$f_i(t) = \mathscr{F}^{-1}[F_i(j\omega)] = \frac{1}{2\pi}\int_{-\infty}^{\infty} F_i(j\omega)e^{j\omega t}\,d\omega$$

$f_o(t)$ can be written as

$$f_o(t) = Kf_i(t - t_0)$$

which shows that the output response is a delayed replica of the input function, with the magnitude of the response changed by the constant factor K.

Causal System

PROBLEM 9-9   The so-called **ideal low-pass filter** is defined as a system for which the system function $H(j\omega)$ is given by

$$H(j\omega) = \begin{cases} e^{-j\omega t_0} & \text{for} \quad |\omega| < \omega_c \\ 0 & \text{for} \quad |\omega| > \omega_c \end{cases}$$

where ω_c is referred to as the cut-off frequency. Find the unit impulse response $h(t)$ of an ideal low-pass filter, and discuss the result.

Solution: Figure 9-8a shows the characteristics of an ideal low-pass filter. From (9.19), the unit impulse response $h(t)$ is obtained by

$$h(t) = \mathscr{F}^{-1}[H(j\omega)] = \frac{1}{2\pi} \int_{-\infty}^{\infty} H(j\omega)e^{j\omega t}\, d\omega$$

$$= \frac{1}{2\pi} \int_{-\omega_c}^{\omega} e^{j\omega(t - t_0)}\, d\omega$$

$$= \frac{1}{\pi(t - t_0)2j} e^{j\omega(t - t_0)} \Big|_{-\omega_c}^{\omega_c}$$

$$= \frac{\omega_c}{\pi} \frac{\sin \omega_c(t - t_0)}{\omega_c(t - t_0)}$$

which is plotted in Figure 9-8b, from which we draw the following conclusions:

1. The applied input is distorted by the system because the filter transmits only a limited range of frequencies.
2. The peak value of the response ω_c/π is proportional to the cut-off frequency ω_c. The width of the main pulse is $2\pi/\omega_c$. We may refer to this quantity as the effective output pulse duration T_d. It is noted that as $\omega_c \to \infty$ (that is, when the filter becomes all-pass), $T_d \to 0$ and the output response peak $\to \infty$; in other words, the response approaches an impulse as it should.
3. The response is not zero before $t = 0$, that is, before the input is impressed. This is characteristic of a physically nonrealizable system. Ideal filters are not physically realizable and therefore are not necessarily causal systems.

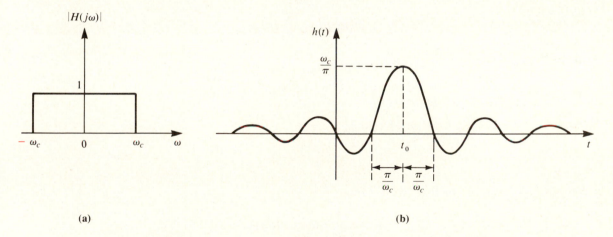

(a)   **(b)**

Figure 9-8   Ideal low-pass filter: (a) system function; (b) unit impulse response.

PROBLEM 9-10   Find the response of the *RC* network of Problem 9-6 to a unit step function $u(t)$ **(a)** by integration of $h(t)$ and **(b)** by convolution.

Solution: From Problem 9-6, we have

$$h(t) = \frac{1}{RC} e^{-t/RC} u(t)$$

(a) By the expression (9.50) of the unit step response when the system is causal, we have

$$v_o(t) = a(t) = \int_0^t h(\tau)\, d\tau = \int_0^t \frac{1}{RC} e^{-\tau/RC}\, d\tau$$

$$= \left[\frac{1}{RC} \int_0^t e^{-\tau/RC}\, d\tau \right] u(t)$$

$$= (1 - e^{-t/RC}) u(t)$$

(b) By the response to an arbitrary input (9.14), we obtain

$$v_o(t) = v_i(t) * h(t) = \int_{-\infty}^{\infty} v_i(\tau)h(t-\tau)\,d\tau$$

$$= \int_{-\infty}^{\infty} u(\tau)\frac{1}{RC}e^{-(t-\tau)/RC}u(t-\tau)\,d\tau$$

$$= \left[\frac{1}{RC}\int_0^t e^{-(t-\tau)/RC}\,d\tau\right]u(t)$$

$$= \left(\frac{1}{RC}e^{-t/RC}\int_0^t e^{\tau/RC}\,d\tau\right)u(t)$$

$$= (1 - e^{-t/RC})u(t)$$

PROBLEM 9-11   Show that the Fourier transform of $a(t)$ is given by

$$A(\omega) = \mathscr{F}[a(t)] = \pi H(0)\delta(\omega) + \frac{1}{j\omega}H(\omega)$$

where $a(t)$ is the unit step response of the system and $H(\omega)$ is the system function of the system.

Solution: From the Fourier transform of a unit step function (6.20), we have

$$\mathscr{F}[f_i(t)] = \mathscr{F}[u(t)] = \pi\delta(\omega) + \frac{1}{j\omega}$$

Now, if $\mathscr{F}[f_0(t)] = \mathscr{F}[a(t)] = A(\omega)$, then from (9.20), we have

$$A(\omega) = \left[\pi\delta(\omega) + \frac{1}{j\omega}\right]H(\omega)$$

$$= \pi\delta(\omega)H(\omega) + \frac{1}{j\omega}H(\omega)$$

$$= \pi H(0)\delta(\omega) + \frac{1}{j\omega}H(\omega)$$

in view of a property of δ-function (4.6).

Alternate Solution: Since from (9.48), that is,

$$a(t) = \int_{-\infty}^{t} h(\tau)\,d\tau$$

it follows from the result of Problem 7-7 that

$$A(\omega) = \frac{1}{j\omega}H(\omega) + \pi H(0)\delta(\omega)$$

PROBLEM 9-12   Let $F(\omega) = R(\omega) + jX(\omega)$ be the Fourier transform of a causal fuction $f(t)$. Then show that $f(t)$ can be expressed in terms of $R(\omega)$ or $X(\omega)$ alone.

Solution: Since $f(t)$ is causal, by definition

$$f(t) = 0 \qquad \text{for} \quad t < 0$$

Accordingly,

$$f(-t) = 0 \qquad \text{for} \quad t > 0$$

Therefore, from the result of Problem 2-1,

$$f(t) = 2f_e(t) = 2f_0(t) \qquad \text{for} \quad t > 0$$

where

$$f(t) = f_e(t) + f_0(t)$$

and $f_e(t)$ and $f_0(t)$ are the even and odd components of $f(t)$, respectively. Then from the results of

Problem 5-11 and Problem 5-12, we obtain

$$f(t) = \frac{2}{\pi} \int_0^\infty R(\omega)\cos \omega t \, d\omega$$

$$= -\frac{2}{\pi} \int_0^\infty X(\omega)\sin \omega t \, d\omega$$

for $t > 0$.

PROBLEM 9-13 Let $F(\omega) = R(\omega) + jX(\omega)$ be the Fourier transform of a causal function $f(t)$. Then prove the following identities:

$$\int_{-\infty}^\infty R^2(\omega) \, d\omega = \int_{-\infty}^\infty X^2(\omega) \, d\omega$$

$$\int_0^\infty f^2(t) \, dt = \frac{2}{\pi} \int_0^\infty R^2(\omega) \, d\omega$$

Solution: With the decomposition of $f(t)$ into its even and odd components,

$$f(t) = f_e(t) + f_o(t)$$

we have from (5.17) and (5.18)

$$\mathscr{F}[f_e(t)] = R(\omega) \quad \text{and} \quad \mathscr{F}[f_o(t)] = jX(\omega)$$

Therefore, from Parseval's theorem (7.18), we have

$$\int_{-\infty}^\infty [f_e(t)]^2 \, dt = \frac{1}{2\pi} \int_{-\infty}^\infty R^2(\omega) \, d\omega$$

$$\int_{-\infty}^\infty [f_o(t)]^2 \, dt = \frac{1}{2\pi} \int_{-\infty}^\infty X^2(\omega) \, d\omega$$

From the causality of $f(t)$ and the result of Problem 9-12,

$$f(t) = 2f_e(t) = 2f_o(t) \qquad \text{for} \quad t > 0$$

Hence,

$$|f_e(t)| = |f_o(t)|$$

Therefore, we conclude that

$$\int_{-\infty}^\infty R^2(\omega) \, d\omega = \int_{-\infty}^\infty X^2(\omega) \, d\omega$$

Since

$$|F(\omega)|^2 = R^2(\omega) + X^2(\omega)$$

and from Parseval's theorem (7.18), we have

$$\int_{-\infty}^\infty f^2(t) \, dt = \frac{1}{2\pi} \int_{-\infty}^\infty |F(\omega)|^2 \, d\omega$$

$$= \frac{1}{2\pi} \int_{-\infty}^\infty |R^2(\omega) + X^2(\omega)| \, d\omega$$

$$= \frac{1}{\pi} \int_{-\infty}^\infty R^2(\omega) \, d\omega$$

$$= \frac{2}{\pi} \int_0^\infty R^2(\omega) \, d\omega$$

in view of $R^2(-\omega) = R^2(\omega)$.

For a causal function $f(t)$, since $f(t) = 0$ for $t < 0$,

$$\int_{-\infty}^\infty f^2(t) \, dt = \int_0^\infty f^2(t) \, dt$$

Hence,

$$\int_0^\infty f^2(t)\, dt = \frac{2}{\pi} \int_0^\infty R^2(\omega)\, d\omega$$

PROBLEM 9-14 If the causal function $f(t)$ contains no impulses at the origin, then show that with $F(\omega) = \mathscr{F}[f(t)] = R(\omega) + jX(\omega)$, $R(\omega)$ and $X(\omega)$ satisfy the following equations:

$$R(\omega) = \frac{1}{\pi} \int_{-\infty}^\infty \frac{X(y)}{\omega - y}\, dy$$

$$X(\omega) = -\frac{1}{\pi} \int_{-\infty}^\infty \frac{R(y)}{\omega - y}\, dy$$

Together, these equations are known as the Hilbert transform pair (see Section 8-7).

Solution: Let

$$f(t) = f_e(t) + f_o(t)$$

where $f_e(t)$ and $f_o(t)$ are the even and odd components of $f(t)$, respectively. Since $f(t)$ is causal, that is, $f(t) = 0$ for $t < 0$, it can be assumed that

$$f_e(t) = -f_o(t) \qquad \text{for} \quad t < 0$$

Also, from the result of Problem 9-12, we have

$$f_e(t) = f_o(t) \qquad \text{for} \quad t > 0$$

Therefore, we may write that

$$f_e(t) = f_o(t)\,\mathrm{sgn}\, t$$
$$f_o(t) = f_e(t)\,\mathrm{sgn}\, t$$

Now, from (5.17), (5.18), and the result of Problem 6-14, we have

$$\mathscr{F}[f_e(t)] = R(\omega)$$
$$\mathscr{F}[f_o(t)] = jX(\omega)$$

$$\mathscr{F}[\mathrm{sgn}\, t] = \frac{2}{j\omega}$$

Hence, from the frequency convolution theorem (7.15), we obtain

$$R(\omega) = \mathscr{F}[f_e(t)] = \mathscr{F}[f_o(t)\,\mathrm{sgn}\, t]$$

$$= \frac{1}{2\pi}\, jX(\omega) * \frac{2}{j\omega}$$

$$= \frac{1}{\pi}\, X(\omega) * \frac{1}{\omega}$$

$$= \frac{1}{\pi} \int_{-\infty}^\infty \frac{X(y)}{\omega - y}\, dy$$

Similarly,

$$jX(\omega) = \mathscr{F}[f_o(t)] = \mathscr{F}[f_e(t)\,\mathrm{sgn}\, t]$$

$$= \frac{1}{2\pi}\, R(\omega) * \frac{2}{j\omega}$$

$$= -j\frac{1}{\pi}\, R(\omega) * \frac{1}{\omega}$$

Hence,

$$X(\omega) = -\frac{1}{\pi}\, R(\omega) * \frac{1}{\omega} = -\frac{1}{\pi} \int_{-\infty}^\infty \frac{R(y)}{\omega - y}\, dy$$

PROBLEM 9-15   The real part of the system function $H(\omega)$ of a causal system is known to be $\pi\delta(\omega)$. Find the system function $H(\omega)$.

Solution: Let

$$H(\omega) = R(\omega) + jX(\omega)$$

Since the system is causal, its impulse response $h(t) = \mathscr{F}^{-1}[H(\omega)]$ must be a causal function. Thus, from the result of Problem 9-14 and $R(\omega) = \pi\delta(\omega)$, we obtain

$$X(\omega) = -\frac{1}{\pi}\int_{-\infty}^{\infty}\frac{\pi\delta(y)}{\omega - y}\,dy = -\int_{-\infty}^{\infty}\delta(y)\frac{1}{\omega - y}\,dy = -\frac{1}{\omega}$$

with the use of definition (6.6) of the δ-function. Hence,

$$H(\omega) = \pi\delta(\omega) - j\frac{1}{\omega} = \pi\delta(\omega) + \frac{1}{j\omega}$$

Response to Random Signals

PROBLEM 9-16   Find the average autocorrelation function of the output of a low-pass RC network (shown in Figure 9-9) when the input is a white noise. Also find the mean-square noise voltage at the output.

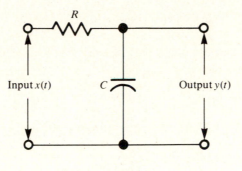

Figure 9-9

Solution: From the result of Problem 9-6, the impulse response $h(t)$ of the network is given by

$$h(t) = \frac{1}{RC}e^{-t/RC}u(t)$$

while from (8.63), the average input (white noise) autocorrelation function is given by

$$\bar{R}_{xx}(\tau) = K\delta(\tau)$$

Then by use of the relation (9.52), we have

$$\bar{R}_{yy}(\tau) = \int_{-\infty}^{\infty}\frac{1}{RC}e^{-\lambda/RC}u(\lambda)\int_{-\infty}^{\infty}\frac{K}{RC}e^{-\sigma/RC}u(\sigma)\delta(\tau + \sigma - \lambda)\,d\sigma\,d\lambda$$

$$= \frac{K}{(RC)^2}\int_{-\infty}^{\infty}e^{-\sigma/RC}u(\sigma)\int_{-\infty}^{\infty}\delta(\tau + \sigma - \lambda)e^{-\lambda/RC}u(\lambda)\,d\lambda\,d\sigma$$

Recalling definition 6.6 of the δ-function, we get

$$\bar{R}_{yy}(\tau) = \frac{K}{(RC)^2}\int_{-\infty}^{\infty}e^{-\sigma/RC}u(\sigma)e^{-(\tau + \sigma)/RC}\,d\sigma$$

$$= \frac{K}{(RC)^2}\int_{0}^{\infty}e^{-\tau/RC}e^{-2\sigma/RC}\,d\sigma$$

since $u(\sigma) = 0$ for $\sigma < 0$ and $u(\sigma) = 1$ for $\sigma > 0$. Hence,

$$\bar{R}_{yy}(\tau) = \frac{K}{(RC)^2}e^{-\tau/RC}\int_{0}^{\infty}e^{-2\sigma/RC}\,d\sigma = \frac{K}{2RC}e^{-\tau/RC} \tag{a}$$

Equation (a) is only valid for τ positive; however, since the autocorrelation function is an even function of

τ (see eq. (7.37)),

$$\bar{R}_{yy}(\tau) = \frac{K}{2RC} e^{-|\tau|/RC}, \qquad -\infty < \tau < \infty \qquad \text{(b)}$$

The mean-square noise voltage at the output is given by

$$\lim_{T \to \infty} \frac{1}{T} \int_{-T/2}^{T/2} [y(t)]^2 \, dt = \bar{R}_{yy}(0) = \frac{K}{2RC} \qquad \text{(c)}$$

PROBLEM 9-17   Find the power spectral density of the output of the RC network of Problem 9-16, when the input is the same white noise. Also check the mean-square noise voltage at the output by eq. (8.59).

Solution: From Problem 9-6, the system function $H(\omega)$ of the RC network is given by

$$H(\omega) = \frac{1/(RC)}{j\omega + 1/(RC)}$$

The power spectral density of the input white noise is given by (8.62),

$$P_i(\omega) = K$$

Thus, from the power spectral densities relationship (9.53), the output power spectral density is given by

$$P_o(\omega) = |H(\omega)|^2 P_i(\omega) = \frac{[1/(RC)]^2}{\omega^2 + [1/(RC)]^2} K$$

From (8.59), the mean-square output voltage may be evaluated from $P_o(\omega)$; thus,

$$\lim_{T \to \infty} \frac{1}{T} \int_{-T/2}^{T/2} [y(t)]^2 \, dt = \frac{1}{2\pi} \int_{-\infty}^{\infty} P_o(\omega) \, d\omega$$

$$= \frac{K}{2\pi(RC)^2} \int_{-\infty}^{\infty} \frac{d\omega}{\omega^2 + [1/(RC)]^2} = \frac{K}{2RC}$$

which agrees with the result of Problem 9-16.

Supplementary Exercises

PROBLEM 9-18   If the input function of the linear system specified by $H(j\omega)$ is a periodic time function with period T, find the output response of the system.

Answer: $f_i(t) = \displaystyle\sum_{n=-\infty}^{\infty} c_n e^{jn\omega_0 t}, \qquad f_o(t) = \displaystyle\sum_{n=-\infty}^{\infty} c_n H(jn\omega_0) e^{jn\omega_0 t}$

PROBLEM 9-19   Show that the steady-state responses of the system specified by $H(j\omega)$ to the input source functions $\cos \omega t$ and $\sin \omega t$ are given by $Re[H(j\omega)e^{j\omega t}]$ and $Im[H(j\omega)e^{j\omega t}]$, respectively, where Re denotes "the real part of" and Im denotes "the imaginary part of."

PROBLEM 9-20   In sinusoidal steady-state analysis, it is customary to use phasor representation for the sinusoidal functions. Thus a cosine function $v(t)$ can be written as

$$v(t) = v_m \cos(\omega t + \beta) = Re[\mathbf{V}_m e^{j\omega t}]$$

where $\mathbf{V}_m = v_m e^{j\beta} = v_m \angle \beta$. The complex quantity $\mathbf{V}_m$ is referred to as the **phasor** representing $v(t)$.

If the system function $H(j\omega)$ is expressed in the phasor form, that is

$$H(j\omega) = |H(j\omega)|e^{j\theta(\omega)} = |H(j\omega)| \angle \theta(\omega)$$

then show that the steady-state responses of the system to input $v_m \cos(\omega t + \beta)$ and $v_m \sin(\omega t + \beta)$ are given by, respectively,

$$Re[H(j\omega)\mathbf{V}_m e^{j\omega t}] = v_m |H(j\omega)| \cos(\omega t + \beta + \theta)$$
$$Im[H(j\omega)\mathbf{V}_m e^{j\omega t}] = v_m |H(j\omega)| \sin(\omega t + \beta + \theta)$$

PROBLEM 9-21   A square-wave voltage source $v(t)$, whose waveform is shown in Figure 9-10a, is applied to the series RL circuit of Figure 9-10b. Find the steady-state response current $i_s(t)$.

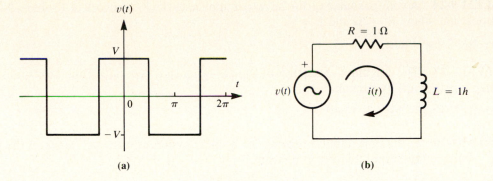

(a)   (b)

Figure 9-10

Answer: $i_s(t) = \dfrac{4V}{\pi}\left[\dfrac{1}{\sqrt{2}}\cos(t - \tan^{-1}1) - \dfrac{1}{3\sqrt{10}}\cos(3t - \tan^{-1}3) + \dfrac{1}{5\sqrt{26}}\cos(5t - \tan^{-1}5) + \cdots\right]$

PROBLEM 9-22   At the terminals a–b of the network in Figure 9-11, the voltage $v_{ab}(t)$ is periodic and defined by the Fourier series

$$v_{ab}(t) = V_0 + \sum_{n=1}^{\infty} V_n\cos(n\omega_0 t + \beta_n)$$

and the steady-state current $i_s(t)$ entering the a terminal is

$$i_s(t) = I_0 + \sum_{n=1}^{\infty} I_n\cos(n\omega_0 t + \alpha_n)$$

Show that the average power input P_{ab} defined by

$$P_{ab} = \frac{1}{T}\int_{-T/2}^{T/2} v_{ab}(t)i_s(t)\,dt$$

is equal to

$$P_{ab} = V_0 I_0 + \frac{1}{2}\sum_{n=1}^{\infty} V_n I_n\cos(\beta_n - \alpha_n)$$

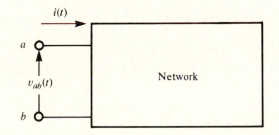

Figure 9-11

PROBLEM 9-23   Find the unit impulse response for the current of the RL network shown in Figure 9-12.

Answer: $h(t) = \dfrac{1}{L}\,e^{-(R/L)t}u(t)$

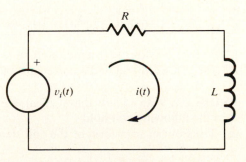

Figure 9-12

PROBLEM 9-24 A voltage source $v_i(t) = 2e^{-t}u(t)$ is applied to the RL network of Problem 9-23. Find the response $i(t)$, where $R = 2\Omega$ and $L = 1h$.

Answer: $2(e^{-t} - e^{-2t})u(t)$

PROBLEM 9-25 The unit impulse response of a linear system is $e^{-t}\cos tu(t)$. Find the response due to a unit step function $u(t)$ by convolution.

Answer: $\frac{1}{2}[e^{-t}(\sin t - \cos t) + 1]u(t)$

PROBLEM 9-26 If the unit impulse response of a linear system is $h(t) = te^{-t}u(t)$ and the input is $f_i(t) = e^{-t}u(t)$, find the output frequency spectrum.

Answer: $1/(1 + j\omega)^3$

PROBLEM 9-27 Show that if the input to a linear system is differentiated, then the response is also differentiated. [*Hint:* Show that $f_i'(t) * h(t) = [f_i(t) * h(t)]' = f_o'(t)$.]

PROBLEM 9-28 Find the unit impulse response $h(t)$ of a linear system whose system function is

$$H(\omega) = \begin{cases} e^{-j\theta_0} & \text{for } \omega > 0 \\ e^{j\theta_0} & \text{for } \omega < 0 \end{cases}$$

[*Hint:* Note that $H(\omega) = \cos\theta_0 - j\sin\theta_0 \text{sgn }\omega$, and use the result of Problem 6-25.]

Answer: $h(t) = \cos\theta_0\delta(t) + \dfrac{\sin\theta_0}{\pi t}$

PROBLEM 9-29 The system of Problem 9-28 is called a **phase shifter**. Show that the response of the system of Problem 9-28 to $\cos\omega_c t$ is $\cos(\omega_c t - \theta_0)$.

PROBLEM 9-30 Find the unit impulse response $h(t)$ of the *ideal high-pass filter* whose system function $H(j\omega)$ is

$$H(j\omega) = \begin{cases} 0 & \text{for } |\omega| < \omega_c \\ e^{-j\omega t_0} & \text{for } |\omega| > \omega_c \end{cases}$$

[*Hint:* Use the result of Problem 9-9, and note that $H(j\omega) = e^{-j\omega t_0} - H_l(j\omega)$, where $H_l(j\omega)$ is the system function of an ideal low-pass filter.]

Answer: $h(t) = \delta(t - t_0) - \dfrac{\omega_c}{\pi}\dfrac{\sin\omega_c(t - t_0)}{\omega_c(t - t_0)}$

PROBLEM 9-31 A **Gaussian filter** is a linear system whose system function is $H(\omega) = e^{-\alpha\omega^2}e^{-j\omega t_0}$. Find its unit impulse response.

Answer: $h(t) = \dfrac{1}{2\sqrt{\pi\alpha}}e^{-(t - t_0)^2/(4\alpha)}$

PROBLEM 9-32 If $H(\omega) = R(\omega) + jX(\omega)$ is the system function of a linear causal system, then show that the unit impulse response $h(t)$ of the system can be expressed as a function of either $R(\omega)$ or $X(\omega)$; that is,

$$h(t) = \frac{2}{\pi}\int_0^\infty R(\omega)\cos\omega t\, d\omega = -\frac{2}{\pi}\int_0^\infty X(\omega)\sin\omega t\, d\omega$$

[*Hint:* $h(t) = 0$ for $t < 0$; hence $h(t)$ can be expressed as $h(t) = 2h_e(t) = 2h_0(t)$ for $t > 0$, where $h_e(t)$ and $h_0(t)$ are the even and odd components of $h(t)$, respectively.]

PROBLEM 9-33 Show that if $H(\omega) = R(\omega) + jX(\omega)$ is the system function of a linear causal system, then the following will hold:
(a) The Fourier transform of the unit step response $a(t)$ of the system is given by

$$\mathscr{F}[a(t)] = \pi R(0)\delta(\omega) + \frac{X(\omega)}{\omega} - j\frac{R(\omega)}{\omega}$$

(b) The unit step response $a(t)$ can be expressed as

$$a(t) = \frac{2}{\pi} \int_0^\infty \frac{R(\omega)}{\omega} \sin \omega t \, d\omega = R(0) + \frac{2}{\pi} \int_0^\infty \frac{X(\omega)}{\omega} \cos \omega t \, d\omega$$

PROBLEM 9-34 Two signals $f_a(t)$ and $f_b(t)$ are applied to two systems, as shown in Figure 9-13. The resultant outputs are $f_1(t)$ and $f_2(t)$. Express the average cross-correlation function $\bar{R}_{12}$ of $f_1(t)$ and $f_2(t)$ in terms of $\bar{R}_{ab}$, $h_1(t)$, and $h_2(t)$, where $h_1(t)$ and $h_2(t)$ are the respective unit impulse responses of the two systems.

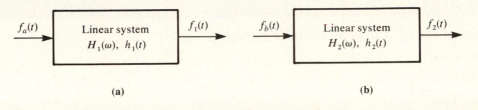

(a) (b)

Figure 9-13

Answer: $\bar{R}_{12}(\tau) = \displaystyle\int_{-\infty}^\infty h_1(\lambda) \int_{-\infty}^\infty \bar{R}_{ab}(\tau + \sigma - \lambda) h_2(\sigma) \, d\sigma \, d\lambda$

PROBLEM 9-35 If the **cross-spectral density** $S_{12}(\omega)$ of two functions $f_1(t)$ and $f_2(t)$ is defined by $S_{12}(\omega) = \mathscr{F}[\bar{R}_{12}(\tau)]$, show that for the two systems of Problem 9-34,

$$S_{12}(\omega) = H_1(\omega) H_2^*(\omega) S_{ab}(\omega)$$

where $S_{ab}(\omega)$ is the cross-spectral density of $f_a(t)$ and $f_b(t)$, and $H_1(\omega)$ and $H_2(\omega)$ are the respective system functions of the two systems.

PROBLEM 9-36 Find the average autocorrelation function of the output of the low-pass network shown in Figure 9-9 when the input has an average autocorrelation function of the form $\bar{R}_{xx}(\tau) = \frac{1}{2}\alpha K e^{-\alpha|\tau|}$.

Answer: $\bar{R}_{yy}(\tau) = \dfrac{b^2 \alpha K}{2(b^2 - \alpha^2)} \left[e^{-\alpha|\tau|} - \dfrac{\alpha}{b} e^{-b|\tau|} \right]$, where $b = \dfrac{1}{RC}$

10 APPLICATIONS TO BOUNDARY-VALUE PROBLEMS

THIS CHAPTER IS ABOUT

☑ **Separation of Variables and Fourier Series**
☑ **Vibration**
☑ **Heat Conduction**
☑ **Potential Theory**

10-1. Separation of Variables and Fourier Series

Many boundary-value problems in mathematical physics can be solved conveniently by a method referred to as the **separation of variables**. We shall illustrate the essence of the method by means of particular examples.

A. The wave equation

Consider the following equation governing small transverse vibrations of an elastic string stretched to length l and then fixed at the end points:

One-dimensional wave equation
$$\frac{\partial^2 u(x,t)}{\partial x^2} - \frac{1}{c^2}\frac{\partial^2 u(x,t)}{\partial t^2} = 0 \qquad (10.1)$$

where $u(x,t)$ is the deflection of the string, and $c^2 = T/\rho$, where ρ is the mass of the string per unit length, and T is the tension of the string. Equation (10.1) is known as the **one-dimensional wave equation**. The boundary conditions are

Boundary conditions
$$u(0,t) = 0 \quad \text{and} \quad u(l,t) = 0 \quad \text{for all} \quad t \qquad (10.2)$$

The initial conditions are

Initial conditions
$$u(x,0) = f(x) \quad \text{and} \quad \left.\frac{\partial u(x,t)}{\partial t}\right|_{t=0} = g(x) \qquad (10.3)$$

EXAMPLE 10-1: Find the solution $u(x,t)$ of the one-dimensional wave equation (10.1) satisfying the boundary conditions (10.2) and the initial conditions (10.3).

Solution: First, assume that the solution $u(x,t)$ of (10.1) will be of the form

Separation of variables
$$u(x,t) = X(x)T(t) \qquad (10.4)$$

which is a product of two functions, each depending on only one of the variables x and t. By differentiating (10.4), we have

$$\frac{\partial^2 u(x,t)}{\partial x^2} = X''(x)T(t) \quad \text{and} \quad \frac{\partial^2 u(x,t)}{\partial t^2} = X(x)T''(t) \qquad (10.5)$$

where primes denote differentiation with respect to the appropriate argument variable of each factor. By substituting (10.5) into (10.1), we obtain

$$X''(x)T(t) = \frac{1}{c^2}X(x)T''(t) \qquad (10.6)$$

Dividing by $X(x)T(t)$, thus separating the variables (one to each side of the equation), we get

$$\frac{X''(x)}{X(x)} = \frac{T''(t)}{c^2 T(t)} \qquad (10.7)$$

Now, the left-hand side of (10.7) is independent of t, and so therefore is the right-hand side. The right-hand side is independent of x, and thus the left-hand side must be also. Hence, the expressions on the left- and right-hand sides of (10.7) must be equal to a constant, which is independent of both x and t. Thus,

Separation constant

$$\frac{X''(x)}{X(x)} = \frac{T''(t)}{c^2 T(t)} = -k^2 \tag{10.8}$$

The constant, which is denoted by $-k^2$, is called the **separation constant**. Equation (10.8) yields the two ordinary linear differential equations

$$X''(x) + k^2 X(x) = 0 \tag{10.9}$$

$$T''(t) + c^2 k^2 T(t) = 0 \tag{10.10}$$

We next determine solutions $X(x)$ and $T(t)$ of (10.9) and (10.10), so that $u(x, t) = X(x)T(t)$ satisfies the conditions (10.2) and (10.3). The general solutions of (10.9) and (10.10) are

$$X(x) = A \cos kx + B \sin kx \tag{10.11}$$

$$T(t) = C \cos kct + D \sin kct \tag{10.12}$$

From the boundary conditions (10.2),

$$u(0, t) = X(0)T(t) = 0$$

Here we have a product of two terms that equals zero. Since $T(t)$ is not identically zero, $X(0)$ must be equal to zero. Similarly, the second condition

$$u(l, t) = X(l)T(t) = 0$$

implies that $X(l) = 0$.
From $X(0) = 0$, we conclude that

$$X(0) = A \cos 0 + B \sin 0 = A = 0 \tag{10.13}$$

hence, by (10.11),

$$X(x) = B \sin kx$$

From the second condition,

$$X(l) = B \sin kl = 0$$

But if $B = 0$, then $X(x) = 0$, and hence $u(x, t) = 0$. This contradicts the initial conditions (10.3) that $u(x, 0) = f(x) \neq 0$. We therefore conclude that

$$\sin kl = 0$$

from which

$$kl = n\pi \quad \text{or} \quad k = \frac{n\pi}{l}, \quad n = 1, 2, \ldots \tag{10.14}$$

Thus, we obtain an infinite set of solutions $X(x) = X_n(x)$, where

$$X_n(x) = B_n \sin \frac{n\pi x}{l}, \quad n = 1, 2, \ldots \tag{10.15}$$

The solution (10.12) now becomes

$$T_n(t) = C_n \cos \frac{cn\pi}{l} t + D_n \sin \frac{cn\pi}{l} t \tag{10.16}$$

Hence, the functions

$$u_n(x, t) = X_n(x)T_n(t) = \sin \frac{n\pi x}{l} \left(E_n \cos \frac{cn\pi}{l} t + F_n \sin \frac{cn\pi}{l} t \right) \tag{10.17}$$

are the solutions of (10.1) satisfying the boundary conditions (10.2).
In (10.17), the coefficients E_n and F_n are as yet undetermined. Note that $B_n C_n = E_n$ and $B_n D_n = F_n$. Clearly, a single solution $u_n(x, t)$ of (10.17) in general will not satisfy the initial conditions (10.3). As (10.1)

is linear, we consider the infinite series

$$u(x,t) = \sum_{n=1}^{\infty} u_n(x,t) = \sum_{n=1}^{\infty} \sin\frac{n\pi x}{l}\left(E_n\cos\frac{cn\pi}{l}t + F_n\sin\frac{cn\pi}{l}t\right) \qquad (10.18)$$

Now, let us require that (10.18) satisfy the initial conditions (10.3). Hence, we find that the coefficients E_n and F_n must satisfy the equations

$$u(x,0) = \sum_{n=1}^{\infty} E_n\sin\frac{n\pi x}{l} = f(x) \qquad (10.19)$$

$$\frac{\partial u(x,t)}{\partial t}\bigg|_{t=0} = \sum_{n=1}^{\infty} F_n\frac{cn\pi}{l}\sin\frac{n\pi x}{l} = g(x) \qquad (10.20)$$

Equation (10.19) shows that the coefficients E_n must be chosen such that $u(x,0)$ becomes the Fourier sine series of $f(x)$ (see Section 2-3); that is,

$$E_n = \frac{2}{l}\int_0^l f(x)\sin\frac{n\pi x}{l}\,dx, \qquad n = 1,2,\dots \qquad (10.21)$$

Similarly, (10.20) indicates that the coefficients F_n must be chosen such that $\partial u(x,t)/\partial t|_{t=0}$ becomes the Fourier sine series of $g(x)$; that is,

$$F_n\frac{cn\pi}{l} = \frac{2}{l}\int_0^l g(x)\sin\frac{n\pi x}{l}\,dx \qquad (10.22)$$

or

$$F_n = \frac{2}{cn\pi}\int_0^l g(x)\sin\frac{n\pi x}{l}\,dx, \qquad n = 1,2,\dots \qquad (10.23)$$

Hence, with the coefficients E_n and F_n given by (10.21) and (10.23), the infinite series (10.18) is the desired solution.

B. Laplace's equation

In stationary heat flow or electrostatic potential problems in a plane (taken as the xy-plane), the temperature distribution function or the electrostatic potential function $u(x, y)$ in a source-free region satisfies the following equation in two-dimensional space:

Laplace's equation
$$\frac{\partial^2 u(x,y)}{\partial x^2} + \frac{\partial^2 u(x,y)}{\partial y^2} = 0 \qquad (10.24)$$

This is known as **Laplace's equation**.

EXAMPLE 10-2: Find the solution of (10.24) with the following boundary conditions:

Boundary conditions
$$u(x,y) = 0 \qquad \text{at} \quad x = 0, y = 0, \text{ and } y = b \qquad (10.25)$$
$$u(x,y) = U_0 \qquad \text{at} \quad x = d \text{ and } 0 < y < b \qquad (10.26)$$

Solution: Assume that the solution of (10.24) is of the form

Separation of variables
$$u(x,y) = X(x)Y(y) \qquad (10.27)$$

where $X(x)$ is a function of x only and $Y(y)$ a function of y only. By substituting (10.27) into (10.24), we have

$$X''(x)Y(y) + X(x)Y''(y) = 0 \qquad (10.28)$$

Dividing by $X(x)Y(y)$, thus separating the variables, we obtain

$$\frac{X''(x)}{X(x)} + \frac{Y''(y)}{Y(y)} = 0 \qquad (10.29)$$

or

$$\frac{X''(x)}{X(x)} = -\frac{Y''(y)}{Y(y)} \qquad (10.30)$$

Now, the left-hand side of (10.30) is independent of y, and therefore so is the right-hand side. The right-hand side is independent of x, and thus the left-hand side must be also. This means that the expressions on the left- and right-hand sides of (10.30) must be independent of both x and y and equal to a constant. Let this separation constant be denoted by k^2; then

Separation constant
$$\frac{X''(x)}{X(x)} = -\frac{Y''(y)}{Y(y)} = k^2 \qquad (10.31)$$

The sign of the separation constant was chosen in such a way that the boundary conditions could be satisfied. Equation (10.31) yields the two ordinary linear differential equations

$$X''(x) - k^2 X(x) = 0 \qquad (10.32)$$
$$Y''(y) + k^2 Y(y) = 0 \qquad (10.33)$$

The general solutions of (10.32) and (10.33) are

$$X(x) = Ae^{kx} + Be^{-kx} \qquad (10.34)$$
$$Y(y) = C\cos ky + D\sin ky \qquad (10.35)$$

From the boundary conditions (10.25), we have

$$X(0) = A + B = 0, \qquad Y(0) = C = 0, \qquad Y(b) = D\sin kb = 0$$

Hence,

$$A = -B$$
$$\sin kb = 0$$

from which

$$kb = n\pi \qquad \text{or} \qquad k = \frac{n\pi}{b}, \qquad n = 1, 2, \dots \qquad (10.36)$$

Thus, we obtain an infinite set of solutions $Y(y) = Y_n(y)$, where

$$Y_n(y) = D_n\sin\frac{n\pi y}{b}, \qquad n = 1, 2, \dots \qquad (10.37)$$

The corresponding general solutions (10.34) now become

$$X_n(x) = A_n(e^{kx} - e^{-kx}) = 2A_n\sinh kx$$

$$= 2A_n\sinh\frac{n\pi x}{b}, \qquad n = 1, 2, \dots \qquad (10.38)$$

Hence, the functions

$$u_n(x, y) = X_n(x)Y_n(y) = E_n\sinh\frac{n\pi x}{b}\sin\frac{n\pi y}{b}, \qquad n = 1, 2, \dots \qquad (10.39)$$

are the solutions of (10.24) satisfying the boundary conditions (10.25). Note that $2A_n D_n$ was replaced by the new arbitrary constant E_n.

Clearly, a single solution $u_n(x, y)$ of (10.39) will not satisfy the other boundary condition of (10.26). Since (10.24) is linear, we consider the infinite series

$$u(x, y) = \sum_{n=1}^{\infty} u_n(x, y) = \sum_{n=1}^{\infty} E_n\sinh\frac{n\pi x}{b}\sin\frac{n\pi y}{b} \qquad (10.40)$$

Applying the boundary condition of (10.26),

$$u(d, y) = U_0 = \sum_{n=1}^{\infty} E_n\sinh\frac{n\pi d}{b}\sin\frac{n\pi y}{b}$$

$$= \sum_{n=1}^{\infty} c_n\sin\frac{n\pi y}{b}, \qquad 0 < y < b \qquad (10.41)$$

where

$$c_n = E_n\sinh\frac{n\pi d}{b}$$

Equation (10.41) is a Fourier sine series and the coefficients c_n may be determined as

$$c_n = \frac{2}{b}\int_0^b U_0 \sin\frac{n\pi y}{b}\,dy = \frac{2U_0}{n\pi}(1 - \cos n\pi) = \begin{cases} \dfrac{4U_0}{n\pi}, & n = 1, 3, \ldots \\ 0, & n = 2, 4, \ldots \end{cases}$$

However,

$$c_n = E_n \sinh\frac{n\pi d}{b}$$

and therefore,

$$E_n = \frac{4U_0}{n\pi\sinh(n\pi d/b)}, \qquad n = 1, 3, 5, \ldots \tag{10.42}$$

which may be substituted into (10.40) to give the desired solution:

$$u(x, y) = \frac{4U_0}{\pi}\sum_{n=\text{odd}}^{\infty}\frac{1}{n}\frac{\sinh(n\pi x/b)}{\sinh(n\pi d/b)}\sin\frac{n\pi y}{b} \tag{10.43}$$

Note: In this section we have obtained formal solutions of certain linear, second-order partial differential equations that satisfy given boundary and initial conditions, but we have *not* shown that any of the solutions are unique. Since the proof of uniqueness is quite involved—and, unfortunately, no general uniqueness theorem exists—we shall not prove the uniqueness of the solutions obtained either in this section or in those to follow.

10-2. Vibration

The vibration of a string and its governing equation, the one-dimensional wave equation, have been discussed in Example 10-1. In the following, we shall apply the Fourier analysis technique to the various problems of vibration.

A. Two-dimensional wave equation—Double Fourier series

The governing equation for the small transverse vibration of a membrane is given by

Two-dimensional wave equation
$$\left(\frac{\partial^2 u}{\partial x^2} + \frac{\partial^2 u}{\partial y^2}\right) - \frac{1}{c^2}\frac{\partial^2 u}{\partial t^2} = 0 \tag{10.44}$$

where $u(x, y, t)$ is the deflection of the membrane, and $c^2 = T/\rho$, where ρ is the mass of the membrane per unit area and T is the tension of the membrane. Equation (10.44) is called the **two-dimensional wave equation**.

EXAMPLE 10-3: Consider the rectangular membrane shown in Figure 10-1 and find the solution of the two-dimensional wave equation (10.44) that satisfies the following boundary condition:

$$u(x, y, t) = 0 \qquad \text{on the boundary of the membrane for all } t$$

that is,

Boundary conditions $u(x, y, t) = 0 \qquad$ for $\quad x = 0, x = a, y = 0, \text{ and } y = b$ $\qquad$ **(10.45)**

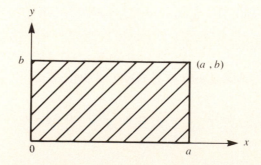

Figure 10-1   A rectangular membrane.

The initial conditions are

$$u(x, y, 0) = f(x, y) \tag{10.46}$$

Initial conditions

$$\left.\frac{\partial u(x, y, t)}{\partial t}\right|_{t=0} = g(x, y) \tag{10.47}$$

where $f(x, y)$ and $g(x, y)$ are the given initial displacement and initial velocity of the membrane, respectively.

Solution: Assume that the solution of (10.44) is of the form

Separation of variables

$$u(x, y, t) = X(x)Y(y)T(t) \tag{10.48}$$

By substituting (10.48) into (10.44), we obtain

$$X''(x)Y(y)T(t) + X(x)Y''(y)T(t) - \frac{1}{c^2}X(x)Y(y)T''(t) = 0 \tag{10.49}$$

where primes denote differentiation with respect to the arguments of each function. Dividing through by $X(x)Y(y)T(t)$ and separating the variables, we obtain

$$\frac{X''(x)}{X(x)} + \frac{Y''(y)}{Y(y)} = \frac{1}{c^2}\frac{T''(t)}{T(t)} \tag{10.50}$$

Since the right-hand side of (10.50) depends only t while the left-hand side does not depend on t, the expressions on both sides must be equal to a constant. Denoting this constant by $-k^2$,

Separation constant

$$\frac{X''(x)}{X(x)} + \frac{Y''(y)}{Y(y)} = \frac{1}{c^2}\frac{T''(t)}{T(t)} = -k^2$$

This yields the two ordinary differential equations

$$T''(t) + c^2 k^2 T(t) = 0 \tag{10.51}$$

$$\frac{X''(x)}{X(x)} + \frac{Y''(y)}{Y(y)} = -k^2$$

or

$$\frac{X''(x)}{X(x)} = -k^2 - \frac{Y''(y)}{Y(y)} \tag{10.52}$$

Again, since the left-hand side of (10.52) depends only on x while the right-hand side depends only on y, the expressions on both sides must be equal to a constant. This constant must be negative (otherwise, the boundary conditions could not be satisfied), say $-k_x^2$. Then

Separation constant

$$\frac{X''(x)}{X(x)} = -k^2 - \frac{Y''(y)}{Y(y)} = -k_x^2$$

This yields the following differential equations:

$$X''(x) + k_x^2 X(x) = 0 \tag{10.53}$$
$$Y''(y) + k_y^2 Y(y) = 0 \tag{10.54}$$

where

$$k_y^2 = k^2 - k_x^2 \quad \text{or} \quad k_x^2 + k_y^2 = k^2 \tag{10.55}$$

The general solutions of (10.51), (10.53), and (10.54) have the forms

$$X(x) = A\cos k_x x + B\sin k_x x \tag{10.56}$$
$$Y(y) = C\cos k_y y + D\sin k_y y \tag{10.57}$$
$$T(t) = E\cos kct + F\sin kct \tag{10.58}$$

From the boundary condition (10.45),

$$X(0) = 0, \quad X(a) = 0, \quad Y(0) = 0, \quad Y(b) = 0$$

Therefore,

$$X(0) = A = 0, \quad X(a) = B\sin k_x a = 0$$

from which

$$k_x a = m\pi \qquad \text{or} \qquad k_x = \frac{m\pi}{a}, \qquad m = 1, 2, \ldots \tag{10.59}$$

Similarly, $Y(0) = C = 0$ and $Y(b) = D \sin k_y b = 0$; hence,

$$k_y b = n\pi \qquad \text{or} \qquad k_y = \frac{n\pi}{b}, \qquad n = 1, 2, \ldots \tag{10.60}$$

In this way, we obtain the solutions

$$X_m(x) = B_m \sin \frac{m\pi x}{a}, \qquad m = 1, 2, \ldots$$

$$Y_n(y) = D_n \sin \frac{n\pi y}{b}, \qquad n = 1, 2, \ldots$$

Since $k^2 = k_x^2 + k_y^2$,

$$k^2 = k_{mn}^2 = \frac{m^2 \pi^2}{a^2} + \frac{n^2 \pi^2}{b^2} \tag{10.61}$$

and the corresponding general solution of (10.51) is

$$T_{mn}(t) = E_{mn} \cos k_{mn} ct + F_{mn} \sin k_{mn} ct$$

It follows that the functions

$$u_{mn}(x, y, t) = X_m(x) Y_n(y) T_{mn}(t)$$

$$= (G_{mn} \cos k_{mn} ct + H_{mn} \sin k_{mn} ct) \sin \frac{m\pi x}{a} \sin \frac{n\pi y}{b} \tag{10.62}$$

where $m = 1, 2, \ldots$, $n = 1, 2, \ldots$, and with k_{mn} given by (10.61), are the solutions of the wave equation (10.44). These are zero on the boundary of the rectangular membrane in Figure 10-1.

We must now evaluate the arbitrary constants G_{mn} and H_{mn}. To obtain the solution that also satisfies the initial conditions (10.46) and (10.47), we proceed in a manner similar to that used in Example 10-1.

Consider the following double series:

$$u(x, y, t) = \sum_{m=1}^{\infty} \sum_{n=1}^{\infty} u_{mn}(x, y, t)$$

$$= \sum_{m=1}^{\infty} \sum_{n=1}^{\infty} (G_{mn} \cos k_{mn} ct + H_{mn} \sin k_{mn} ct) \sin \frac{m\pi x}{a} \sin \frac{n\pi y}{b} \tag{10.63}$$

From (10.63) and (10.46), we have

**Double Fourier
series**
$$u(x, y, 0) = f(x, y) = \sum_{m=1}^{\infty} \sum_{n=1}^{\infty} G_{mn} \sin \frac{m\pi x}{a} \sin \frac{n\pi y}{b} \tag{10.64}$$

The series (10.64) is called the **double Fourier series** representing $f(x, y)$ in the region $0 < x < a$ and $0 < y < b$.

The Fourier coefficients G_{mn} of $f(x, y)$ in (10.64) can be determined by setting

$$J_m(y) = \sum_{n=1}^{\infty} G_{mn} \sin \frac{n\pi y}{b} \tag{10.65}$$

Then we may write (10.64) in the form

$$f(x, y) = \sum_{m=1}^{\infty} J_m(y) \sin \frac{m\pi x}{a} \tag{10.66}$$

For fixed y, (10.66) is the Fourier sine series of $f(x, y)$ and is considered a function of x. It follows from (2.19) that the coefficients of this expansion are given by

$$J_m(y) = \frac{2}{a} \int_0^a f(x, y) \sin \frac{m\pi x}{a} \, dx \tag{10.67}$$

Now, (10.65) is the Fourier sine series of $J_m(y)$, and hence the coefficients G_{mn} are given by

$$G_{mn} = \frac{2}{b} \int_0^b J_m(y)\sin\frac{n\pi y}{b}\, dy \tag{10.68}$$

Substituting (10.67) into (10.68), we obtain

$$G_{mn} = \frac{4}{ab} \int_0^b \int_0^a f(x, y)\sin\frac{m\pi x}{a}\sin\frac{n\pi y}{b}\, dx\, dy \tag{10.69}$$

where $m = 1, 2, \ldots, n = 1, 2, \ldots$.
To determine the H_{mn} of (10.63), we differentiate (10.63) term-by-term with respect to t. Then, using (10.47), we have

$$\left.\frac{\partial u}{\partial t}\right|_{t=0} = g(x, y) = \sum_{m=1}^{\infty} \sum_{n=1}^{\infty} H_{mn} c k_{mn} \sin\frac{m\pi x}{a}\sin\frac{n\pi y}{b} \tag{10.70}$$

Proceeding as before,

$$H_{mn} = \frac{4}{abck_{mn}} \int_0^b \int_0^a g(x, y)\sin\frac{m\pi x}{a}\sin\frac{n\pi y}{b}\, dx\, dy \tag{10.71}$$

where $m = 1, 2, \ldots, n = 1, 2, \ldots$.
Hence, (10.63), with the coefficients given by (10.69) and (10.71), is the desired solution.

B. Vibration of an infinite string—Fourier integral

In Example 10-4 we shall consider the vibration of an infinite string. In this case we do not have boundary conditions, but only the initial conditions.

EXAMPLE 10-4: Determine the displacement $u(x, t)$ of an infinite string with zero initial velocity. The initial displacement is given by $f(x)$ for $-\infty < x < \infty$.

Solution: The function $u(x, t)$ satisfies the one-dimensional wave equation (10.1):

Wave equation
$$\frac{\partial^2 u(x,t)}{\partial x^2} - \frac{1}{c^2}\frac{\partial^2 u(x,t)}{\partial t^2} = 0$$

The initial conditions are

$$u(x, 0) = f(x), \qquad -\infty < x < \infty \tag{10.72}$$

Initial conditions
$$\left.\frac{\partial u(x,t)}{\partial t}\right|_{t=0} = 0 \tag{10.73}$$

Proceeding as in Example 10-1, we substitute

Separation of variables
$$u(x, t) = X(x)T(t)$$

into (10.1). This yields two ordinary differential equation:

$$X''(x) + k^2 X(x) = 0 \tag{10.74}$$
$$T''(t) + c^2 k^2 T(t) = 0 \tag{10.75}$$

The functions

$$X(x) = A\cos kx + B\sin kx$$
$$T(t) = C\cos kct + D\sin kct$$

are solutions of (10.74) and (10.75), respectively.
By using the initial condition (10.73),

$$T'(0) = kcD = 0$$

Hence, $D = 0$, and

$$u(x, t; k) = (F\cos kx + G\sin kx)\cos kct \tag{10.76}$$

is a solution of (10.1) satisfying (10.73).

Any series of functions (10.76), found in the usual manner by taking k as multiples of a fixed number, would lead to a function that is periodic in x when $t = 0$. However, since $f(x)$ in (10.72) is not assumed periodic, it is natural to use the Fourier integral instead of the Fourier series in the present case.

Since F and G in (10.76) are arbitrary, we may consider them as functions of k and write $F = F(k)$ and $G = G(k)$. Since the wave equation (10.1) is linear and homogeneous, the function

$$u(x, t) = \int_0^\infty u(x, t; k)\, dk = \int_0^\infty [F(k)\cos kx + G(k)\sin kx]\cos kct\, dk \tag{10.77}$$

is also a solution of (10.1).

From (10.72),

$$u(x, 0) = f(x) = \int_0^\infty [F(k)\cos kx + G(k)\sin kx]\, dk \tag{10.78}$$

Now, from Fourier's integral theorem (see Problem 6-19),

$$f(t) = \frac{1}{\pi} \int_0^\infty \left[\int_{-\infty}^\infty f(x)\cos \omega(t - x)\, dx \right] d\omega$$

we may write

$$f(x) = \frac{1}{\pi} \int_0^\infty \left[\int_{-\infty}^\infty f(y)\cos k(x - y)\, dy \right] dk$$

$$= \frac{1}{\pi} \int_0^\infty \left[\int_{-\infty}^\infty f(y)(\cos kx \cos ky + \sin kx \sin ky)\, dy \right] dk$$

$$= \frac{1}{\pi} \int_0^\infty \left[\cos kx \int_{-\infty}^\infty f(y)\cos ky\, dy + \sin kx \int_{-\infty}^\infty f(y)\sin ky\, dy \right] dk \tag{10.79}$$

If we set

$$F(k) = \frac{1}{\pi} \int_{-\infty}^\infty f(y)\cos ky\, dy, \qquad G(k) = \frac{1}{\pi} \int_{-\infty}^\infty f(y)\sin ky\, dy$$

then (10.79) can be written in the form

$$f(x) = \int_0^\infty [F(k)\cos kx + G(k)\sin kx]\, dk \tag{10.80}$$

Comparing (10.80) and (10.78), we may write (10.78) as

$$u(x, 0) = f(x) = \frac{1}{\pi} \int_0^\infty \left[\int_{-\infty}^\infty f(y)\cos k(x - y)\, dy \right] dk \tag{10.81}$$

Then, from (10.77),

$$u(x, t) = \frac{1}{\pi} \int_0^\infty \left[\int_{-\infty}^\infty f(y)\cos k(x - y)\cos kct\, dy \right] dk \tag{10.82}$$

By the use of the trigonometric identity

$$\cos k(x - y)\cos kct = \frac{1}{2} [\cos k(x + ct - y) + \cos k(x - ct - y)]$$

(10.82) becomes

$$u(x, t) = \frac{1}{2} \frac{1}{\pi} \int_0^\infty \left[\int_{-\infty}^\infty f(y)\cos k(x + ct - y)\, dy \right] dk$$

$$+ \frac{1}{2} \frac{1}{\pi} \int_0^\infty \left[\int_{-\infty}^\infty f(y)\cos k(x - ct - y)\, dy \right] dk \tag{10.83}$$

If we replace x by $x \pm ct$ in the first equation of (10.79), we have

$$f(x \pm ct) = \frac{1}{\pi} \int_0^\infty \left[\int_{-\infty}^\infty f(y)\cos k(x \pm ct - y)\, dy \right] dk$$

and comparing this with (10.83), we conclude that

$$u(x, t) = \frac{1}{2} f(x + ct) + \frac{1}{2} f(x - ct) \qquad (10.84)$$

which is the familiar equation for **traveling waves** (see Problem 10-2).

C. Fourier transform technique

In the previous chapters, we have dealing with the Fourier transform pair $f(t)$ and $F(\omega)$, in which the first term denotes a function of time, and the second, a function of frequency. The use of the Fourier transform is by no means restricted to time–frequency domains. If the functions $f(x)$ and $F(s)$ form a Fourier transform pair, then

Fourier transform pair

$$F(s) = \mathscr{F}[f(x)] = \int_{-\infty}^{\infty} f(x)e^{-jsx}\, dx \qquad (10.85)$$

$$f(x) = \mathscr{F}^{-1}[F(s)] = \frac{1}{2\pi} \int_{-\infty}^{\infty} F(s)e^{jsx}\, ds \qquad (10.86)$$

In Example 10-5, we shall apply the Fourier transform technique to solve the initial-value problem.

EXAMPLE 10-5: Using the Fourier transform technique, rework Example 10-4.

Solution: Let the Fourier transform of the solution $u(x, t)$ with respect to x be

$$U(s, t) = \mathscr{F}[u(x, t)] = \int_{-\infty}^{\infty} u(x, t)e^{-jsx}\, dx \qquad (10.87)$$

then,

$$u(x, t) = \mathscr{F}^{-1}[U(s, t)] = \frac{1}{2\pi} \int_{-\infty}^{\infty} U(s, t)e^{jsx}\, ds \qquad (10.88)$$

We shall assume that the solutions $u(x, t)$ and $\partial u(x, t)/\partial x$ are small for large x and approach zero as $x \to \pm\infty$.

Let

$$u_{xx}(x, t) = \frac{\partial^2 u(x, t)}{\partial x^2}, \qquad u_x(x, t) = \frac{\partial u(x, t)}{\partial x}$$

$$u_{tt}(x, t) = \frac{\partial^2 u(x, t)}{\partial t^2}, \qquad u_t(x, t) = \frac{\partial u(x, t)}{\partial t}$$

By applying successive partial integration, the Fourier transform of $u_{xx}(x, t)$ is

$$\mathscr{F}[u_{xx}(x, t)] = \int_{-\infty}^{\infty} u_{xx}(x, t)e^{-jsx}\, dx$$

$$= \int_{(x=-\infty)}^{(x=\infty)} e^{-jsx}\, du_x(x, t)$$

$$= e^{-jsx} u_x(x, t)\Big|_{-\infty}^{\infty} + js \int_{-\infty}^{\infty} u_x(x, t)e^{-jsx}\, dx$$

$$= js \int_{(x=-\infty)}^{(x=\infty)} e^{-jsx}\, du(x, t)$$

$$= jse^{-jsx}u(x, t)\Big|_{-\infty}^{\infty} - js(-js) \int_{-\infty}^{\infty} u(x, t)e^{-jsx}\, dx$$

$$= -s^2 U(s, t) \qquad (10.89)$$

since $u_x(\pm\infty, t) = u(\pm\infty, t) = 0$.

Since we are taking the transform with respect to x, the Fourier transform of $u_{tt}(x, t)$ is

$$\mathscr{F}[u_{tt}(x, t)] = \int_{-\infty}^{\infty} u_{tt}(x, t)e^{-jsx}\, dx = \frac{\partial^2}{\partial t^2} \int_{-\infty}^{\infty} u(x, t)e^{-jsx}\, dx = U_{tt}(s, t) \qquad (10.90)$$

Now, applying the Fourier transform to the wave equation (10.1) and from (10.89) and (10.90), we have

$$-s^2 U(s,t) - \frac{1}{c^2} U_{tt}(s,t) = 0$$

or

$$\frac{\partial^2 U(s,t)}{\partial t^2} + s^2 c^2 U(s,t) = 0 \tag{10.91}$$

which is the equation for the transform $U(s,t)$.

The general solution of (10.91) is

$$U(s,t) = A(s)e^{jsct} + B(s)e^{-jsct} \tag{10.92}$$

where $A(s)$ and $B(s)$ are constants with respect to t. Applying the Fourier transform to the initial conditions (10.72) and (10.73), we have

$$U(s,0) = \mathscr{F}[u(x,0)] = \int_{-\infty}^{\infty} u(x,0)e^{-jsx}\,dx$$

$$= \int_{-\infty}^{\infty} f(x)e^{-jsx}\,dx$$

$$= F(s) \tag{10.93}$$

$$U_t(s,0) = \mathscr{F}[u_t(x,t)|_{t=0}] = 0 \tag{10.94}$$

From (10.93) and (10.94), we can now evaluate $A(s)$ and $B(s)$ of (10.92); thus,

$$F(s) = U(s,0) = A(s) + B(s), \qquad 0 = U_t(s,0) = jsc[A(s) - B(s)]$$

Solving $A(s)$ and $B(s)$ from these two algebraic equations, we obtain

$$A(s) = B(s) = \frac{1}{2}F(s)$$

Hence, from (10.92),

$$U(s,t) = \frac{1}{2}F(s)e^{jsct} + \frac{1}{2}F(s)e^{-jsct} \tag{10.95}$$

The desired solution $u(x,t)$ is the inverse Fourier transform of $U(s,t)$, namely,

$$u(x,t) = \mathscr{F}^{-1}[U(s,t)] = \frac{1}{2}\mathscr{F}^{-1}[F(s)e^{jsct}] + \frac{1}{2}\mathscr{F}^{-1}[F(s)e^{-jsct}] \tag{10.96}$$

By means of the space-shifting property (see the time-shifting property (5.6)), we obtain

$$\mathscr{F}^{-1}[F(s)e^{jsct}] = f(x + ct) \tag{10.97}$$

$$\mathscr{F}^{-1}[F(s)e^{-jsct}] = f(x - ct) \tag{10.98}$$

Thus,

$$u(x,t) = \frac{1}{2}f(x + ct) + \frac{1}{2}f(x - ct)$$

which is exactly the same result as (10.84).

10-3. Heat Conduction

A. Heat equation

The heat flow in a body of homogeneous material is governed by the **heat equation**

Heat equation
$$\nabla^2 u(x,y,z,t) - \frac{1}{c^2}\frac{\partial u(x,y,z,t)}{\partial t} = 0 \tag{10.99}$$

where $u(x,y,z,t)$ is the temperature in the body, and $c^2 = K/(\rho\sigma)$, where K is the thermal conductivity, σ is the specific heat, and ρ is the density of material of the body. The **Laplacian** of u is

$\nabla^2 u$, and in rectangular coordinates, it can be expressed as

Laplacian
$$\nabla^2 u = \frac{\partial^2 u}{\partial x^2} + \frac{\partial^2 u}{\partial y^2} + \frac{\partial^2 u}{\partial z^2} \qquad (10.100)$$

B. One-dimensional heat equation

EXAMPLE 10-6: Consider the temperature in a uniform bar of length l which is oriented along the x-axis. Both ends of the bar are held at zero temperature. If the initial temperature in the bar is given by

$$f(x) = \begin{cases} x & \text{for} \quad 0 < x < \frac{1}{2}l \\ l - x & \text{for} \quad \frac{1}{2}l < x < l \end{cases}$$

where x is the distance measured from one end, find the temperature distribution after time t.

Solution: Since the temperature $u(x, t)$ depends only on x and t, the heat equation (10.99) becomes the so-called **one-dimensional heat equation**:

One-dimensional
heat equation
$$\frac{\partial^2 u(x,t)}{\partial x^2} - \frac{1}{c^2} \frac{\partial u(x,t)}{\partial t} = 0 \qquad (10.101)$$

The boundary conditions are

$$u(0, t) = 0, \qquad u(l, t) = 0 \qquad (10.102)$$

and the initial condition is

$$u(x, 0) = f(x) = \begin{cases} x & \text{for} \quad 0 < x < \frac{1}{2}l \\ l - x & \text{for} \quad \frac{1}{2}l < x < l \end{cases} \qquad (10.103)$$

Once again, assume the solution to be in the form of the product

$$u(x, t) = X(x)T(t) \qquad (10.104)$$

Substituting into the one-dimensional heat equation (10.101), we obtain

$$X''(x)T(t) - \frac{1}{c^2} X(x)T'(t) = 0 \qquad (10.105)$$

Dividing by $X(x)T(t)$ and separating variables, we have

$$\frac{X''(x)}{X(x)} = \frac{1}{c^2} \frac{T'(t)}{T(t)} \qquad (10.106)$$

The expression on the left-hand side depends only on x, while the right-hand side depends only on t; therefore, we conclude that both expressions must be equal to a constant. This constant, say K, must be negative since if $K > 0$, the only solution $u(x, t) = X(x)T(t)$ that satisfies the boundary conditions (10.102) is $u(x, t) = 0$. This is shown as follows: If

$$\frac{X''(x)}{X(x)} = K = k^2$$

then

$$X''(x) - k^2 X(x) = 0$$

and the general solution will be

$$X(x) = Ae^{kx} + Be^{-kx}$$

Applying the boundary conditions (10.102),

$$A + B = 0 \qquad \text{and} \qquad Ae^{kl} + Be^{-kl} = 0$$

Solving for A and B, we have $A = -B = 0$. Thus, $X(x) = 0$, and consequently $u(x, t) = 0$. This gives only a trivial solution. Hence, letting $K = -k^2$,

$$\frac{X''(x)}{X(x)} = \frac{1}{c^2} \frac{T'(t)}{T(t)} = -k^2 \tag{10.107}$$

and from this we get the two ordinary differential equations

$$X''(x) + k^2 X(x) = 0 \tag{10.108}$$
$$T'(t) + c^2 k^2 T(t) = 0 \tag{10.109}$$

The general solutions of (10.108) and (10.109) are

$$X(x) = A \cos kx + B \sin kx \tag{10.110}$$
$$T(t) = Ce^{-c^2 k^2 t} \tag{10.111}$$

From the boundary conditions (10.102), we get

$$X(0) = A = 0$$
$$X(l) = B \sin kl = 0$$

Thus,

$$kl = n\pi \quad \text{or} \quad k = \frac{n\pi}{l}, \quad n = 1, 2, \ldots \tag{10.112}$$

We thus obtain the solution of (10.108) that satisfies (10.102) as

$$X_n(x) = B_n \sin \frac{n\pi x}{l}, \quad n = 1, 2, \ldots \tag{10.113}$$

The corresponding solutions of (10.109) are

$$T_n(t) = C_n e^{-c^2 k^2 t} = C_n e^{-c^2 n^2 \pi^2 t / l^2} = C_n e^{-\lambda_n^2 t}, \quad n = 1, 2, \ldots \tag{10.114}$$

where

$$\lambda_n = \frac{cn\pi}{l}$$

Hence, the functions

$$u_n(x, t) = X_n(t) T_n(t) = b_n e^{-\lambda_n^2 t} \sin \frac{n\pi x}{l}, \quad n = 1, 2 \ldots \tag{10.115}$$

where $b_n = B_n C_n$, are the solutions of the heat equation (10.101) satisfying the boundary conditions (10.102).

To find a solution that also satisfies the initial condition (10.103), we consider the series

$$u(x, t) = \sum_{n=1}^{\infty} u_n(x, t) = \sum_{n=1}^{\infty} b_n e^{-\lambda_n^2 t} \sin \frac{n\pi x}{l} \tag{10.116}$$

From the initial condition (10.103) and the series (10.116), we have

$$u(x, 0) = f(x) = \sum_{n=1}^{\infty} b_n \sin \frac{n\pi x}{l} \tag{10.117}$$

Hence, for (10.116) to satisfy (10.103), the coefficients b_n must be chosen such that (10.117) becomes the Fourier sine series of $f(x)$; that is,

$$\begin{aligned} b_n &= \frac{2}{l} \int_0^l f(x) \sin \frac{n\pi x}{l} \, dx \\ &= \frac{2}{l} \left[\int_0^{l/2} x \sin \frac{n\pi x}{l} \, dx + \int_{l/2}^l (l - x) \sin \frac{n\pi x}{l} \, dx \right] \\ &= \begin{cases} 0 & \text{for } n \text{ even} \\ \dfrac{4l}{n^2 \pi^2} & \text{for } n = 1, 5, 9, \ldots \\ -\dfrac{4l}{n^2 \pi^2} & \text{for } n = 3, 7, 11, \ldots \end{cases} \end{aligned} \tag{10.118}$$

Hence, the solution is

$$u(x,t) = \frac{4l}{\pi^2}\left[\sin\frac{\pi x}{l}e^{-(c\pi/l)^2t} - \frac{1}{9}\sin\frac{3\pi x}{l}e^{-(3c\pi/l)^2t} + \cdots\right] \tag{10.119}$$

Note that the solution $u(x,t)$ of (10.119) becomes small after a long period of time; that is, it tends to zero as $t \to \infty$.

10-4. Potential Theory

In this section we shall apply Fourier analysis to potential theory. **Potential theory** is the theory of the solution of Laplace's equation

Laplace's equation $$\nabla^2 u = 0 \tag{10.120}$$

where $\nabla^2 u$ is the Laplacian of u. Laplace's equation occurs in connection with gravitational potentials, electrostatic potentials, stationary heat problems, potentials of incompressible inviscid fluid flow, etc.

In rectangular coordinates, the Laplacian of a function u in three-dimensional space is expressed as

$$\nabla^2 u = \frac{\partial^2 u}{\partial x^2} + \frac{\partial^2 u}{\partial y^2} + \frac{\partial^2 u}{\partial z^2} \tag{10.121}$$

In cylindrical coordinates (ρ, ϕ, z), as shown in Figure 10-2,

$$\nabla^2 u = \frac{\partial^2 u}{\partial \rho^2} + \frac{1}{\rho}\frac{\partial u}{\partial \rho} + \frac{1}{\rho^2}\frac{\partial^2 u}{\partial \phi^2} + \frac{\partial^2 u}{\partial z^2} \tag{10.122}$$

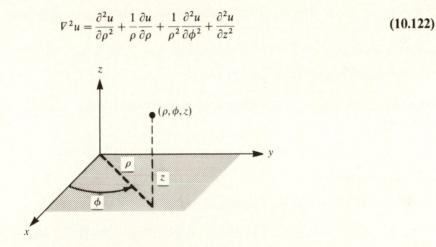

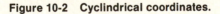

Figure 10-2   Cyclindrical coordinates.

In spherical coordinates (r, θ, ϕ), as shown in Figure 10-3,

$$\nabla^2 u = \frac{1}{r^2}\frac{\partial}{\partial r}\left(r^2\frac{\partial u}{\partial r}\right) + \frac{1}{\sin\theta}\frac{\partial}{\partial\theta}\left(\sin\theta\frac{\partial u}{\partial\theta}\right) + \frac{1}{\sin^2\theta}\frac{\partial^2 u}{\partial\phi^2} \tag{10.123}$$

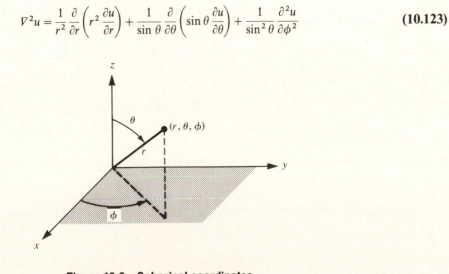

Figure 10-3   Spherical coordinates.

EXAMPLE 10-7: Consider the rectangular box shown in Figure 10-4. Find the potential distribution if the potential is zero on all sides and the bottom and is $f(x, y)$ on the top.

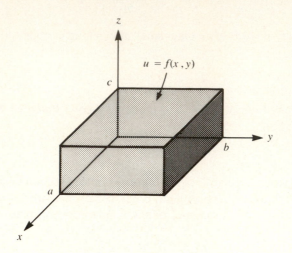

Figure 10-4   A rectangular box.

Solution: Let $u(x, y, z)$ be the potential distribution in the rectangular box shown in Figure 10-4. Then, $u(x, y, z)$ satisfies

Laplace's equation $\qquad\qquad V^2 u = u_{xx} + u_{yy} + u_{zz} = 0$ $\qquad\qquad\qquad$ **(10.124)**

and the boundary conditions are

Boundary $\qquad\quad u(0, y, z) = u(a, y, z) = u(x, 0, z) = u(x, b, z) = u(x, y, 0) = 0$ $\qquad$ **(10.125)**
conditions $\qquad\qquad\qquad\qquad u(x, y, c) = f(x, y)$ $\qquad\qquad\qquad\qquad\qquad$ **(10.126)**

The method of separation of variables suggests assuming a solution of the form

Separation of variables $\qquad\qquad u(x, y, z) = X(x)Y(y)Z(z)$ $\qquad\qquad\qquad$ **(10.127)**

Substituting this into Laplace's equation (10.124) reduces it to

$$X''(x)Y(y)Z(z) + X(x)Y''(y)Z(z) + X(x)Y(y)Z''(z) = 0 \qquad\qquad \textbf{(10.128)}$$

Dividing by $X(x)Y(y)Z(z)$ and separating the variables, we obtain

$$-\frac{X''(x)}{X(x)} = \frac{Y''(y)}{Y(y)} + \frac{Z''(z)}{Z(z)} = k_x^2 \qquad\qquad \textbf{(10.129)}$$

where k_x^2 is the separation constant. The separation here depends upon the fact that the left-hand side is independent of both y and z and the right-hand side is independent of x. Hence,

$$X''(x) + k_x^2 X(x) = 0 \qquad\qquad \textbf{(10.130)}$$

After a second separation,

$$-\frac{Y''(y)}{Y(y)} = \frac{Z''(z)}{Z(z)} - k_x^2 = k_y^2 \qquad\qquad \textbf{(10.131)}$$

This yields the following differential equations:

$$Y''(y) + k_y^2 Y(y) = 0 \qquad\qquad \textbf{(10.132)}$$
$$Z''(z) - k_z^2 Z(z) = 0 \qquad\qquad \textbf{(10.133)}$$

where $k_z^2 = k_x^2 + k_y^2$. The general solutions of (10.130), (10.132), and (10.133) are

$$X(x) = A \cos k_x x + B \sin k_x x \qquad\qquad \textbf{(10.134)}$$
$$Y(y) = C \cos k_y y + D \sin k_y y \qquad\qquad \textbf{(10.135)}$$
$$Z(z) = E \cosh k_z Z + F \sinh k_z z \qquad\qquad \textbf{(10.136)}$$

From the boundary conditions (10.125),

$$X(0) = X(a) = 0$$
$$Y(0) = Y(b) = 0$$
$$Z(0) = 0$$

Therefore,

$$X(0) = A = 0$$
$$X(a) = B \sin k_x a = 0$$

hence,

$$k_x a = m\pi \quad \text{or} \quad k_x = \frac{m\pi}{a}, \quad m = 1, 2, \ldots \qquad (10.137)$$

Similarly,

$$Y(0) = C = 0$$
$$Y(b) = D \sin k_y b = 0$$

hence,

$$k_y b = n\pi \quad \text{or} \quad k_y = \frac{n\pi}{b}, \quad n = 1, 2, \ldots \qquad (10.138)$$

Also

$$Z(0) = E = 0$$

If we write further

$$k_z^2 = k_x^2 + k_y^2 = \pi^2 \left(\frac{m^2}{a^2} + \frac{n^2}{b^2} \right) = k_{mn}^2$$

or

$$k_z = k_{mn} = \pi \sqrt{\frac{m^2}{a^2} + \frac{n^2}{b^2}} \qquad (10.139)$$

we obtain the solutions

$$X(x) = X_m(x) = B_m \sin \frac{m\pi x}{a}, \quad m = 1, 2, \ldots$$

$$Y(y) = Y_n(y) = D_n \sin \frac{n\pi y}{b}, \quad n = 1, 2, \ldots$$

$$Z(z) = Z_{mn}(z) = F_{mn} \sinh k_{mn} z$$

Thus, writing $b_{mn} = B_m D_n F_{mn}$, it follows that the functions

$$u_{mn}(x, y, z) = X_m(x) Y_n(y) Z_{mn}(z)$$

$$= b_{mn} \sin \frac{m\pi x}{a} \sin \frac{n\pi y}{b} \sinh k_{mn} z \qquad (10.140)$$

where $m = 1, 2, \ldots, n = 1, 2, \ldots$, with k_{mn} defined by (10.139), are solutions of Laplace's equation (10.124) satisfying the boundary condition (10.125).

In order to satisfy the boundary condition (10.126), we assume the desired solution in the form

$$u(x, y, z) = \sum_{m=1}^{\infty} \sum_{n=1}^{\infty} u_{mn}(x, y, z)$$

$$= \sum_{m=1}^{\infty} \sum_{n=1}^{\infty} b_{mn} \sin \frac{m\pi x}{a} \sin \frac{n\pi y}{b} \sinh k_{mn} z \qquad (10.141)$$

If we let

$$c_{mn} = b_{mn} \sinh k_{mn} c \qquad (10.142)$$

the boundary condition (10.126) takes the form

$$f(x, y) = \sum_{m=1}^{\infty} \sum_{n=1}^{\infty} c_{mn} \sin \frac{m\pi x}{a} \sin \frac{n\pi y}{b}, \qquad 0 < x < a, 0 < y < b \tag{10.143}$$

Thus, the coefficients c_{mn} are the coefficients of the double Fourier sine series expansion of $f(x, y)$ over the indicated rectangle. From (10.69), these coefficients are readily determined as

$$c_{mn} = \frac{4}{ab} \int_0^b \int_0^a f(x, y) \sin \frac{m\pi x}{a} \sin \frac{n\pi y}{b} \, dx \, dy \tag{10.144}$$

With these values of c_{mn} and with the notation of (10.142), the solution (10.141) becomes

$$u(x, y, z) = \sum_{m=1}^{\infty} \sum_{n=1}^{\infty} c_{mn} \sin \frac{m\pi x}{a} \sin \frac{n\pi y}{b} \frac{\sinh k_{mn} z}{\sinh k_{mn} c} \tag{10.145}$$

where k_{mn} is defined by (10.139).

SUMMARY

1. The "separation of variables" method is used to solve many boundary-value and initial-value problems.
2. The Fourier series representation technique is used to obtain solutions for bounded-region boundary-value problems.
3. The Fourier transform technique is used to obtain solutions for unbounded-region boundary-value problems.

RAISE YOUR GRADES

Can you explain ...?

☑ how to select the separation constants in the "separation of variables" method
☑ how the Fourier sine series expansion is effectively used in matching the boundary conditions
☑ under what conditions the Fourier transform technique should be used instead of the Fourier series representation

SOLVED PROBLEMS

Separation of Variables and Fourier Series

PROBLEM 10-1 Find the solution of the one-dimensional wave equation (10.1) with the same boundary conditions as (10.2) ($u(0, t) = 0$ and $u(l, t) = 0$ for all t), but with the triangular initial deflection shown in Figure 10-5 and zero initial velocity; that is,

$$u(x, 0) = f(x) = \begin{cases} \dfrac{2k}{l} x & \text{for } 0 < x < \dfrac{1}{2} l \\[2mm] \dfrac{2k}{l}(l - x) & \text{for } \dfrac{1}{2} l < x < l \end{cases}$$

$$\frac{\partial u(x, t)}{\partial t}\bigg|_{t=0} = g(x) = 0$$

Solution: Since $g(x) = 0$, from (10.23), $F_n = 0$. From the result of Problem 2-9 (expanding in a Fourier sine series), we see that the coefficients E_n of (10.21) are given by

$$E_n = \frac{8k}{n^2 \pi^2} \sin \frac{n\pi}{2}$$

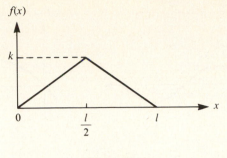

Figure 10-5

Hence, the Fourier sine series of $f(x)$ is given by

$$u(x,0) = \frac{8k}{\pi^2}\left(\sin\frac{\pi x}{l} - \frac{1}{3^2}\sin\frac{3\pi x}{l} + \frac{1}{5^2}\sin\frac{5\pi x}{l} - \cdots\right)$$

Thus, from (10.18), we obtain

$$u(x,t) = \frac{8k}{\pi^2}\left(\sin\frac{\pi x}{l}\cos\frac{c\pi t}{l} - \frac{1}{3^2}\sin\frac{3\pi x}{l}\cos\frac{3c\pi t}{l} + \cdots\right)$$

PROBLEM 10-2 In Example 10-1, if $u(x,0) = f(x)$ but $\partial u(x,t)/\partial t|_{t=0} = g(x) = 0$, show that the solution of the one-dimensional wave equation (10.1) can be expressed as

$$u(x,t) = \frac{1}{2}f_1(x-ct) + \frac{1}{2}f_1(x+ct)$$

where $f_1(x)$ is the odd periodic extension of $f(x)$ with the period $2l$. Also give the physical interpretation of this expression.

Solution: The general solution of (10.1) is given by (10.18); that is,

$$u(x,t) = \sum_{n=1}^{\infty}\sin\frac{n\pi x}{l}\left(E_n\cos\frac{cn\pi t}{l} + F_n\sin\frac{cn\pi t}{l}\right)$$

Since the initial velocity $g(x)$ is identically zero, from (10.22), the coefficients F_n are zero, and (10.18) reduces to

$$u(x,t) = \sum_{n=1}^{\infty}E_n\sin\frac{n\pi x}{l}\cos\frac{cn\pi t}{l} \qquad \textbf{(a)}$$

Using the trigonometric identity

$$\sin A\cos B = \frac{1}{2}[\sin(A-B) + \sin(A+B)]$$

it follows that

$$\sin\frac{n\pi x}{l}\cos\frac{cn\pi t}{l} = \frac{1}{2}\left[\sin\frac{n\pi}{l}(x-ct) + \sin\frac{n\pi}{l}(x+ct)\right]$$

Hence, we can rewrite eq. (a) in the form

$$u(x,t) = \frac{1}{2}\sum_{n=1}^{\infty}E_n\sin\frac{n\pi}{l}(x-ct) + \frac{1}{2}\sum_{n=1}^{\infty}E_n\sin\frac{n\pi}{l}(x+ct)$$

Comparing with (10.19), we conclude that the above two series are those obtained by substituting $(x-ct)$ and $(x+ct)$, respectively, for the variable x in the Fourier sine series (10.19) for $f(x)$. Therefore,

$$u(x,t) = \frac{1}{2}f_1(x-ct) + \frac{1}{2}f_1(x+ct)$$

where $f_1(x)$ is the odd periodic extension of $f(x)$ with the period $2l$ shown in Figure 10-6(a).

Now, the graph of $f_1(x-ct)$ is obtained from the graph of $f_1(x)$ by shifting ct units to the right (see Figure 10-6b). Also, we recognize that we may stay on a particular reference value of the function by

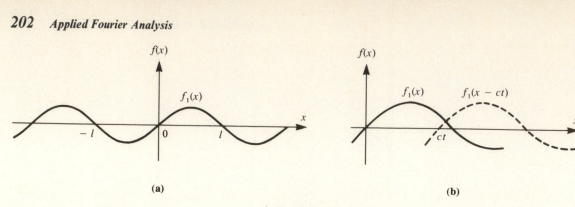

Figure 10-6

keeping the argument, $x - ct$, a constant, that is, by moving in the positive x direction with velocity c as time increases. This means that $f_1(x - ct)$, $(c > 0)$, represents a wave that is traveling to the right as t increases. Similarly, $f_1(x + ct)$ represents a wave traveling to the left with velocity c. Hence, the solution $u(x, t)$ is the superposition of these two waves.

Vibration

PROBLEM 10-3 Find the solution of the two-dimensional wave equation (10.44) with the following boundary and initial conditions:

$$u(x, y, t) = 0 \quad \text{for} \quad x = 0, x = a, y = 0, \text{ and } y = b$$

$$u(x, y, 0) = xy(x - a)(y - b).$$

$$\frac{\partial u}{\partial t}\bigg|_{t=0} = 0$$

Solution: From the double series (10.63), we have

$$u(x, y, t) = \sum_{m=1}^{\infty} \sum_{n=1}^{\infty} (G_{mn}\cos k_{mn}ct + H_{mn}\sin k_{mn}ct)\sin \frac{m\pi x}{a} \sin \frac{n\pi y}{b}$$

Letting $t = 0$, we get the double Fourier series

$$u(x, y, 0) = \sum_{m=1}^{\infty} \sum_{n=1}^{\infty} G_{mn}\sin \frac{m\pi x}{a} \sin \frac{n\pi y}{b}$$

According to (10.69),

$$G_{mn} = \frac{4}{ab} \int_0^b \int_0^a u(x, y, 0)\sin \frac{m\pi x}{a} \sin \frac{n\pi y}{b} dx\, dy$$

$$= \frac{4}{ab} \int_0^a x(x - a)\sin \frac{m\pi x}{a} dx \int_0^b y(y - b)\sin \frac{n\pi y}{b} dy$$

$$= \frac{4}{ab} \frac{2a^3}{m^3\pi^3} [(-1)^m - 1] \frac{2b^3}{n^3\pi^3} [(-1)^n - 1]$$

$$= \begin{cases} \dfrac{64a^2b^2}{\pi^6 m^3 n^3} & \text{if } n \text{ and } m \text{ are odd} \\ 0 & \text{otherwise} \end{cases}$$

Since $\partial u/\partial t|_{t=0} = 0$ and, according to (10.71), $H_{mn} = 0$, the final solution is

$$u(x, y, t) = \frac{64a^2b^2}{\pi^6} \sum_{m=\text{odd}}^{\infty} \sum_{n=\text{odd}}^{\infty} \frac{1}{m^3 n^3} \cos k_{mn}ct \sin \frac{m\pi x}{a} \sin \frac{n\pi y}{b}$$

where

$$k_{mn}^2 = \left(\frac{m\pi}{a}\right)^2 + \left(\frac{n\pi}{b}\right)^2$$

PROBLEM 10-4 The small, free transverse vibrations of a uniform cantilever beam lying along the x-axis are governed by the fourth-order equation

$$\frac{\partial^4 u(x,t)}{\partial x^4} + \frac{1}{c^2}\frac{\partial^2 u(x,t)}{\partial t^2} = 0 \tag{a}$$

where $c^2 = EI/(\rho A)$; E = Young's modulus of elasticity, I = moment of inertia of the cross-section, ρ = density, and A = cross-sectional area. Find the solution of eq. (a) that satisfies the following conditions:

$$u(0,t) = 0, \qquad u(l,t) = 0 \tag{b}$$

$$\left.\frac{\partial^2 u}{\partial x^2}\right|_{x=0} = \left.\frac{\partial^2 u}{\partial x^2}\right|_{x=l} = 0 \tag{c}$$

$$u(x,0) = x(l - x) \tag{d}$$

$$\left.\frac{\partial u}{\partial t}\right|_{t=0} = 0 \tag{e}$$

Solution: Assume that the solution of eq. (a) will be of the form

$$u(x,t) = X(x)T(t) \tag{f}$$

Substituting eq. (f) into eq. (a), we obtain

$$x^{(4)}(x)T(t) + \frac{1}{c^2}X(x)T''(t) = 0$$

Dividing by $X(x)T(t)$ and separating the variables, we obtain

$$\frac{X^{(4)}(x)}{X(x)} = -\frac{1}{c^2}\frac{T''(t)}{T(t)} \tag{g}$$

Since the left-hand side of eq. (g) depends only on x and the right-hand side depends only on t, the expressions on both sides must be equal to a constant. The constant, say k^4, must be positive from physical considerations; namely, to make $T(t)$ oscillatory. Thus, we obtain the two ordinary differential equations

$$X^{(4)}(x) - k^4 X(x) = 0 \tag{h}$$
$$T''(t) + c^2 k^4 T(t) = 0 \tag{i}$$

The general solutions of eq. (h) and eq. (i) are

$$X(x) = A\cos kx + B\sin kx + C\cosh kx + D\sinh kx \tag{j}$$
$$T(t) = E\cos k^2 ct + F\sin k^2 ct \tag{k}$$

Now, from the boundary conditions (b),

$$X(0) = A + C = 0 \tag{l}$$
$$X(l) = A\cos kl + B\sin kl + C\cosh kl + D\sinh kl = 0 \tag{m}$$

Since

$$X''(x) = -k^2(A\cos kx + B\sin kx - C\cosh kx - D\sinh kx)$$

using the boundary conditions (c),

$$X''(0) = -k^2(A - C) = 0 \tag{n}$$
$$X''(l) = -k^2(A\cos kl + B\sin kl - C\cosh kl - D\sinh kl) = 0 \tag{o}$$

From eqs. (l) and (n), we see that $A + C = 0$ and $A - C = 0$; hence $A = C = 0$. Then from eqs. (m) and (o),

$$B\sin kl + D\sinh kl = 0$$
$$B\sin kl - D\sinh kl = 0$$

and hence,

$$B\sin kl = 0, \qquad D\sinh kl = 0$$

The second condition gives $D = 0$ since if $\sinh kl = 0$, then $k = 0$ and hence $X(x) = 0$, which would give a trivial solution. Then from the first condition,

$$\sin kl = 0$$

that is,

$$kl = n\pi \quad \text{or} \quad k = \frac{n\pi}{l}, \quad n = 1, 2, \ldots$$

Thus, we obtain the infinite set of solutions $X(x) = X_n(x)$; that is,

$$X_n(x) = B_n \sin \frac{n\pi x}{l}, \quad n = 1, 2, \ldots$$

Next, since

$$T'(t) = k^2 c(-E \sin k^2 ct + F \cos k^2 ct)$$

from the initial condition (e),

$$T'(0) = k^2 cF = 0$$

Hence, $F = 0$, and the corresponding solutions $T_n(t)$ become

$$T_n(t) = E_n \cos \frac{n^2 \pi^2 ct}{l^2}$$

Therefore, the functions

$$u_n(x, t) = X_n(x) T_n(t) = b_n \sin \frac{n\pi x}{l} \cos \frac{n^2 \pi^2 ct}{l^2}$$

where $b_n = B_n E_n$, are the solutions of eq. (a) satisfying the boundary conditions (b) and (c) and the zero initial velocity condition (e).

In order to satisfy the initial condition (d), we consider

$$u(x, t) = \sum_{n=1}^{\infty} u_n(x, t)$$

$$= \sum_{n=1}^{\infty} b_n \sin \frac{n\pi x}{l} \cos \frac{n^2 \pi^2 ct}{l^2}$$

Hence, from condition (d), we have

$$u(x, 0) = x(l - x) = \sum_{n=1}^{\infty} b_n \sin \frac{n\pi x}{l}$$

Thus, the coefficients b_n are the sine Fourier coefficients of $x(l - x)$ and are given by

$$b_n = \frac{2}{l} \int_0^l x(l - x) \sin \frac{n\pi x}{l} \, dx$$

$$= \begin{cases} \dfrac{8l^2}{n^3 \pi^3} & \text{for } n \text{ odd} \\ 0 & \text{for } n \text{ even} \end{cases}$$

The final solution is therefore

$$u(x, t) = \frac{8l^2}{\pi^3} \sum_{n=\text{odd}}^{\infty} \frac{1}{n^3} \sin \frac{n\pi x}{l} \cos \frac{n^2 \pi^2 ct}{l^2}$$

Heat Conduction

PROBLEM 10-5   Find the temperature distribution $u(x, t)$ in the case of an infinite bar. The initial temperature distribution is given by $f(x)$ for $-\infty < x < \infty$.

Solution: The function $u(x, t)$ satisfies the one-dimensional heat equation

$$\frac{\partial^2 u(x, t)}{\partial x^2} - \frac{1}{c^2} \frac{\partial u(x, t)}{\partial t} = 0 \tag{a}$$

and the initial condition is

$$u(x,0) = f(x) \qquad \text{for} \quad -\infty < x < \infty \tag{b}$$

Proceeding as in Example 10-6, we substitute

$$u(x,t) = X(x)T(t)$$

into eq. (a). This yields two ordinary differential equations

$$X''(x) + k^2 X(x) = 0 \tag{c}$$

$$T'(t) + c^2 k^2 T(t) = 0 \tag{d}$$

The general solutions of eqs. (c) and (d) are

$$X(x) = A\cos kx + B\sin kx$$

$$T(t) = Ce^{-c^2 k^2 t}$$

Hence,

$$u(x,t;k) = X(x)T(t) = (D\cos kx + E\sin kx)e^{-c^2 k^2 t} \tag{e}$$

is a solution of eq. (a), where D and E are the arbitrary constants. Since $f(x)$ in eq. (b) is in general not periodic, following the similar argument for the case of an infinite string vibration (see Example 10-4), we may consider D and E as functions of k. Then the function

$$u(x,t) = \int_0^\infty u(x,t;k)\,dk$$

$$= \int_0^\infty [D(k)\cos kx + E(k)\sin kx]e^{-c^2 k^2 t}\,dk \tag{f}$$

is also a solution of eq. (a).

From initial condition (b),

$$u(x,0) = f(x) = \int_0^\infty [D(k)\cos kx + E(k)\sin kx]\,dk \tag{g}$$

Now if

$$D(k) = \frac{1}{\pi}\int_{-\infty}^\infty f(y)\cos ky\,dy$$

$$E(k) = \frac{1}{\pi}\int_{-\infty}^\infty f(y)\sin ky\,dy$$

then with the Fourier integral theorem (see Problem 6-19), we may write eq. (g) as

$$u(x,0) = \frac{1}{\pi}\int_0^\infty \left[\int_{-\infty}^\infty f(y)\cos k(x-y)\,dy\right]dk \tag{h}$$

Thus, from eq. (f),

$$u(x,t) = \frac{1}{\pi}\int_0^\infty \left[\int_{-\infty}^\infty f(y)\cos k(x-y)e^{-c^2 k^2 t}\,dy\right]dk \tag{i}$$

Assuming that we can interchange the order of integration,

$$u(x,t) = \frac{1}{\pi}\int_{-\infty}^\infty f(y)\left[\int_0^\infty e^{-c^2 k^2 t}\cos k(x-y)\,dk\right]dy \tag{j}$$

In order to evaluate the inner integral, we proceed as follows.

From a table of integral formulas,

$$\int_0^\infty e^{-s^2}\cos 2bs\,ds = \frac{\sqrt{\pi}}{2}e^{-b^2} \tag{k}$$

Introducing a new variable of integration k by setting $s = ck\sqrt{t}$ and choosing

$$b = \frac{x-y}{2c\sqrt{t}}$$

the formula (k) becomes

$$\int_0^\infty e^{-c^2 k^2 t} \cos k(x-y)\, dk = \frac{\sqrt{\pi}}{2c\sqrt{t}} e^{-(x-y)^2/(4c^2 t)}$$

(l)

Substituting eq. (l) into eq. (j), we obtain

$$u(x,t) = \frac{1}{2c\sqrt{\pi t}} \int_{-\infty}^\infty f(y) e^{-(x-y)^2/(4c^2 t)}\, dy$$

(m)

Introducing a new variable of integration, $q = (x-y)/(2c\sqrt{t})$, eq. (m) can be written as

$$u(x,t) = \frac{1}{\sqrt{\pi}} \int_{-\infty}^\infty f(x - 2cq\sqrt{t}) e^{-q^2}\, dq$$

(n)

PROBLEM 10-6 Using the Fourier transform technique, rework Problem 10-5.

Solution: Let the Fourier transform of the solution $u(x,t)$ with respect to x be

$$U(s,t) = \mathscr{F}[u(x,t)] = \int_{-\infty}^\infty u(x,t) e^{-jsx}\, dx$$

(a)

then,

$$u(x,t) = \mathscr{F}^{-1}[U(s,t)] = \frac{1}{2\pi} \int_{-\infty}^\infty U(s,t) e^{jsx}\, ds$$

(b)

We shall assume that the solution $u(x,t)$ and $\partial u(x,t)/\partial t$ are small for large $|x|$ and approach zero as $x \to +\infty$.

From (10.89), the Fourier transform of $u_{xx}(x,t)$ is

$$\mathscr{F}[u_{xx}(x,t)] = \int_{-\infty}^\infty u_{xx}(x,t) e^{-jsx}\, dx = -s^2 U(s,t)$$

(c)

The Fourier transform of $u_t(x,t)$ is

$$\mathscr{F}[u_t(x,t)] = \int_{-\infty}^\infty u_t(x,t) e^{-jsx}\, dx = \frac{\partial}{\partial t} U(s,t) = U_t(s,t)$$

(d)

Now, applying the Fourier transform to the heat equation (10.101), we have

$$-s^2 U(s,t) - \frac{1}{c^2} U_t(s,t) = 0$$

or

$$\frac{\partial U(s,t)}{\partial t} + c^2 s^2 U(s,t) = 0$$

(e)

The solution of eq. (e) is

$$U(s,t) = U(s,0) e^{-c^2 s^2 t}$$

(f)

But by applying the Fourier transform to the initial condition (b) of Problem 10-5, we have

$$U(s,0) = \int_{-\infty}^\infty u(x,0) e^{-jsx}\, dx$$

$$= \int_{-\infty}^\infty f(x) e^{-jsx}\, dx$$

$$= \int_{-\infty}^\infty f(y) e^{-jsy}\, dy$$

(g)

Substituting eq. (g) into solution (f), we have

$$U(s,t) = e^{-c^2 s^2 t} \int_{-\infty}^\infty f(y) e^{-jsy}\, dy$$

(h)

Now, the solution $u(x, t)$ can be obtained by taking the inverse Fourier transform of eq. (h); that is,

$$u(x, t) = \frac{1}{2\pi} \int_{-\infty}^{\infty} U(s, t) e^{jsx} \, dx$$

$$= \frac{1}{2\pi} \int_{-\infty}^{\infty} e^{(jsx - c^2 s^2 t)} \left[\int_{-\infty}^{\infty} f(y) e^{-jsy} \, dy \right] ds \tag{i}$$

Assuming that we can interchange the order of integration,

$$u(x, t) = \frac{1}{2\pi} \int_{-\infty}^{\infty} f(y) \left\{ \int_{-\infty}^{\infty} e^{[js(x-y) - c^2 s^2 t]} \, ds \right\} dy \tag{j}$$

In order to evaluate the inner integral, we proceed as follows.
From the integral table,

$$\int_{-\infty}^{\infty} e^{-w^2} \, dw = \sqrt{\pi} \tag{k}$$

Now,

$$\int_{-\infty}^{\infty} e^{[js(x-y) - c^2 s^2 t]} \, ds = \int_{-\infty}^{\infty} \exp \left[\left(\frac{x-y}{2c\sqrt{t}} + jcs\sqrt{t} \right)^2 - \left(\frac{x-y}{2c\sqrt{t}} \right)^2 \right] ds$$

$$= e^{-(x-y)^2/(4c^2 t)} \int_{-\infty}^{\infty} \exp \left(\frac{x-y}{2c\sqrt{t}} + jcs\sqrt{t} \right)^2 ds$$

Introducing a new variable of integration w by setting

$$\frac{x-y}{2c\sqrt{t}} + jcs\sqrt{t} = jw$$

we have

$$\int_{-\infty}^{\infty} e^{[js(x-y) - c^2 s^2 t]} \, ds = \frac{1}{c\sqrt{t}} e^{-(x-y)^2/(4c^2 t)} \int_{-\infty}^{\infty} e^{-w^2} \, dw$$

$$= \frac{1}{c} \sqrt{\frac{\pi}{t}} e^{-(x-y)^2/(4c^2 t)} \tag{1}$$

in view of eq. (k).
Substituting eq. (l) into eq. (j), we finally obtain

$$u(x, t) = \frac{1}{2c\sqrt{\pi t}} \int_{-\infty}^{\infty} f(y) e^{-(x-y)^2/(4c^2 t)} \, dy \tag{m}$$

which is exactly eq. (m) of Problem 10-5.
Now eqs. (j) or (m) can be expressed as

$$u(x, t) = \int_{-\infty}^{\infty} f(y) G(x - y, t) \, dy$$

where

Green's
function
$$G(x - y, t) = \frac{1}{2\pi} \int_{-\infty}^{\infty} e^{[js(x-y) - c^2 s^2 t]} \, ds = \frac{1}{2c\sqrt{\pi t}} e^{-(x-y)^2/(4c^2 t)}$$

which is called **Green's function** of the heat equation (10.101) for the infinite interval.

PROBLEM 10-7 Assume a bar to be semi-infinite, extending from 0 to ∞. The end at $x = 0$ is held at zero temperature, and the initial temperature distribution is $f(x)$ for $0 < x < \infty$. Find the temperature $u(x, t)$ in the bar. It is assumed that the condition at the infinite end is such that $u(x, t) \to 0$ as $x \to \infty$.

Solution: There are several ways to solve this problem, but we shall use the *method of images* here.
Since the temperature at $x = 0$ is held at 0, we extend the given initial function $f(x)$, $x > 0$, to be an odd function for $-\infty < x < \infty$. Then this becomes an infinite bar problem (see Problem 10-5).
From eq. (m) of Problem 10-5, we have

$$u(x, t) = \frac{1}{2c\sqrt{\pi t}} \int_{-\infty}^{\infty} f(y) e^{-(x-y)^2/(4c^2 t)} \, dy$$

Using the fact that $f(-y) = -f(y)$,

$$u(x,t) = \frac{1}{2c\sqrt{\pi t}} \int_0^\infty f(y)e^{-(x-y)^2/(4c^2t)}\,dy + \frac{1}{2c\sqrt{\pi t}} \int_0^\infty f(-y)e^{-(x+y)^2/(4c^2t)}\,dy$$

$$= \frac{1}{2c\sqrt{\pi t}} \int_0^\infty f(y)[e^{-(x-y)^2/(4c^2t)} - e^{-(x+y)^2/(4c^2t)}]\,dy$$

which is the desired solution.

Potential Theory

PROBLEM 10-8 Rework Example 10-7 when $f(x, y) = U_0$, a constant.

Solution: From (10.144),

$$c_{mn} = \frac{4}{ab} \int_0^b \int_0^a U_0 \sin\frac{m\pi x}{a} \sin\frac{n\pi y}{b}\,dx\,dy$$

$$= \frac{4U_0}{ab} \int_0^a \sin\frac{m\pi x}{a}\,dx \int_0^b \sin\frac{n\pi y}{b}\,dy$$

$$= \begin{cases} \dfrac{16U_0}{mn\pi^2} & \text{for} \quad m, n \text{ odd} \\ 0 & \text{for} \quad m, n \text{ even} \end{cases}$$

Hence, from (10.145), we obtain

$$u(x, y, z) = \frac{16U_0}{\pi^2} \sum_{m=\text{odd}}^\infty \sum_{n=\text{odd}}^\infty \frac{1}{mn} \sin\frac{m\pi x}{a} \sin\frac{n\pi y}{b} \frac{\sinh k_{mn}z}{\sinh k_{mn}c}$$

where $k_{mn} = \pi[(m^2/a^2) + (n^2/b^2)]^{1/2}$.

PROBLEM 10-9 Find the steady-state temperature distribution in a semicircular plate of radius a, insulated on both faces, with its curved boundary kept at a constant temperature U_0, and its bounding diameter kept at zero temperature (see Figure 10-7).

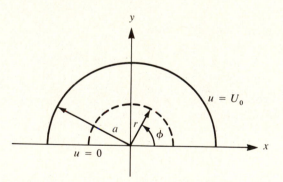

Figure 10-7   A semicircular plate.

Solution: In Section 10-3 (see eq. 10.99), the heat flow equation is written as

$$\nabla^2 u - \frac{1}{c^2}\frac{\partial u}{\partial t} = 0$$

In the steady state, the temperature u is independent of time; hence $\partial u/\partial t = 0$ and u satisfies Laplace's equation, that is,

$$\nabla^2 u = 0$$

Since in this problem, the space of heat flow is two-dimensional and the boundaries are cylindrical, the two-dimensional Laplacian of u in cylindrical coordinates (or polar coordinates) will be used. Hence, from (10.122),

$$\nabla^2 u(r, \phi) = \frac{\partial^2 u(r, \phi)}{\partial r^2} + \frac{1}{r}\frac{\partial u(r, \phi)}{\partial r} + \frac{1}{r^2}\frac{\partial^2 u(r, \phi)}{\partial \phi^2} = 0 \qquad \textbf{(a)}$$

The temperature $u(r, \phi)$ considered as a function of r and ϕ satisfies eq. (a) and the boundary conditions

$$u(a, \phi) = U_0 \tag{b}$$
$$u(r, 0) = 0 \tag{c}$$
$$u(r, \pi) = 0 \tag{d}$$

The method of separation of variables suggests assuming a solution of eq. (a) of the form

$$u(r, \phi) = R(r)\Phi(\phi) \tag{e}$$

Substituting eq. (e) into eq. (a), we have

$$R''(r)\Phi(\phi) + \frac{1}{r} R'(r)\Phi(\phi) + \frac{1}{r^2} R(r)\Phi''(\phi) = 0$$

or

$$r^2 R''(r)\Phi(\phi) + rR'(r)\Phi(\phi) + R(r)\Phi''(\phi) = 0 \tag{f}$$

Dividing eq. (f) by $R(r)\Phi(\phi)$ and separating the variables, we obtain

$$r^2 \frac{R''(r)}{R(r)} + r \frac{R'(r)}{R(r)} = -\frac{\Phi''(\phi)}{\Phi(\phi)} = k^2 \tag{g}$$

where k^2 is the separation constant. The separation here results from the fact that the left-hand side is independent of ϕ and the right-hand side is independent of r. The sign of the separation constant was chosen in such a way that sine and cosine functions, rather than exponential functions, will be introduced in $\Phi(\phi)$. Equation (g) then yields the following two equations:

$$r^2 R''(r) + rR'(r) - k^2 R(t) = 0 \tag{h}$$
$$\Phi''(\phi) + k^2 \Phi(\phi) = 0 \tag{i}$$

The general solution of eq. (i) is

$$\Phi(\phi) = A \cos k\phi + B \sin k\phi \tag{j}$$

In order to solve eq. (h), we make the transformation

$$r = e^s$$

Then,

$$R'(r) = \frac{dR}{dr} = \frac{dR}{ds}\frac{ds}{dr} = \frac{1}{r}\frac{dR}{ds}$$

$$R''(r) = \frac{1}{r^2}\frac{d^2 R}{ds^2} - \frac{1}{r^2}\frac{dR}{ds}$$

and eq. (h) reduces to

$$\frac{d^2 R}{ds^2} - k^2 R = 0$$

The general solution of this equation is

$$R = Ce^{ks} + De^{-ks}$$

Since $e^s = r$,

$$R(r) = Cr^k + Dr^{-k} \tag{k}$$

From the boundary conditions (c) and (d),

$$\Phi(0) = \Phi(\pi) = 0$$

Hence,

$$\Phi(0) = A = 0 \quad \text{and} \quad \Phi(\pi) = B \sin k\pi = 0$$

Since a trivial solution results if $B = 0$, we must have $\sin k\pi = 0$, from which

$$k\pi = n\pi \quad \text{or} \quad k = n, \quad n = 1, 2, \dots$$

Hence, we find the solutions

$$\Phi(\phi) = \Phi_n(\phi) = B_n \sin n\phi, \qquad n = 1, 2, \dots \tag{l}$$

In eq. (k), we see that as $r \to 0$, the term $r^{-k} \to \infty$, since $k = n > 0$. Since at $r = 0$, $R(0) = 0$, D must be equal to zero. Thus,

$$R(r) = R_n(r) = C_n r^n, \qquad n = 1, 2, \dots \tag{m}$$

Then, it follows that the functions

$$u_n(r, \phi) = R_n(r)\Phi_n(\phi) = b_n r^n \sin n\phi, \qquad n = 1, 2, \dots \tag{n}$$

where $b_n = B_n C_n$, satisfy eq. (a) as well as the boundary conditions (c) and (d).

In order to satisfy the boundary condition (b), we assume the desired solution in the form

$$u(r, \phi) = \sum_{n=1}^{\infty} u_n(r, \phi) = \sum_{n=1}^{\infty} b_n r^n \sin n\phi \tag{o}$$

From condition (b),

$$u(a, \phi) = U_0 = \sum_{n=1}^{\infty} b_n a^n \sin n\phi \tag{p}$$

Thus, the coefficients $b_n a^n$ are the sine Fourier coefficients of U_0, and

$$b_n a^n = \frac{2}{\pi} \int_0^\pi U_0 \sin n\phi \, d\phi$$

$$= \begin{cases} \dfrac{4U_0}{n\pi} & \text{for} \quad n = 1, 3, \dots \\ 0 & \text{for} \quad n = 2, 4, \dots \end{cases}$$

Hence,

$$b_n = \frac{4U_0}{\pi n a^n}, \qquad n = 1, 3, \dots$$

With these values of b_n, the solution (o) becomes

$$u(r, \phi) = \frac{4U_0}{\pi} \sum_{n=\text{odd}}^{\infty} \frac{1}{n} \left(\frac{r}{a}\right)^n \sin n\phi$$

PROBLEM 10-10   Find the solution $u(x, y)$ of the Laplace's equation for the half-plane $y > 0$, when $u(x, 0) = f(x)$ for $-\infty < x < \infty$. (See Figure 10-8.)

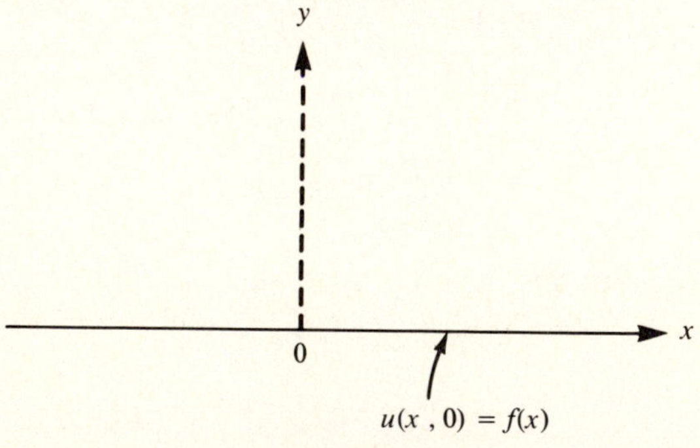

$$u(x, 0) = f(x)$$

Figure 10-8

Solution: To the Laplace's equation

$$u_{xx}(x, y) + u_{yy}(x, y) = 0$$

we apply the Fourier transform with respect to the variable x, namely

$$U(s, y) = \mathscr{F}[u(x, y)] = \int_{-\infty}^{\infty} u(x, y) e^{-jsx} \, dx$$

Assuming that $u(x, y)$ and $u_x(x, y)$ vanish for $x \to \pm\infty$, we obtain the equation for $U(s, y)$ as (see eq. (10.89))

$$\frac{\partial^2 U(s, y)}{\partial y^2} - s^2 U(s, y) = 0 \qquad \text{(a)}$$

The general solution of eq. (a) is

$$U(s, y) = A(s)e^{sy} + B(s)e^{-sy} \qquad \text{(b)}$$

We shall also assume that $u(x, y)$ is bounded as $y \to +\infty$. Hence, for $s > 0$, we set $A(s) = 0$, and

$$U(s, y) = B(s)e^{-sy} \qquad \text{for} \quad s > 0 \qquad \text{(c)}$$

Since $U(s, 0) = B(s)$, we can rewrite eq. (c) as

$$U(s, y) = U(s, 0)e^{-sy} \qquad \text{for} \quad s > 0 \qquad \text{(d)}$$

Similarly, for $s < 0$, we set $B(s) = 0$ in solution (b) and write

$$U(s, y) = A(s)e^{sy} \qquad \text{for} \quad s < 0 \qquad \text{(e)}$$

Again, since $U(s, 0) = A(s)$, we can rewrite eq. (e) as

$$U(s, y) = U(s, 0)e^{sy} \qquad \text{for} \quad s < 0 \qquad \text{(f)}$$

The two equations (d) and (f) can be combined as

$$U(s, y) = U(s, 0)e^{-|s|y} \qquad \text{(g)}$$

Since $u(x, 0) = f(x)$,

$$U(s, 0) = \mathscr{F}[u(x, 0)] = \int_{-\infty}^{\infty} f(x')e^{-jsx'} dx' \qquad \text{(h)}$$

From eq. (g), we have

$$U(s, y) = \left[\int_{-\infty}^{\infty} f(x')e^{-jsx'} dx' \right] e^{-|s|y} \qquad \text{(i)}$$

The desired solution $u(x, y)$ is the inverse Fourier transform of eq. (i); that is,

$$u(x, y) = \mathscr{F}^{-1}[U(s, y)] = \frac{1}{2\pi} \int_{-\infty}^{\infty} U(s, y)e^{jsx} ds$$

$$= \frac{1}{2\pi} \int_{-\infty}^{\infty} e^{jsx} \left[\int_{-\infty}^{\infty} f(x')e^{-jsx'} dx' \right] e^{-|s|y} dx \qquad \text{(j)}$$

Interchanging the order of integration,

$$u(x, y) = \frac{1}{2\pi} \int_{-\infty}^{\infty} f(x') \left\{ \int_{-\infty}^{\infty} e^{[js(x-x') - |s|y]} ds \right\} dx' \qquad \text{(k)}$$

Now,

$$\int_{-\infty}^{\infty} e^{[js(x-x') - |s|y]} ds = \int_{-\infty}^{0} e^{[js(x-x') + sy]} ds + \int_{0}^{\infty} e^{[js(x-x') - sy]} ds$$

$$= \frac{e^{js(x-x') + sy}}{j(x - x') + y}\bigg|_{-\infty}^{0} + \frac{e^{js(x-x') - sy}}{j(x - x') - y}\bigg|_{0}^{\infty}$$

$$= \frac{1}{j(x - x') + y} - \frac{1}{j(x - x') - y}$$

$$= \frac{2y}{(x - x')^2 + y^2} \qquad \text{(l)}$$

Substituting eq. (l) into eq. (k), we finally obtain

$$u(x, y) = \frac{y}{\pi} \int_{-\infty}^{\infty} \frac{f(x') dx'}{(x - x')^2 + y^2}, \qquad y > 0$$

Supplementary Exercises

PROBLEM 10-11 Solve (10.1) using the boundary conditions of (10.2) and with the initial conditions

$$u(x,0) = f(x) = \begin{cases} \dfrac{k}{a}x & \text{for} \quad 0 < x < a, \\[2ex] \dfrac{k}{l-a}(l-x) & \text{for} \quad a < x < l, \end{cases} \qquad \text{and} \quad \left.\frac{\partial u(x,t)}{\partial t}\right|_{t=0} = 0$$

(See Figure 10-9.)

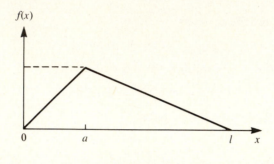

Figure 10-9

Answer: $u(x,t) = \dfrac{2kl^2}{\pi^2 a(l-a)} \displaystyle\sum_{n=1}^{\infty} \dfrac{1}{n^2} \sin\left(\dfrac{n\pi a}{l}\right)\sin\left(\dfrac{n\pi x}{l}\right)\cos\left(\dfrac{n\pi ct}{l}\right)$

PROBLEM 10-12 Prove that the function $u(x,t) = f(x - ct) + g(x + ct)$ is a solution of the one-dimensional wave equation (10.1) whenever f and g are twice differentiable functions of a single variable.

PROBLEM 10-13 The temperature in a uniformly insulated bar of length l satisfies the endpoint conditions $u(0,t) = 0$, $u(l,t) = 1$, and the initial condition $u(x,0) = \sin(\pi x/l)$. Find **(a)** the temperature distribution after time t and **(b)** the steady-state temperature, that is, the temperature as $t \to \infty$, in the bar.

Answer:

(a) $u(x,t) = \dfrac{x}{l} + \dfrac{2}{\pi}\displaystyle\sum_{n=1}^{\infty}\dfrac{(-1)^n}{n}e^{-\lambda_n^2 t}\sin\left(\dfrac{n\pi x}{l}\right), \qquad \lambda_n = \dfrac{cn\pi}{l}$

(b) $\left. u(x,t)\right|_{t=\infty} = \dfrac{x}{l}$

PROBLEM 10-14 Solve

$$\frac{\partial^2 u}{\partial x^2} - \frac{1}{c^2}\frac{\partial u}{\partial t} = 0 \qquad \text{for } 0 < x < \pi, \quad t > 0$$

with

$$\frac{\partial u}{\partial x}(0,t) = 0, \qquad \frac{\partial u}{\partial x}(\pi,t) = 0, \qquad \text{and} \qquad u(x,0) = \sin x$$

Answer: $u(x,t) = \dfrac{2}{\pi} - \dfrac{4}{\pi}\displaystyle\sum_{n=1}^{\infty}\dfrac{1}{(4n^2-1)}e^{-4n^2c^2 t}\cos 2nx$

PROBLEM 10-15 Solve

$$\frac{\partial^2 u}{\partial x^2} + \frac{\partial^2 u}{\partial y^2} = 0 \qquad \text{for } 0 < x < a, \quad 0 < y < b$$

with the boundary and initial conditions $u(0,y) = u(a,y) = u(x,b) = 0$ and $u(x,0) = f(x)$.

Answer: $u(x,y) = \displaystyle\sum_{n=1}^{\infty} b_n \dfrac{\sinh[n\pi(b-y)/a]}{\sinh(n\pi b/a)}\sin\left(\dfrac{n\pi x}{a}\right),$

where

$$b_n = \frac{2}{a} \int_0^a f(x) \sin\left(\frac{n\pi x}{a}\right) dx$$

PROBLEM 10-16   Solve

$$\frac{\partial^2 u}{\partial x^2} + \frac{\partial^2 u}{\partial y^2} = 0 \qquad \text{for} \quad 0 < x < a, \quad 0 < y < \infty$$

with $u(x, y) \to 0$ when $y \to \infty$, $u(0, y) = 0$, $u(a, y) = 0$, and $u(x, 0) = x(a - x)$.

Answer: $u(x, y) = \dfrac{4a^2}{\pi^3} \displaystyle\sum_{n=1}^{\infty} \frac{(1 - \cos n\pi)}{n^3} e^{-n\pi y/a} \sin\left(\frac{n\pi x}{a}\right)$

PROBLEM 10-17   Solve

$$\frac{\partial^2 u}{\partial r^2} + \frac{1}{r}\frac{\partial u}{\partial r} + \frac{1}{r^2}\frac{\partial^2 u}{\partial \phi^2} = 0 \qquad \text{for } r < 1, \quad 0 < \phi < \pi$$

with $u(r, 0) = u(r, \pi) = 0$ and $u(1, \phi) = \phi(\pi - \phi)$.

Answer: $u(r, \phi) = \dfrac{8}{\pi} \displaystyle\sum_{n=1}^{\infty} \frac{1}{(2n-1)^3} r^{2n-1} \sin(2n-1)\phi$

PROBLEM 10-18   Solve

$$\frac{\partial^2 u}{\partial r^2} + \frac{1}{r}\frac{\partial u}{\partial r} + \frac{1}{r^2}\frac{\partial^2 u}{\partial \phi^2} = 0 \qquad \text{for } r < 1, \quad 0 < \phi < \frac{\pi}{2}$$

with $u(r, 0) = 0$, $\dfrac{\partial u}{\partial \phi}\left(r, \frac{1}{2}\pi\right) = 0$, and $u(1, \phi) = \phi$.

Answer: $u(r, \phi) = \dfrac{4}{\pi} \displaystyle\sum_{n=1}^{\infty} (-1)^{n-1} \frac{1}{(2n-1)^2} r^{2n-1} \sin(2n-1)\phi$

PROBLEM 10-19   Find the temperature distribution $u(x, t)$ for an infinite bar. The initial temperature distribution is

$$f(x) = \begin{cases} 0 & \text{for} \quad x < 0 \\ T & \text{for} \quad x > 0 \end{cases}$$

where T is a constant.

Answer: $u(x, t) = \dfrac{T}{2}\left[1 + \operatorname{erf}\left(\dfrac{x}{2c\sqrt{t}}\right)\right]$

where

$$\operatorname{erf} y = \frac{2}{\sqrt{\pi}} \int_0^y e^{-\xi^2} d\xi$$

PROBLEM 10-20   Using the Fourier transform, solve

$$\frac{\partial^2 u}{\partial x^2} - \frac{\partial u}{\partial t} = f(x, t) \qquad \text{for} \quad -\infty < x < \infty, \quad t > 0$$

with the initial condition $u(x, 0) = 0$ for $t > 0$.

Answer: $u(x, t) = \dfrac{1}{2\sqrt{\pi}} \displaystyle\int_{-\infty}^{\infty} \int_{-\infty}^{\infty} \frac{e^{-(x-\xi)^2/4(t-\tau)}}{\sqrt{t-\tau}} H(t - \tau) f(\xi, \tau) \, d\xi \, d\tau$

where

$$H(\lambda) = \begin{cases} 1 & \text{for} \quad \lambda > 0 \\ 0 & \text{for} \quad \lambda < 0 \end{cases}$$

FINAL EXAMINATION

1. Suppose an electric circuit is excited by a voltage $v(t)$ given by

$$v(t) = V_0 + \sum_{n=1}^{\infty} V_n \cos(n\omega_0 t + \theta_n), \qquad \omega_0 = \frac{2\pi}{T}$$

The corresponding steady-state current $i(t)$ is given by

$$i(t) = I_0 + \sum_{n=1}^{\infty} I_n \cos(n\omega_0 t + \phi_n)$$

Show that the average input power P at the input terminals defined by

$$P = \frac{1}{T} \int_{-T/2}^{T/2} v(t)i(t)\, dt$$

is equal to

$$P = V_0 I_0 + \frac{1}{2} \sum_{n=1}^{\infty} V_n I_n \cos(\theta_n - \phi_n)$$

2. Using the differentiation technique, find the complex Fourier series for the sawtooth wave function $f(t)$ shown in the figure.

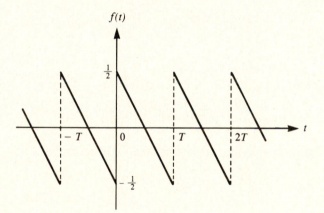

3. If $f(t)$ is a solution of the differential equation

$$\frac{d^2 f(t)}{dt^2} - t^2 f(t) = f(t)$$

show that its Fourier transform is a solution of the same equation.

4. Use convolution to find

$$\mathscr{F}^{-1}\left[\frac{1}{(1 + j\omega)(2 + j\omega)} \right]$$

Solutions

1. Using the trigonometric identity

$$\cos A \cos B = \frac{1}{2}\cos(A+B) + \frac{1}{2}\cos(A-B)$$

and the result

$$\int_{-T/2}^{T/2} \cos(m\omega_0 t + \alpha)\, dt = 0 \qquad \text{for } m \neq 0$$

we obtain

$$P = \frac{1}{T}\int_{-T/2}^{T/2}\left[V_0 + \sum_{n=1}^{\infty} V_n\cos(n\omega_0 t + \theta_n)\right]\left[I_0 + \sum_{k=1}^{\infty} I_k\cos(k\omega_0 t + \phi_k)\right]dt$$

$$= \frac{1}{T}\int_{-T/2}^{T/2}\left[V_0 I_0 + V_0\sum_{k=1}^{\infty} I_k\cos(k\omega_0 t + \phi_k) + I_0\sum_{n=1}^{\infty} V_n\cos(n\omega_0 t + \theta_n)\right.$$

$$\left. + \sum_{n=1}^{\infty}\sum_{k=1}^{\infty} V_n I_k\cos(n\omega_0 t + \theta_n)\cos(k\omega_0 t + \phi_k)\right]dt$$

$$= V_0 I_0 \frac{1}{T}\int_{-T/2}^{T/2} dt + V_0\sum_{k=1}^{\infty} I_k \frac{1}{T}\int_{-T/2}^{T/2}\cos(k\omega_0 t + \phi_k)\, dt + I_0\sum_{n=1}^{\infty} V_n \frac{1}{T}\int_{-T/2}^{T/2}\cos(n\omega_0 t + \theta_n)\, dt$$

$$+ \sum_{n=1}^{\infty} V_n \sum_{k=1}^{\infty} I_k \frac{1}{T}\int_{-T/2}^{T/2}\cos(n\omega_0 t + \theta)\cos(k\omega_0 t + \phi_k)\, dt$$

$$= V_0 I_0 + \sum_{n=1}^{\infty}\sum_{k=1}^{\infty}\frac{1}{2}V_n I_k\left\{\frac{1}{T}\int_{-T/2}^{T/2}\cos[(n+k)\omega_0 t + \theta_n + \phi_k]\, dt + \frac{1}{T}\int_{-T/2}^{T/2}\cos[(n-k)\omega_0 t + \theta_n - \phi_k]\, dt\right\}$$

The first integral is zero for all $n, k > 0$. The second integral is zero unless $n = k$, in which case we obtain

$$\frac{1}{T}\int_{-T/2}^{T/2}\cos(\theta_n - \phi_n)\, dt = \cos(\theta_n - \phi_n)$$

Thus,

$$P = V_0 I_0 + \frac{1}{2}\sum_{n=1}^{\infty} V_n I_n \cos(\theta_n - \phi_n)$$

2. Assume that

$$f(t) = \sum_{n=-\infty}^{\infty} c_n e^{jn\omega_0 t}, \qquad \omega_0 = 2\pi/T$$

Using term-by-term differentiation, we have

$$f'(t) = \sum_{n=-\infty}^{\infty}(jn\omega_0)c_n e^{jn\omega_0 t}$$

Now from the figure, we can express $f'(t)$ as

$$f'(t) = -\frac{1}{T} + \sum_{n=-\infty}^{\infty}\delta(t - nT) = -\frac{1}{T} + \frac{1}{T}\sum_{n=-\infty}^{\infty} e^{jn\omega_0 t}$$

since

$$\sum_{n=-\infty}^{\infty} \delta(t - nT) = \frac{1}{T} \sum_{n=-\infty}^{\infty} e^{jn\omega_0 t}, \qquad \omega_0 = 2\pi/T$$

Therefore, we obtain

$$(jn\omega_0)c_n = \frac{1}{T} \qquad \text{for } n \neq 0$$

Hence,

$$c_n = \frac{1}{jn\omega_0 T} = \frac{1}{j2\pi n}$$

Now

$$c_0 = \frac{1}{T} \int_{-T/2}^{T/2} f(t)\, dt = 0$$

since $f(t)$ is odd.

3. Because

$$\frac{d^n f(t)}{dt^n} \leftrightarrow (j\omega)^n F(\omega)$$

$$(-jt)^n f(t) \leftrightarrow \frac{d^n F(\omega)}{d\omega^n}$$

we can take the Fourier transforms of both sides, to obtain

$$(j\omega)^2 F(\omega) + \frac{d^2 F(\omega)}{d\omega^2} = F(\omega)$$

or

$$\frac{d^2 F(\omega)}{d\omega^2} - \omega^2 F(\omega) = F(\omega)$$

4. Since

$$\mathscr{F}[e^{-at} u(t)] = \frac{1}{a + j\omega}$$

we have

$$\mathscr{F}^{-1}\left[\frac{1}{1 + j\omega}\right] = e^{-t} u(t) \qquad \text{and} \qquad \mathscr{F}^{-1}\left[\frac{1}{2 + j\omega}\right] = e^{-2t} u(t)$$

Now from the convolution theorem

$$\mathscr{F}^{-1}[F_1(\omega) F_2(\omega)] = f_1(t) * f_2(t)$$

we obtain

$$\mathscr{F}^{-1}\left[\frac{1}{(1 + j\omega)(2 + j\omega)}\right] = e^{-t} u(t) * e^{-2t} u(t)$$

$$= \int_{-\infty}^{\infty} e^{-\tau} u(\tau) e^{-2(t-\tau)} u(t - \tau)\, d\tau$$

$$= e^{-2t} \int_{-\infty}^{\infty} e^{\tau} u(\tau) u(t - \tau)\, d\tau$$

$$= e^{-2t} \int_{0}^{t} e^{\tau}\, d\tau$$

$$= e^{-2t}(e^t - 1) u(t) = (e^{-t} - e^{-2t}) u(t)$$

since

$$u(\tau) u(t - \tau) = \begin{cases} 1, & 0 < \tau < t \\ 0, & \text{otherwise} \end{cases}$$

APPENDIX A: *Three Forms of Fourier Series*

Form 1: Trigonometric $\quad f(t) = \dfrac{a_0}{2} + \displaystyle\sum_{n=1}^{\infty} (a_n \cos n\omega_0 t + b_n \sin n\omega_0 t)$

Form 2: Trigonometric $\quad f(t) = C_0 + \displaystyle\sum_{n=1}^{\infty} C_n \cos(n\omega_0 t - \theta_n)$

Form 3: Complex Exponential $\quad f(t) = \displaystyle\sum_{n=-\infty}^{\infty} c_n e^{jn\omega_0 t}$

For all of the above $\quad f(t + T) = f(t), \quad \omega_0 = \dfrac{2\pi}{T}$

Conversion Formulas:

For $n \neq 0$

$$c_n = \tfrac{1}{2}(a_n - jb_n), \qquad c_{-n} = \tfrac{1}{2}(a_n + jb_n) = c_n^*$$

$$c_n = |c_n| e^{j\phi_n}, \qquad |c_n| = \tfrac{1}{2}\sqrt{a_n^2 + b_n^2}, \qquad \phi_n = \tan^{-1}\left(-\frac{b_n}{a_n}\right)$$

$$a_n = 2\,Re[c_n], \qquad b_n = -2\,Im[c_n]$$

$$C_n = 2|c_n| = \sqrt{a_n^2 + b_n^2}, \qquad \theta_n = \tan^{-1}\left(\frac{b_n}{a_n}\right) = -\phi_n$$

For $n = 0$, $\qquad\qquad\qquad\qquad c_0 = \tfrac{1}{2}a_0 = C_0$

APPENDIX B: *Summary of Symmetry Conditions*

Summary of Symmetry Conditions for Periodic Waveforms and Fourier Coefficients

$$f(t + T) = f(t), \qquad \omega_0 = 2\pi/T$$

Type of symmetry	Conditions	Form of the Fourier series	Formulas for the Fourier coefficients
Even	$f(t) = f(-t)$	$f(t) = \dfrac{a_0}{2} + \displaystyle\sum_{n=1}^{\infty} a_n \cos n\omega_0 t$	$a_n = \dfrac{4}{T}\displaystyle\int_0^{T/2} f(t)\cos(n\omega_0 t)\,dt$
Odd	$f(t) = f(-t)$	$f(t) = \displaystyle\sum_{n=1}^{\infty} b_n \sin n\omega_0 t$	$b_n = \dfrac{4}{T}\displaystyle\int_0^{T/2} f(t)\sin(n\omega_0 t)\,dt$
Half-wave	$f(t) = -f\left(t + \dfrac{T}{2}\right)$	$f(t) = \displaystyle\sum_{n=1}^{\infty} [a_{2n-1}\cos(2n-1)\omega_0 t$ $+\, b_{2n-1}\sin(2n-1)\omega_0 t]$	$\left.\begin{array}{c} a_{2n-1} \\ b_{2n-1} \end{array}\right\} = \dfrac{4}{T}\displaystyle\int_0^{T/2} f(t)\begin{Bmatrix}\cos \\ \sin\end{Bmatrix}[(2n-1)\omega_0 t]\,dt$
Even quarter-wave	$f(t) = f(-t)$ $f(t) = -f\left(t + \dfrac{T}{2}\right)$	$f(t) = \displaystyle\sum_{n=1}^{\infty} a_{2n-1}\cos(2n-1)\omega_0 t$	$a_{2n-1} = \dfrac{8}{T}\displaystyle\int_0^{T/4} f(t)\cos[(2n-1)\omega_0 t]\,dt$
Odd quarter-wave	$f(t) = -f(t)$ $f(t) = -f\left(t + \dfrac{T}{2}\right)$	$f(t) = \displaystyle\sum_{n=1}^{\infty} b_{2n-1}\sin(2n-1)\omega_0 t$	$b_{2n-1} = \dfrac{8}{T}\displaystyle\int_0^{T/4} f(t)\sin[(2n-1)\omega_0 t]\,dt$

APPENDIX C:
Properties of the Fourier Transform

$f(t)$	$F(\omega)$
$a_1 f_1(t) + a_2 f_2(t)$	$a_1 F_1(\omega) + a_2 F_2(\omega)$
$f(at)$	$\dfrac{1}{\|a\|} F\left(\dfrac{\omega}{a}\right)$
$f(-t)$	$F(-\omega)$
$f(t - t_0)$	$F(\omega)e^{-j\omega t_0}$
$f(t)e^{j\omega_0 t}$	$F(\omega - \omega_0)$
$f(t)\cos \omega_0 t$	$\dfrac{1}{2} F(\omega - \omega_0) + \dfrac{1}{2} F(\omega + \omega_0)$
$f(t)\sin \omega_0 t$	$\dfrac{1}{2j} F(\omega - \omega_0) - \dfrac{1}{2j} F(\omega + \omega_0)$
$f(t) = f_e(t) + f_o(t)$	$F(\omega) = R(\omega) + jX(\omega)$
$f_e(t) = \dfrac{1}{2}\left[f(t) + f(-t)\right]$	$R(\omega)$
$f_o(t) = \dfrac{1}{2}\left[f(t) - f(-t)\right]$	$jX(\omega)$
$F(t)$	$2\pi f(-\omega)$
$f'(t)$	$j\omega F(\omega)$
$f^{(n)}(t)$	$(j\omega)^n F(\omega)$
$\displaystyle\int_{-\infty}^{t} f(x)\, dx$	$\dfrac{1}{j\omega} F(\omega) + \pi F(0)\delta(\omega)$
$-jtf(t)$	$F'(\omega)$
$(-jt)^n f(t)$	$F^{(n)}(\omega)$
$f_1(t) * f_2(t) = \displaystyle\int_{-\infty}^{\infty} f_1(x)f_2(t - x)\, dx$	$F_1(\omega)F_2(\omega)$
$f_1(t)f_2(t)$	$\dfrac{1}{2\pi} F_1(\omega) * F_2(\omega) = \dfrac{1}{2\pi} \displaystyle\int_{-\infty}^{\infty} F_1(y)F_2(\omega - y)\, dy$
$e^{-at}u(t)$	$\dfrac{1}{j\omega + a}$
$e^{-a\|t\|}$	$\dfrac{2a}{a^2 + \omega^2}$
e^{-at^2}	$\sqrt{\dfrac{\pi}{a}}\, e^{-\omega^2/(4a)}$
$p_a(t) = \begin{cases} 1 & \text{for} \quad \|t\| < a/2 \\ 0 & \text{for} \quad \|t\| > a/2 \end{cases}$	$a\dfrac{\sin(\omega a/2)}{(\omega a/2)}$
$\dfrac{\sin at}{\pi t}$	$p_{2a}(\omega) = \begin{cases} 1 & \text{for} \quad \|\omega\| < a \\ 0 & \text{for} \quad \|\omega\| > a \end{cases}$
$te^{-at}u(t)$	$\dfrac{1}{(j\omega + a)^2}$
$\dfrac{t^{n-1}}{(n-1)!}\, e^{-at}u(t)$	$\dfrac{1}{(j\omega + a)^n}$

$f(t)$	$F(\omega)$
$e^{-at}\sin bt\,u(t)$	$\dfrac{b}{(j\omega + a)^2 + b^2}$
$e^{-at}\cos bt\,u(t)$	$\dfrac{j\omega + a}{(j\omega + a)^2 + b^2}$
$\dfrac{1}{a^2 + t^2}$	$\dfrac{\pi}{a}\,e^{-a|\omega|}$
$\dfrac{\cos bt}{a^2 + t^2}$	$\dfrac{\pi}{2a}\left[e^{-a|\omega - b|} + e^{-a|\omega + b|}\right]$
$\dfrac{\sin bt}{a^2 + b^2}$	$\dfrac{\pi}{2aj}\left[e^{-a|\omega - b|} - e^{-a|\omega + b|}\right]$
$\delta(t)$	1
$\delta(t - t_0)$	$e^{-j\omega t_0}$
$\delta'(t)$	$j\omega$
$\delta^{(n)}(t)$	$(j\omega)^n$
$u(t)$	$\pi\delta(\omega) + \dfrac{1}{j\omega}$
$u(t - t_0)$	$\pi\delta(\omega) + \dfrac{1}{j\omega}\,e^{-j\omega t_0}$
1	$2\pi\delta(\omega)$
t	$2\pi j\delta'(\omega)$
t^n	$2\pi j^n \delta^{(n)}(\omega)$
$e^{j\omega_0 t}$	$2\pi\delta(\omega - \omega_0)$
$\cos\omega_0 t$	$\pi[\delta(\omega - \omega_0) + \delta(\omega + \omega_0)]$
$\sin\omega_0 t$	$-j\pi[\delta(\omega - \omega_0) - \delta(\omega + \omega_0)]$
$\sin\omega_0 t\,u(t)$	$\dfrac{\omega_0}{\omega_0^2 - \omega^2} + \dfrac{\pi}{2j}[\delta(\omega - \omega_0) - \delta(\omega + \omega_0)]$
$\cos\omega_0 t\,u(t)$	$\dfrac{j\omega}{\omega_0^2 - \omega^2} + \dfrac{\pi}{2}[\delta(\omega - \omega_0) + \delta(\omega + \omega_0)]$
$t\,u(t)$	$j\pi\delta'(\omega) - \dfrac{1}{\omega^2}$
$\dfrac{1}{t}$	$\pi j - 2\pi j u(\omega)$
$\dfrac{1}{t^n}$	$\dfrac{(-j\omega)^{n-1}}{(n-1)!}[\pi j - 2\pi j u(\omega)]$
$\operatorname{sgn} t$	$\dfrac{2}{j\omega}$
$\delta_T(t) = \displaystyle\sum_{n=-\infty}^{\infty} \delta(t - nT)$	$\omega_0\delta_{\omega_0}(\omega) = \omega_0 \displaystyle\sum_{n=-\infty}^{\infty} \delta(\omega - n\omega_0)$

Other properties:

$$\int_{-\infty}^{\infty} f_1(t)f_2(t)\,dt = \frac{1}{2\pi}\int_{-\infty}^{\infty} F_1(\omega)F_2^*(\omega)\,d\omega$$

$$\int_{-\infty}^{\infty} |f(t)|^2\,dt = \frac{1}{2\pi}\int_{-\infty}^{\infty} |F(\omega)|^2\,d\omega$$

$$\int_{-\infty}^{\infty} f(x)G(x)\,dx = \int_{-\infty}^{\infty} F(x)g(x)\,dx$$

APPENDIX D: *List of Symbols*

$\left.\begin{array}{lll} a_n & A_n & b_{mn} \\ b_n & B_n & C_{mn} \\ c_n & C_n & G_{mn} \\ D_n & H_{mn} \\ E_n & F_n \end{array}\right\}$ Fourier coefficients

C_n	Harmonic amplitude
$a(t)$	Unit step response
E	Energy content
E_k	Mean-square error
f	Frequency, function
$f(t)$	Time function
$f_e(t)$	Even function
$f_i(t)$	Input function
$f_o(t)$	Odd function
$f_o(t)$	Output function
$f^{(n)}(t)$	nth derivative of $f(t)$
$f_+(t)$	Analytical signal
$\hat{f}(t)$	Hilbert transform of $f(t)$
$F(\omega), F(j\omega)$	Fourier transform of $f(t)$
$F(s)$	Fourier transform of $f(x)$
$|F(\omega)|$	Magnitude spectrum (absolute value of $F(\omega)$)
$\hat{F}(\omega)$	Fourier transform of $\hat{f}(t)$
$g(t)$	Time function, generalized function
$G(\omega)$	Fourier transform of $g(t)$
$h(t)$	Unit impulse response
$H(p)$	Operational system function
$H(\omega), H(j\omega)$	Fourier transform of $h(t)$, system function
j	$\sqrt{-1}$
$k^2, k_x, \ldots, k_{mn}$	Separation constant
L	Linear operator
$m(t)$	Signal
$n(t)$	Noise
p	Operator d/dt
$p_d(t), p_{2a}(\omega)$	Rectangular pulse with unit amplitude
P	Power
$P(\omega)$	Power spectral density, power spectrum
$P_i(\omega)$	Input power spectral density
$P_o(\omega)$	Output power spectral density

$R(\omega)$	Real part of $F(\omega)$
$R_{11}, R_{22}, R_{ff}, \ldots$	Autocorrelation functions
$R_{12}, R_{21}, R_{xy}, \ldots$	Cross-correlation functions
$\bar{R}_{11}, \bar{R}_{22}, \ldots$	Average autocorrelation functions
$\bar{R}_{12}, \bar{R}_{21}, \ldots$	Average cross-correlation functions
sgn t	signum t
Si	Sine-integral function
Sa(t)	Sampling function
$S_k(t)$	Sum of the first $(2k + 1)$ terms of a Fourier series of $f(t)$
T	Period of a periodic function, operator
T_D	Equivalent pulse duration
$u(t)$	Unit step function
u_t	Partial derivative of u with respect to t
u_x	Partial derivative of u with respect to x
$u_{tt}, \ldots$	Second partial derivative of u with respect to $t, \ldots$
W_B	Equivalent spectral bandwidth
$X(\omega)$	Imaginary part of $F(\omega)$
α_n, β_n	Fourier coefficients
γ	Autocorrelation (normalized)
$\delta(t)$	Delta function or unit impulse function
$\delta_T(t), \delta_{\omega_0}(\omega)$	Periodic train of unit impulses
ε_k	Error between $f(t)$ and $S_k(t)$
θ, θ_n, ϕ_n	Phase angle
$\phi(t)$	Testing function
$\phi(\omega)$	Phase spectrum
$\phi_n(t)$	Orthogonal function
ω	Radian frequency
ω_c	Carrier frequency
ω_0	Fundamental radian frequency
$\mathscr{F}$	Fourier transform of
$\mathscr{F}^{-1}$	Inverse Fourier transform of
$*$	Convolution, conjugate of (superscript)

220

INDEX